Advances in Air Quality Monitoring and Assessment

Advances in Air Quality Monitoring and Assessment

Editor

Thomas Maggos

MDPI • Basel • Beijing • Wuhan • Barcelona • Belgrade • Manchester • Tokyo • Cluj • Tianjin

Editor
Thomas Maggos
Atmospheric Chemistry and
Innovative Technologies
Laboratory
NCSR DEMOKRITOS
Ag. Paraskevi
Greece

Editorial Office
MDPI
St. Alban-Anlage 66
4052 Basel, Switzerland

This is a reprint of articles from the Special Issue published online in the open access journal *Applied Sciences* (ISSN 2076-3417) (available at: www.mdpi.com/journal/applsci/special_issues/air_quality_monitoring_assessment).

For citation purposes, cite each article independently as indicated on the article page online and as indicated below:

LastName, A.A.; LastName, B.B.; LastName, C.C. Article Title. *Journal Name* **Year**, *Volume Number*, Page Range.

ISBN 978-3-0365-2140-4 (Hbk)
ISBN 978-3-0365-2139-8 (PDF)

© 2021 by the authors. Articles in this book are Open Access and distributed under the Creative Commons Attribution (CC BY) license, which allows users to download, copy and build upon published articles, as long as the author and publisher are properly credited, which ensures maximum dissemination and a wider impact of our publications.

The book as a whole is distributed by MDPI under the terms and conditions of the Creative Commons license CC BY-NC-ND.

Contents

About the Editor vii

Thomas Maggos
Advances in Air Quality Monitoring and Assessment
Reprinted from: *Appl. Sci.* **2021**, *11*, 5817, doi:10.3390/app11135817 1

Athanasios Paralikis, Emmaouil Gagaoudakis, Viktoras Kampitakis, Elias Aperathitis, George Kiriakidis and Vassilios Binas
Study on the Ozone Gas Sensing Properties of rf-Sputtered Al-Doped NiO Films
Reprinted from: *Appl. Sci.* **2021**, *11*, 3104, doi:10.3390/app11073104 5

Miloš Davidović, Sonja Dmitrašinović, Maja Jovanović, Jelena Radonić and Milena Jovašević-Stojanović
Diurnal, Temporal and Spatial Variations of Main Air Pollutants Before and during Emergency Lockdown in the City of Novi Sad (Serbia)
Reprinted from: *Appl. Sci.* **2021**, *11*, 1212, doi:10.3390/app11031212 17

Yiniva Camargo-Caicedo, Laura C. Mantilla-Romo and Tomás R. Bolaño-Ortiz
Emissions Reduction of Greenhouse Gases, Ozone Precursors, Aerosols and Acidifying Gases from Road Transportation during the COVID-19 Lockdown in Colombia
Reprinted from: *Appl. Sci.* **2021**, *11*, 1458, doi:10.3390/app11041458 37

Peng Li and Lin Lü
Evaluating the Real-World NOx Emission from a China VI Heavy-Duty Diesel Vehicle
Reprinted from: *Appl. Sci.* **2021**, *11*, 1335, doi:10.3390/app11031335 55

Athita Onuean, Hanmin Jung and Krisana Chinnasarn
Finding Optimal Stations Using Euclidean Distance and Adjustable Surrounding Sphere
Reprinted from: *Appl. Sci.* **2021**, *11*, 848, doi:10.3390/app11020848 75

Afifa Aslam, Muhammad Ibrahim, Imran Shahid, Abid Mahmood, Muhammad Kashif Irshad, Muhammad Yamin, Ghazala, Muhammad Tariq and Redmond R. Shamshiri
Pollution Characteristics of Particulate Matter ($PM_{2.5}$ and PM_{10}) and Constituent Carbonaceous Aerosols in a South Asian Future Megacity
Reprinted from: *Appl. Sci.* **2020**, *10*, 8864, doi:10.3390/app10248864 93

Arthur K. Cho, Yasuhiro Shinkai, Debra A. Schmitz, Emma Di Stefano, Arantza Eiguren-Fernandez, Aline Lefol Nani Guarieiro, Erika M. Salinas, John R. Froines and William P. Melega
Chemical and Biological Characterization of Particulate Matter (PM 2.5) and Volatile Organic Compounds Collected at Different Sites in the Los Angeles Basin
Reprinted from: *Appl. Sci.* **2020**, *10*, 3245, doi:10.3390/app10093245 111

Juanming Zhan, Minyi Wang, Yonghong Liu, Chunming Feng, Ting Gan, Li Li, Ruiwen Ou and Hui Ding
Impact of the '13th Five-Year Plan' Policy on Air Quality in Pearl River Delta, China: A Case Study of Haizhu District in Guangzhou City Using WRF-Chem
Reprinted from: *Appl. Sci.* **2020**, *10*, 5276, doi:10.3390/app10155276 125

Chengming Li, Zhaoxin Dai, Xiaoli Liu and Pengda Wu
Transport Pathways and Potential Source Region Contributions of $PM_{2.5}$ in Weifang: Seasonal Variations
Reprinted from: *Appl. Sci.* **2020**, *10*, 2835, doi:10.3390/app10082835 . . . . **147**

JinSoo Park and Sungroul Kim
Improved Interpolation and Anomaly Detection for Personal $PM_{2.5}$ Measurement
Reprinted from: *Appl. Sci.* **2020**, *10*, 543, doi:10.3390/app10020543 . . . . **165**

A. Stamatelopoulou, M. Dasopoulou, K. Bairachtari, S. Karavoltsos, A. Sakellari and T. Maggos
Contamination and Potential Risk Assessment of Polycyclic Aromatic Hydrocarbons (PAHs) and Heavy Metals in House Settled Dust Collected from Residences of Young Children
Reprinted from: *Appl. Sci.* **2021**, *11*, 1479, doi:10.3390/app11041479 . . . . **179**

Konstantinos G. Koukoulakis, Panagiotis George Kanellopoulos, Eirini Chrysochou, Danae Costopoulou, Irene Vassiliadou, Leondios Leondiadis and Evangelos Bakeas
Atmospheric Concentrations and Health Implications of PAHs, PCBs and PCDD/Fs in the Vicinity of a Heavily Industrialized Site in Greece
Reprinted from: *Appl. Sci.* **2020**, *10*, 9023, doi:10.3390/app10249023 . . . . **195**

John F. Joseph, Chad Furl, Hatim O. Sharif, Thankam Sunil and Charles G. Macias
Towards Improving Transparency of Count Data Regression Models for Health Impacts of Air Pollution
Reprinted from: *Appl. Sci.* **2021**, *11*, 3375, doi:10.3390/app11083375 . . . . **219**

About the Editor

Thomas Maggos

Thomas Maggos M.Sc., Ph.D. is a research director and Head of the Atmospheric Chemistry and Innovative Technologies Lab (AirTech-Lab) at the National Centre for Scientific Research "Demokritos". Following his basic studies in chemistry at the University of Crete, he joined the University of Athens where he received his M.Sc in environmental chemistry and technology. He received his Ph.D. in the Mechanical Engineering department of the University of West Macedonia. In 1998, he joined NCSRD, working on innovative technological systems for air quality assessments and on photocatalytic processes for abatement of air pollutants. He is a coordinator and principal investigator in European and National R&D projects and a member of CEN/TC working groups. He is in charge of the accreditation of AirTech-Lab under the terms of ISO 17025:2017, the deputy coordinator to the Indoor TG of ERNCIP and a deputy member of the management board of COST17136. He has more than 80 publications in journals and more than 120 in conferences.

Editorial

Advances in Air Quality Monitoring and Assessment

Thomas Maggos

Atmospheric Chemistry and Innovative Technologies Lab., NCSR "Demokritos", 15310 Athens, Greece; tmaggos@ipta.demokritos.gr; Tel.: +30-210-6503716

1. Introduction

Air quality monitoring is a long-term assessment of pollutant levels that helps to assess the extent of pollution and provide information about air quality trends. Furthermore, an air quality monitoring system (AQMS) supports research by providing the information necessary for scientists to perform long-term studies of population exposure to various atmospheric substances and generally estimate the health effects of air pollution. In addition, an advanced AQMS could make useful information available to policy makers and planners in order to help make informed decisions about managing and improving air quality by better understanding the sources of air pollution. The purpose of this Special Issue is to provide an overview of recent advances in environmental monitoring and assessment, which includes the design, development and application of advanced monitoring systems based on cutting edge scientific knowledge.

2. Current and Future Challenges in Air Quality Monitoring

In this Special Issue, 23 papers were submitted and 13 were accepted for publication (57% acceptance rate). Forty-six percent of the published studies originated from Asian countries, 31% of them were conducted in European countries and 23% were done in the USA and South America. Various topics were addressed in the contributed articles, which can be distinguished into two main groups: (i) development of emerging AQ monitoring systems and methodologies and (ii) evaluation and modeling of AQMN data in terms of AQ and health impact assessment. A quite smaller group included studies on the improvement of methodological approaches to analyzing AQ data.

In the first group, Paralikis et al. [1] developed Al-doped NiO films which can potentially be used as a sensing element for ozone gas sensors. The gas-sensing performance of the film for ozone was studied at different operating temperatures and was able to detect ozone at an ultra-low concentration of 10 ppb. Davidović et al. [2] evaluated the changes in air pollution in Serbia due to the COVID-19 pandemic using data from permanently operating air quality monitoring stations as well as by deploying low-cost particulate matter (PM) sensors. Beyond the useful outcomes for the improvement of air quality due to the reduction of transport and industrial activities, the study confirmed the low-cost PM sensors' usefulness in air quality assessment, as they increase spatial resolution. It also pointed out the necessity to calibrate them and follow the QA/QC protocols in order to verify their reliability. Following a similar methodological approach, Yiniva Camargo-Caicedo et al. [3] observed the changes in air quality using data from an air quality network and from the Ozone Monitoring Instrument (OMI) satellite in order to estimate improvements in air quality in Colombia due to COVID-19 pandemic lockdown. Furthermore, emissions from road transportation of four groups of pollutants (greenhouse gases, ozone precursor gases, aerosols and acidifying gases) before and during the lockdown were estimated and compared. The results could serve decision makers in adopting strategies to improve air quality related to the transportation sector. Regarding this sector, Li et al. [4] presented and analyzed data from four portable emissions measurement system (PEMS) tests of heavy-duty vehicles. More specifically, they analyzed the NOx emission of urban, rural and motorway sections and calculated the moving averaging window (MAW) NOx emission under the required boundary conditions. Finally, they explored the proper methods to evaluate real-world NOx emission based on the MAW method. It is worth noting that the study pointed out the insufficiency of the current evaluation method for real-world NOx emission of heavy-duty vehicles, indicating where the research community should focus future studies. It is obvious from the previous studies that an air quality monitoring network (AQMN) plays an important role in air pollution management. However, setting up an initial network in a city often lacks the necessary information, such as historical pollution and geographical data, which makes establishing an effective network challenging. Meanwhile, cities with an existing one do not adequately represent spatial coverage of air pollution issues or face rapid urbanization, where additional stations are needed. To resolve the two cases, Athita Onuean et al. [5] proposed four methods for finding stations and constructing a network. They introduced and applied a coverage percentage and weighted coverage degree for evaluating the results from the proposed methods that will be implemented as a guide for establishing a new network and can be a tool for improving spatial coverage of an existing network for future expansions in air monitoring.

In the second group, Afifa Aslam et al. [6] investigated the concentration level of PM2.5 and PM10 as well as their carbonaceous fraction, including organic carbon (OC), elemental carbon (EC) and total carbon (TC) from samples collected from five different sectors in Pakistan. It is well-known that studying the chemical composition of particulate matter (PM) provides an opportunity to conduct additional studies on source identification, impact assessment and trend analysis. Furthermore, Cho et al. [7] have shown that quantitative assessments of chemical and biological properties of ambient PM2.5 and VOCs can be used effectively to characterize, compare and contrast air pollution across different geographical regions (Los Angeles basin) to account for effects of atmospheric modifications on air mass and to evaluate exposure proximity to an emission source. Observational data from city AQ monitoring stations are usually analyzed and used for numerical simulations in order to evaluate the impact of emission control scenarios. To that end, Zhan et al. [8] analyzed the air quality observational data of major air pollutants in 2015 and during pollution episodes in Haizhu district, China, and the impacts of emission control scenarios on air quality by the year 2020 were evaluated using a WRF-Chem numerical simulation. In the same direction, Li et al. [9] investigated the pollution characteristics, transport pathways and potential sources of PM2.5 based on PM2.5 monitoring data from 2015 to 2016 in Weifang, China. For that purpose, they used three methods: Hybrid Single-Particle Lagrangian Integrated Trajectory (HYSPLIT), the potential source contribution function (PSCF) and concentration weighted trajectory (CWT). Nowadays, fine dust data acquired by various personal monitoring devices is of great value as training data for predicting future fine dust concentrations and innovatively alerting people of potential danger. However, most of the fine dust data obtained from these devices include either missing or abnormal data caused by various factors such as sensor malfunction, transmission errors or storage errors. Park et al. [10] presented methods for interpolating the missing data and detecting anomalies in PM2.5 time series data. These methods are expected to contribute greatly to improving the reliability of data.

Air pollution data obtained from various monitoring campaigns are usually used for health impact assessment. Stamatelopoulou et al. [11] examined the concentrations and sources of PAHs and trace metals in indoor dust and, more specifically, focused on residences with infants and young children. Exposure to toxicants contained in house-settled dust is of paramount concern, especially in the case of young children, due to their particular behavioral characteristics. In this context, extracts of sieved vacuum cleaner dust from 20 residences with young children in Athens, Greece were examined for the presence of PAHs and trace metals. Outdoor environment and, more specifically, industrialized areas also play a significant role in citizens' health. To that end, Koukoulakis et al. [12] simultaneously monitored PAHs, PCDD/Fs, dlPCBs and indPCBs bonded to particulate matter (PM10) for the estimation of their health risks to nearby citizens. SPSS statistical package was employed for statistical analysis and source apportionment purposes. Cancer risk was also estimated from total persistent organic pollutants (POPs) dataset according to the available literature. Specific attention should be given to studies on the health impacts of air pollution where regression analysis is used. The complexity of count data regression models can lead to false inference and overfitting. Joseph et al. [13] presented a simple histogram of predicted and observed count values (POCH) which, while rarely found in the environmental literature but presented in authoritative statistical texts, can dramatically reduce the risk of accepting untrue hypotheses.

3. Conclusions

In conclusion, the papers in this Special Issue have highlighted two thematic areas of AQ monitoring that researchers currently focus on: (i) the improvement of AQ monitoring methods and (ii) the use of AQ data in order to better assess the impact of air pollution to the environment and health. While this Special Issue has been closed, further research towards these directions is expected shortly, as there are still several challenging research questions to be answered.

Funding: This research received no external funding.

Acknowledgments: This issue would not be possible without the contributions of various talented authors, hardworking and professional reviewers and the dedicated editorial team of Applied Sciences. Congratulations to all authors—no matter what the final decisions of the submitted manuscripts were, the feedback, comments and suggestions from the reviewers and editors substantially helped the authors to improve their papers. Finally, I place on record my gratitude to the editorial team of Applied Sciences, and special thanks to Sara Zhan, Assistant Editor from MDPI Branch Office, Wuhan.

Conflicts of Interest: The author declares no conflict of interest.

References

1. Paralikis, A.; Gagaoudakis, E.; Kampitakis, V.; Aperathitis, E.; Kiriakidis, G.; Binas, V. Study on the Ozone Gas Sensing Properties of rf-Sputtered Al-Doped NiO Films. *Appl. Sci.* **2021**, *11*, 3104. [CrossRef]
2. Davidović, M.; Dmitrašinović, S.; Jovanović, M.; Radonić, J.; Jovašević-Stojanović, M. Diurnal, Temporal and Spatial Variations of Main Air Pollutants Before and during Emergency Lockdown in the City of Novi Sad (Serbia). *Appl. Sci.* **2021**, *11*, 1212. [CrossRef]
3. Camargo-Caicedo, Y.; Mantilla-Romo, L.; Bolaño-Ortiz, T. Emissions Reduction of Greenhouse Gases, Ozone Precursors, Aerosols and Acidifying Gases from Road Transportation during the COVID-19 Lockdown in Colombia. *Appl. Sci.* **2021**, *11*, 1458. [CrossRef]
4. Li, P.; Lü, L. Evaluating the Real-World NOx Emission from a China VI Heavy-Duty Diesel Vehicle. *Appl. Sci.* **2021**, *11*, 1335. [CrossRef]
5. Onuean, A.; Jung, H.; Chinnasarn, K. Finding Optimal Stations Using Euclidean Distance and Adjustable Surrounding Sphere. *Appl. Sci.* **2021**, *11*, 848. [CrossRef]
6. Aslam, A.; Ibrahim, M.; Shahid, I.; Mahmood, A.; Irshad, M.; Yamin, M.; Ghazala; Tariq, M.; Shamshiri, R. Pollution Characteristics of Particulate Matter (PM2.5 and PM10) and Constituent Carbonaceous Aerosols in a South Asian Future Megacity. *Appl. Sci.* **2020**, *10*, 8864. [CrossRef]
7. Cho, A.; Shinkai, Y.; Schmitz, D.; Di Stefano, E.; Eiguren-Fernandez, A.; Guarieiro, A.; Salinas, E.; Froines, J.; Melega, W. Chemical and Biological Characterization of Particulate Matter (PM 2.5) and Volatile Organic Compounds Collected at Different Sites in the Los Angeles Basin. *Appl. Sci.* **2020**, *10*, 3245. [CrossRef]
8. Zhan, J.; Wang, M.; Liu, Y.; Feng, C.; Gan, T.; Li, L.; Ou, R.; Ding, H. Impact of the '13th Five-Year Plan' Policy on Air Quality in Pearl River Delta, China: A Case Study of Haizhu District in Guangzhou City Using WRF-Chem. *Appl. Sci.* **2020**, *10*, 5276. [CrossRef]
9. Li, C.; Dai, Z.; Liu, X.; Wu, P. Transport Pathways and Potential Source Region Contributions of PM2.5 in Weifang: Seasonal Variations. *Appl. Sci.* **2020**, *10*, 2835. [CrossRef]
10. Park, J.; Kim, S. Improved Interpolation and Anomaly Detection for Personal PM2.5 Measurement. *Appl. Sci.* **2020**, *10*, 543. [CrossRef]
11. Stamatelopoulou, A.; Dasopoulou, M.; Bairachtari, K.; Karavoltsos, S.; Sakellari, A.; Maggos, T. Contamination and Potential Risk Assessment of Polycyclic Aromatic Hydrocarbons (PAHs) and Heavy Metals in House Settled Dust Collected from Residences of Young Children. *Appl. Sci.* **2021**, *11*, 1479. [CrossRef]

12. Koukoulakis, K.; Kanellopoulos, P.; Chrysochou, E.; Costopoulou, D.; Vassiliadou, I.; Leondiadis, L.; Bakeas, E. Atmospheric Concentrations and Health Implications of PAHs, PCBs and PCDD/Fs in the Vicinity of a Heavily Industrialized Site in Greece. *Appl. Sci.* **2020**, *10*, 9023. [CrossRef]
13. Joseph, J.; Furl, C.; Sharif, H.; Sunil, T.; Macias, C. Towards Improving Transparency of Count Data Regression Models for Health Impacts of Air Pollution. *Appl. Sci.* **2021**, *11*, 3375. [CrossRef]

… Article

Study on the Ozone Gas Sensing Properties of rf-Sputtered Al-Doped NiO Films

Athanasios Paralikis [1,2], Emmaouil Gagaoudakis [1,*], Viktoras Kampitakis [1,2], Elias Aperathitis [1], George Kiriakidis [1] and Vassilios Binas [1,2,*]

1. Institute of Electronic Structure & Laser, Foundation for Research and Technology (FORTH-IESL), 71100 Herakleion, Greece; ph4888@edu.physics.uoc.gr (A.P.); vkabitakis@physics.uoc.gr (V.K.); eaper@iesl.forth.gr (E.A.); kiriakid@iesl.forth.gr (G.K.)
2. Department of Physics, University of Crete, 70013 Heraklion, Greece
* Correspondence: mgagas@iesl.forth.gr (E.G.); binasbill@iesl.forth.gr (V.B.); Tel.: +30-2810391272 (E.G.); +30-2810391269 (V.B.)

Featured Application: Authors are encouraged to provide a concise description of the specific application or a potential application of the work. This section is not mandatory.

Abstract: Al-doped NiO (NiO:Al) has attracted the interest of researchers due to its excellent optical and electrical properties. In this work, NiO:Al films were deposited on glass substrates by the radio frequencies (rf) sputtering technique at room temperature and they were tested against ozone gas. The Oxygen content in (Ar-O_2) plasma was varied from 2% to 4% in order to examine its effect on the gas sensing performance of the films. The thickness of the films was between 160.3 nm and 167.5 nm, while the Al content was found to be between 5.3 at% and 6.7 at%, depending on the oxygen content in plasma. It was found that NiO:Al films grown with 4% O_2 in plasma were able to detect 60 ppb of ozone with a sensitivity of 3.18% at room temperature, while the detection limit was further decreased to 10 ppb, with a sensitivity of 2.54%, at 80 °C, which was the optimum operating temperature for these films. In addition, the films prepared in 4% O_2 in plasma had lower response and recovery time compared to those grown with lower O_2 content in plasma. Finally, the role of the operating temperature on the gas sensing properties of the NiO:Al films was investigated.

Keywords: p-type sensor; Al-doped NiO; rf sputtering; ozone gas sensing

1. Introduction

Ozone (O_3) is a well-known harmful gas existing in the atmosphere as a product of photochemical reactions of Nitrogen dioxide (NO_2) and/or Volatile Organic Compounds (VOCs), which are very common environmental pollutants that come from industrial activity, cars, etc. [1]. As a result, over 80% of people live in cities where air pollution is higher than the safety limits of the World Health Organization (WHO) [2]. Taking into account that ozone is associated with various respiratory symptoms, including dyspnea, upper airway irritation, coughing, and chest tightness [1], the need of its detection in outdoor as well as indoor environments becomes of paramount importance.

Various kinds of materials, such as Metal Oxide Semiconductors (MOS) [3–5], inorganic perovskites [6], as well as hybrid perovskites [7] have been examined as gas sensing elements for ozone detection during the last decades. Among them, MOS are by far the most well studied materials for ozone gas sensing applications due to their excellent electrical and optical properties as well as the fact that they can be grown by a number of methods even at large scale [8,9]. Nickel oxide (NiO) is a p-type metal oxide semiconductor with a wide energy band gap of 3.4–3.8 eV that has interesting optoelectronic properties [10]. Thus, it can be used in UV photo-detectors [11] as an active material in photovoltaics, such as perovskite or dye-sensitized solar cells [12–16], as well as in electrochromic [17], thermoelectric [18], and gas sensing devices for the detection of gases, such as H_2 [19,20], CH_4 [21], NO_2 [22], ethanol [23], O_3 [24], etc. Furthermore, the introduction of Aluminum (Al) atoms in the NiO structure seems to enhance its electrical as well as optical properties [25], while not affecting its gas sensing performance. As a result, Al-doped NiO films have been already tested against NO_2 [26,27], ethanol [28], CO [29], H_2, and CH_4 [30]. However, NiO:Al films have not been examined for ozone detection, according to our literature research.

In this work, rf-sputtered Al-doped NiO films were tested, for first time, against ozone. The effect of Oxygen content in Ar-O_2 plasma as well as the operating temperature on the gas sensing performance were studied. It was found that NiO:Al films have the ability to detect ozone even at room temperature, while those prepared with 4% O_2 in plasma successfully detected ozone at an ultra-low concentration of 10 ppb at 80 °C. The latter is one of the lower operating temperatures for metal oxide gas sensors.

2. Materials and Methods

2.1. Deposition Conditions

A Nordiko RFG2500 rf sputtering system was employed to grow NiO:Al films using a Ni metal target and Ar-O_2 air mixture in plasma. Two (2) Al pellets were placed on the Ni target surface for Al-doping. Both pressure and sputtering power were kept constant during deposition, being 5 mTorr and 300 W, respectively. The films were deposited on three different substrates, namely the commercial (Metrohm/DropSens) InterDigitated Transducers (glass substrate, Pt electrodes, bands/gaps = 5 μm) for sensing measurements, a piece of silicon wafer for structural characterization, and microscope glass for optical measurements. All depositions were done at room temperature (RT). The O_2 content in plasma was varied, being 2.0%, 2.8%, and 4.0% in order to examine its effect on the gas sensing characteristics of the films. After deposition, the films underwent thermal annealing at 400 °C under air for 24 h, in order to improve both their crystallinity and optical properties.

2.2. Characterization

The structure of the Al-doped NiO films was examined by the X-ray Diffraction (XRD) technique, using a PANalyticalEmpyrean diffractometer equipped with Cu-LFF as an X-Ray source, at λ = 0.15406 nm. The measurement mode was 2θ/θ varying from 20° to 90°, with a step of 0.013°/s. Using the XRD pattern the crystallite size (D) of the films was calculated according to Scherrer's Equation (1)

$$D(nm) = (0.9 \cdot \lambda)/(B \cdot \cos\theta) \tag{1}$$

where λ is the X-ray wavelength which equals to 0.154 nm. B is the Full Width at Half Maximum (FWHM) of the corresponding peak at an angle 2θ. Moreover, the lattice constant (a_0) for the cubic structure of NiO was calculated according to Equation (2)

$$a = (h^2 + k^2 + \ell^2)^{1/2}/d \tag{2}$$

where d is the distance between the adjacent planes in the set (hkℓ). Field Emission Surface Electron Microscopy (FESEM) was used to investigate the surface morphology of the films using a Hitachi S570 microscope, equipped with an Energy Dispersive X-ray (EDX) spectrometry system, which was employed to determine the Al at% of the NiO:Al films. Optical properties of the films were studied with a Perkin Elmer Lambda 950 UV/Vis/NIR spectrophotometer in the wavelength range of 250–2500 nm. The transmittance spectra and the corresponding Tauc plot were used in order to calculate the optical energy bandgap of the films assuming that direct transitions are permitted according to Equation (3)

$$(\alpha h \nu) = A (h\nu - E_g)^{1/2} \tag{3}$$

where α is the absorption coefficient, hv the photon energy, A is a constant and E_g the optical energy bandgap. The thickness of the films was measured by using a Veeco Dektak 150 stylus profilometer.

2.3. Gas Sensing

In order to investigate the gas sensing performance in ozone, the NiO:Al films deposited on IDTs were placed in a homemade stainless steel test chamber. A mechanical pump was used to initially evacuate the chamber, while an FP-400 temperature controller was employed to regulate the operating temperature. Ozone of different concentrations was produced by an ozone analyzer (Thermo, Model 49i, Thermofisher scientific, Waltham, MA, USA). A Keithley 6517A electrometer was used to apply a constant voltage on the films and measure the electrical current variations upon the interaction between the film and the target gas. The chamber was initially evacuated at a pressure of 1 mbar. Prior to the ozone exposure, the films were treated with synthetic (dry) air for 45 min in order to be stabilized in the testing environment. After that, the films were exposed to ozone of specific concentrations for a constant time duration in which the electrical current was increased, while the recovery of the films in its initial state was done by exposing them in synthetic air for a specific time period. Both ozone and synthetic air were introduced in the chamber with a flow of 200 sccm (standard cubic centimeters per minute) regulated by mass flow controllers, while the pressure in the chamber was kept constant of 700 mbar. The whole procedure was monitored by a PC through an appropriate LabVIEW program. The sensitivity (S) of the sensor is defined by the following Equation (4)

$$S(\%) = [(I_{gas} - I_{air})/I_{air}] \cdot 100\% \qquad (4)$$

where I_{gas} is the maximum value of electrical current in the presence of ozone, while I_{air} is the minimum value of electrical current in the presence of synthetic air (absence of ozone), as depicted in Figure 1. Moreover, the response time (t_{resp}) is defined as the time that is needed for the electrical current to become equal to the 90% of the value of I_{gas} in the presence of ozone, while the recovery time (t_{rec}) is defined as the required time for the electrical current to be equal to 10% of the value of I_{gas} in the absence of ozone (in the presence of synthetic air).

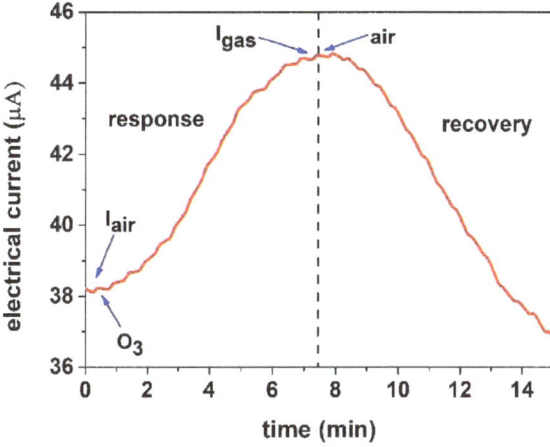

Figure 1. Typical sensing measurement of an O_3-synthetic air cycle.

All measurements were taken under the room relative humidity (RH) which was about 60%.

3. Results and Discussion

3.1. Characterization

The XRD patterns of rf-sputtered Al-doped NiO films deposited on Si substrates at room temperature with different O_2 content in plasma that underwent thermal annealing at 400 °C for 24 h are presented in Figure 2. It can be seen that all films had a polycrystalline structure with a peak of high intensity at $2\theta = 43.3°$ corresponding to a reflection from the (200) planes and two low intensity peaks at $2\theta = 36.5°$ and $62.9°$ corresponding to reflections from the (111) and (220) planes, respectively, according to JCPDS Card No. 047-1049. No peaks corresponding to other materials were found, indicating that all Al atoms were introduced in the NiO structure. Using Equation (1) for the diffraction peak with the highest intensity (200), the crystallite size of the films was calculated (Table 1) and found to be between 8.68 nm for films prepared with 2% and 2.8% O_2 in plasma and 7.89 nm for those prepared with 4% O_2 in plasma. In the case of the films prepared with 4% O_2, the decrease of the crystallite size may lead to a further increase of defect states inside the crystal structure, which can also be observed through the intensity of the main peak. In addition, the lattice constant (a_0) of the films was calculated from Equation (2) and was varied between 0.417 nm and 0.421 nm (Table 1). Both the crystallite size and lattice constant are in agreement with the values reported in the literature for Al-doped NiO films prepared by various techniques [25,31].

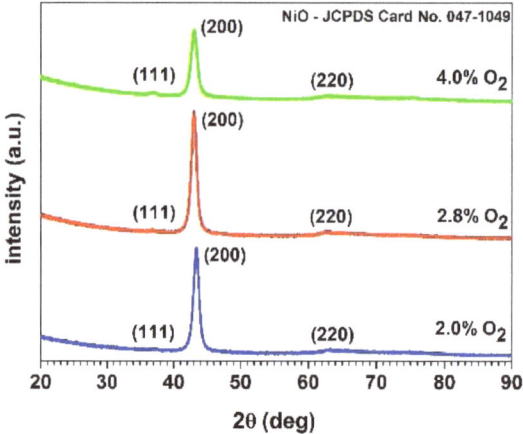

Figure 2. X-ray Diffraction (XRD) patterns of rf-sputtered NiO:Al films grown at room temperature with different O_2 content in plasma that underwent thermal annealing at 400 °C for 24 h.

Table 1. Deposition parameters, thickness (t), crystallite size (D), lattice constant (a_0), and optical energy bandgap (E_g) of rf-sputtered Al-doped NiO films.

Sample No.	% O_2 in Plasma	%at. Al	t (nm)	D (nm)	a_0 (nm)	E_g (eV)
S1749	2.0	5.3	160.3	8.68	0.417	3.66
S1746	2.8	6.5	168.7	8.68	0.421	3.65
S1743	4.0	6.7	167.5	7.89	0.420	3.66

The surface morphology of the NiO:Al films is presented in Figure 3a–c. It can be seen that all films had a smooth and uniform surface consisting of grains with a diameter of around 40 nm The increased defects on the surface of the film prepared with 4% O_2 can be also be observed.

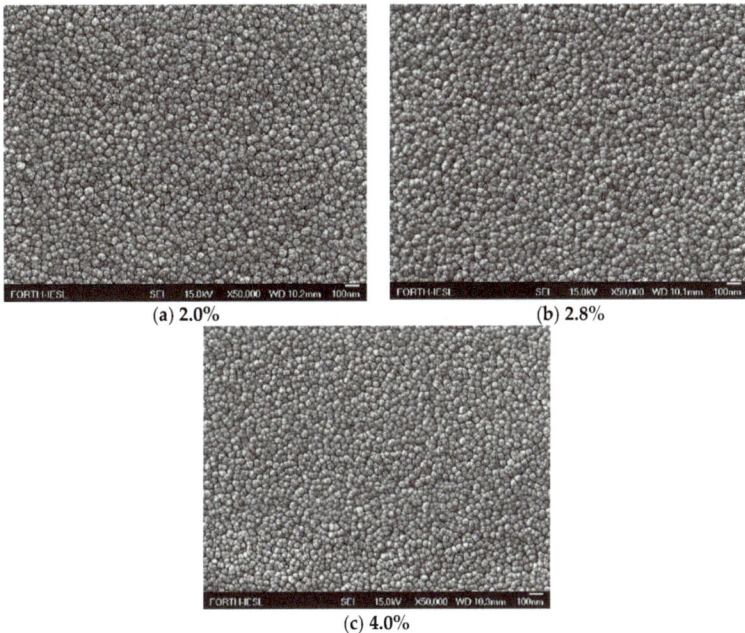

Figure 3. SEM images of rf-sputtered NiO:Al films grown at room temperature with (**a**) 2.0%, (**b**) 2.8%, and (**c**) 4.0% O_2 content in plasma that underwent thermal annealing at 400 °C for 24 h.

The transmittance spectra of the Al-doped NiO films are shown in Figure 4. All films were highly transparent (>70%) in the visible region as well as in the near infrared one. Using Equation (3) and the Tauc plot (inset of Figure 4), the optical energy bandgap was calculated and found to be around 3.66 eV (Table 1), comparable with the values reported by Chattopadhyay et al. for Al-doped NiO films grown by the rf sputtering technique [25].

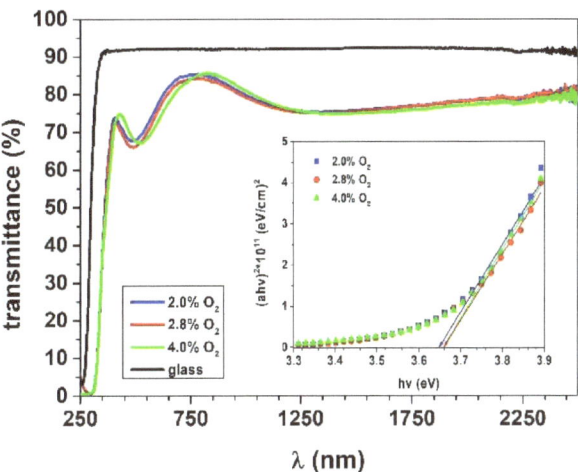

Figure 4. Transmittance spectra of rf-sputtered NiO:Al films grown at room temperature with different O_2 content in plasma that underwent thermal annealing at 400 °C for 24 h. Inset: Tauc plot to determine the optical energy band gap of the films.

3.2. Gas Sensing

The gas sensing performance of rf-sputtered Al-doped NiO films was investigated by exposing them to ozone gas with a concentration varying from 10 ppb to 3000 ppb, at operating temperatures between RT and 150 °C. The duration in ozone exposure was 5 min for films prepared with 4% O_2 in plasma and 10 min for the films prepared with 2% and 2.8% O_2 in plasma, while the recovery of the sensors was carried out by exposing them to synthetic air for 10 min and 15 min, respectively. The applied voltage was 10 V when the operating temperature equals to RT, while for the rest of the operating temperatures the applied voltage was 1 V.

A typical measurement of electrical current variation for different O_3 concentrations is presented in Figure 5. In the presence of ozone the electrical current of the film is increased reaching a maximum value, while in the absence of the target gas the electrical current of the film is decreased, as a result of its interaction with synthetic air, reaching its initial value. Moreover, it is noted that the sensor cannot be fully recovered to its initial electrical current value, especially at high concentrations (>600 ppb), probably due to residual oxidation. The gas sensing mechanism for Al-doped NiO films is the typical one for a p-type semiconductor. Thus, upon the exposure to synthetic air, Oxygen species (O$^-$, O^{2-}, etc.) are formed as a result of the O_2 adsorption on the films' surface. This is described through the following Equations (5)–(7).

$$O_{2(gas)} + e^- \rightarrow O_{2(adsorbed)}^- \tag{5}$$

$$O_{2(gas)} + 2e^- \rightarrow 2O_{(adsorbed)}^- \tag{6}$$

$$O_{2(adsorbed)}^- + e^- \rightarrow O_{(adsorbed)}^{2-} \tag{7}$$

where the kind of oxygen ions is dependent on the operating temperature (T_{oper}), being O_2^- for T_{oper} < 100 °C, O^- for 100 °C < T_{oper} < 300 °C, and O^{2-} for T_{oper} > 300 °C. The Oxygen species take out electrons from NiO, increasing the number of holes, leading to an increase of electrical conductivity, thus a depletion layer is formed. As a result, the height of the potential barrier at the grain boundaries is decreased and the electrical current is slightly increased. When NiO:Al films interact with ozone, which is a strongly oxidizing gas, the ozone molecules are adsorbed on the films' surface trapping the free electrons, resulting in an increase of the hole concentration as well as in a decrease of the thickness of the depletion layer. Thus, an extra increase of the electrical current is occurred [32,33].

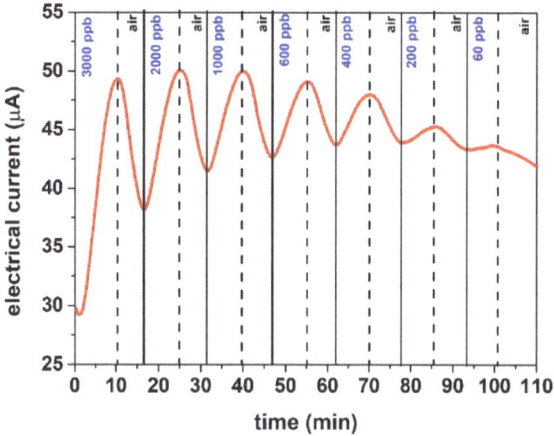

Figure 5. Electrical current variation of a NiO:Al film prepared with 4% O_2 in plasma, as a function of time for different O_3 concentrations, at 110 °C.

In Figure 6a–c the variation in sensitivity of the NiO:Al films grown with different O_2 content in plasma are shown. This was calculated by Equation (4) as a function of ozone concentration for various operating temperatures and plotted in logarithmic scale. In addition, a linear fitting has been applied confirming that the sensitivity is increased with O_3 concentration, independent of the Oxygen content in the plasma or the operating temperature. From Figure 6, it becomes clear that all films can detect the ozone gas even at an ultra-low concentration of 10 ppb. Specifically, the films prepared with 2% O_2 in plasma had a lowest detection limit of around 200 ppb with a sensitivity 1.88%, 11.31% and 38.66% at RT, 110 °C and 150 °C, respectively, while at 80 °C the lowest detection limit was 400 ppb with a sensitivity of 5.17%. Moreover, films synthesized with 2.8% O_2 in plasma had a lowest detection limit of 400 ppb [27] (sensitivity 0.84%), 200 ppb (sensitivity 6.25%), 200 ppb (sensitivity 6.94%) and 60 ppb (sensitivity 6.22%) when the operating temperature equals to RT, 80 °C, 110 °C, and 150 °C, respectively. Finally, the NiO:Al prepared with 4% O_2 in plasma appeared to have a lowest detection limit of 60 ppb (sensitivity 3.18%), 10 ppb (sensitivity 2.54%), 60 ppb (sensitivity 2.44%) and 400 ppb (sensitivity 9.02%) when the operating temperature equals to RT, 80 °C, 110 °C, and 150 °C. The detection limit of 10 ppb at 80 °C is one order of magnitude less than the one that has been reported [32] for Co_3O_4 nanobricks at almost the same working temperature (85 °C). The above mentioned values for all NiO:Al films are presented in Table 2.

Furthermore, in Figure 7 the effect of the operating temperature on gas sensing performance of the rf-sputtered NiO:Al films is shown. The films tested against 400 ppb ozone at various operating temperatures, namely RT, 80 °C, 110 °C, and 150 °C. It can be seen that films prepared with 4% O_2 in plasma appeared to have the best sensitivity at 80 °C, while the rest of them showed a maximum sensitivity at 150 °C. This can be attributed to both less Al content and smaller crystallite size of the films grown with 4% O_2 in plasma compared to the others, which also leads to a further decrease of the defect states that can reduce the adsorption energy of the O_3 gas molecules [34]. In addition, the operating temperature of 80 °C is one of the lowest that has been reported for ultra-low ozone concentration (10 ppb) [4,32].

Finally, the response and recovery time of the NiO:Al films for each operating temperature were calculated and presented in Figure 8a–c. Films grown with 2% and 2.8% O_2 in plasma had a response time between 400 s and 800 s, depending on both operating temperature and ozone concentration, while for those prepared with 4% O_2 in plasma the response time was ranging from 150 s to 500 s. In the same way, the recovery time for the former films was between 400 s and 900 s, while it was varied from 200 s to 600 s for films grown with 4% O_2 in plasma. It should be noticed that the Al: NiO films grown with 4% O_2 in plasma showed the best response and the recovery time equals to 189.6 s and 243.6 s, respectively, against 10 ppb ozone at an operating temperature of 80 °C. The response time values of Al: NiO films are higher than those in other reported works, while the recovery time values are comparable to those reported in the literature [4,32,34–38]. However, most of these sensors work at higher temperatures and/or have a higher concentration detection limit than those presented in this work.

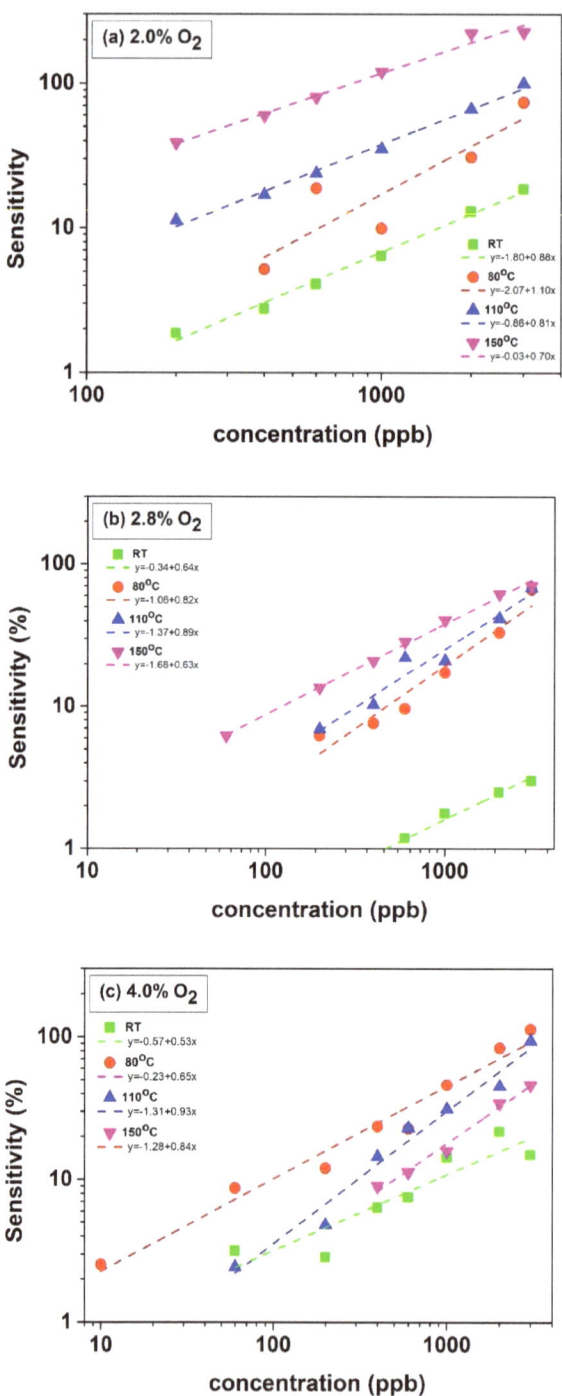

Figure 6. Sensitivity on different ozone concentrations at various operating temperatures of rf-sputtered NiO:Al with (**a**) 2.0%, (**b**) 2.8%, and (**c**) 4.0% O_2 in plasma. A linear fitting plot has been applied.

Table 2. Deposition parameters, temperature of operation (°C), minimum detection concentration (ppb), and sensitivity (S) of rf-sputtered Al-doped NiO films.

Sample No.	% O_2 in Plasma	T_{oper} (°C)	Minimum Detected Concentration (ppb)	S (%)
S1749	2.0	RT	213	1.88
		80	393	5.17
		110	213	11.31
		150	207	38.66
S1746	2.8	RT	393	0.84
		80	212	6.25
		110	215	6.94
		150	67	6.22
S1743	4.0	RT	58	3.18
		80	13	2.54
		110	59	2.44
		150	330	9.02

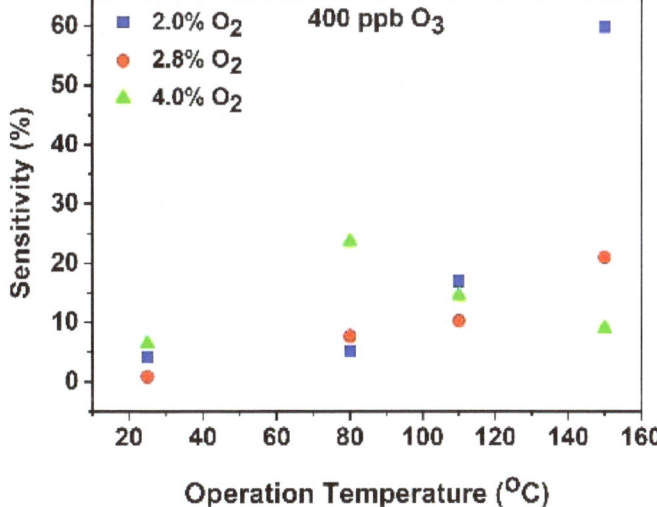

Figure 7. Sensitivity of rf-sputtered NiO:Al films grown with different O_2 content in plasma as a function of operating temperature.

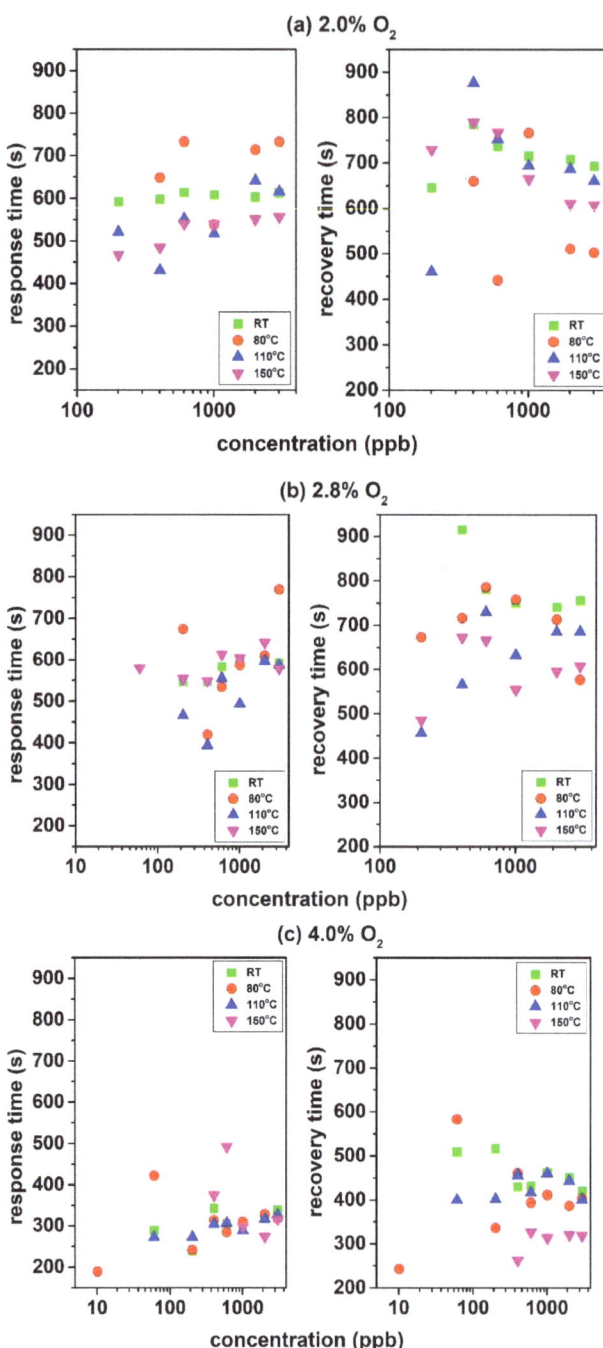

Figure 8. Response and recovery times at different operating temperatures of rf-sputtered NiO:Al with (**a**) 2.0%, (**b**) 2.8%, and (**c**) 4.0% O_2 in plasma as a function of ozone concentrations.

4. Conclusions

Al-doped NiO films were grown by the rf sputtering technique with different oxygen contents in Ar-O$_2$ plasma, i.e., 2%, 2.8%, and 4%, at room temperature. All films were polycrystalline as revealed by the XRD technique, while their surface was smooth as was confirmed by Scanning Electron Microscopy. Moreover, all films were highly transparent, showing a transmittance of more than 70% in the visible spectrum. The gas sensing performance of the films against ozone was studied at different operating temperatures, namely 25 °C, 80 °C, 110 °C, and 150 °C. All films were sensitive to ozone of different concentrations. Specifically, films grown with 4% O$_2$ in plasma were able to detect ozone at an ultra-low concentration of 10 ppb at a low operating temperature of 80 °C. Thus NiO:Al films can be potentially used as a sensing element for ozone gas sensors.

Author Contributions: Conceptualization, A.P. and V.K.; methodology, E.G.; data curation, E.G.; writing—original draft preparation, E.G.; writing—review and editing, V.B., G.K. and E.A.; supervision, E.G.; project administration, V.B. All authors have read and agreed to the published version of the manuscript.

Funding: Part of this work supported by the project "Quality of Life" (MIS 5002464) which is implemented under the "Action for the Strategic Development on the Research and Technological Sector", funded by the Operational Programme "Competitiveness, Entrepreneurship and Innovation" (NSRF 2014-2020) and co-financed by Greece and the European Union (European Regional Development Fund).

Institutional Review Board Statement: Not applicable.

Informed Consent Statement: Not applicable.

Data Availability Statement: Not applicable.

Conflicts of Interest: The authors declare no conflict of interest.

References

1. Huangfu, P.; Atkinson, R. Long-term exposure to NO$_2$ and O$_3$ and all-cause and respiratory mortality: A systematic review and meta-analysis. *Environ. Int.* **2020**, *144*, 105998. [CrossRef]
2. Barkjohn, K.K.; Norris, C.; Cui, X.; Fang, L.; Zheng, T.; Schauer, J.J.; Li, Z.; Zhang, Y.; Black, M.; Zhang, J.; et al. Real-time measurements of PM 2.5 and ozone to assess the effectiveness of residential indoor air filtration in Shanghai homes. *Indoor Air* **2021**, *31*, 74–87. [CrossRef]
3. Bejaoui, A.; Guerin, J.; Zapien, J.; Aguir, K. Theoretical and experimental study of the response of CuO gas sensor under ozone. *Sens. Actuators B Chem.* **2014**, *190*, 8–15. [CrossRef]
4. Rocha, L.; Foschini, C.; Silva, C.; Longo, E.; Simões, A. Novel ozone gas sensor based on ZnO nanostructures grown by the microwave-assisted hydrothermal route. *Ceram. Int.* **2016**, *42*, 4539–4545. [CrossRef]
5. Petromichelaki, E.; Gagaoudakis, E.; Moschovis, K.; Tsetseris, L.; Anthopoulos, T.D.; Kiriakidis, G.; Binas, V. Highly sensitive and room temperature detection of ultra-low concentrations of O$_3$ using self-powered sensing elements of Cu$_2$O nanocubes. *Nanoscale Adv.* **2019**, *1*, 2009–2017. [CrossRef]
6. Brintakis, K.; Gagaoudakis, E.; Kostopoulou, A.; Faka, V.; Argyrou, A.; Binas, V.; Kiriakidis, G.; Stratakis, E. Ligand-free all-inorganic metal halide nanocubes for fast, ultra-sensitive and self-powered ozone sensors. *Nanoscale Adv.* **2019**, *1*, 2699–2706. [CrossRef]
7. Kakavelakis, G.; Gagaoudakis, E.; Petridis, K.; Petromichelaki, V.; Binas, V.; Kiriakidis, G.; Kymakis, E. Solution Processed CH3NH3PbI3–xClxPerovskite Based Self-Powered Ozone Sensing Element Operated at Room Temperature. *ACS Sens.* **2018**, *3*, 135–142. [CrossRef] [PubMed]
8. Wetchakun, K.; Samerjai, T.; Tamaekong, N.; Liewhiran, C.; Siriwong, C.; Kruefu, V.; Wisitsoraat, A.; Tuantranont, A.; Phanichphant, S. Semiconducting metal oxides as sensors for environmentally hazardous gases. *Sens. Actuators B Chem.* **2011**, *160*, 580–591. [CrossRef]
9. Moseley, P.T. Progress in the development of semiconducting metal oxide gas sensors: A review. *Meas. Sci. Technol.* **2017**, *28*, 082001. [CrossRef]
10. Reddy, A.M.; Reddy, A.S.; Lee, K.-S.; Reddy, P.S. Effect of oxygen partial pressure on the structural, optical and electrical properties of sputtered NiO films. *Ceram. Int.* **2011**, *37*, 2837–2843. [CrossRef]
11. Echresh, A.; Chey, C.O.; Shoushtari, M.Z.; Khranovskyy, V.; Nour, O.; Willander, M. UV photo-detector based on p-NiO thin film/n-ZnO nanorods heterojunction prepared by a simple process. *J. Alloys Compd.* **2015**, *632*, 165–171. [CrossRef]

12. Bonomo, M.; Naponiello, G.; Venditti, I.; Zardetto, V.; Di Carlo, A.; Dini, D. Electrochemical and Photoelectrochemical Properties of Screen-Printed Nickel Oxide Thin Films Obtained from Precursor Pastes with Different Compositions. *J. Electrochem. Soc.* **2016**, *164*, H137–H147. [CrossRef]
13. Di Girolamo, D.; Matteocci, F.; Piccinni, M.; Di Carlo, A.; Dini, D. Anodically electrodeposited NiO nanoflakes as hole selective contact in efficient air processed p-i-n perovskite solar cells. *Sol. Energy Mater. Sol. Cells* **2020**, *205*, 110288. [CrossRef]
14. Di Girolamo, D.; Di Giacomo, F.; Matteocci, F.; Marrani, A.G.; Dini, D.; Abate, A. Progress, highlights and perspectives on NiO in perovskite photovoltaics. *Chem. Sci.* **2020**, *11*, 7746–7759. [CrossRef]
15. Bonomo, M.; Dini, D.; Decker, F. Electrochemical and Photoelectrochemical Properties of Nickel Oxide (NiO) With Nanostructured Morphology for Photoconversion Applications. *Front. Chem.* **2018**, *6*, 601. [CrossRef] [PubMed]
16. Bonomo, M.; Sheehan, S.; Dowling, D.P.; Gontrani, L.; Dini, D. First Evidence of Electrode Reconstruction in Mesoporous NiO After Operation as Photocathode of Dye-Sensitized Solar Cells. *ChemistrySelect* **2018**, *3*, 6729–6736. [CrossRef]
17. Sahu, D.; Lee, Y.-H.; Wu, T.-J.; Wang, S.-C.; Huang, J.-L. Synthesis and electrochromic property improvement of NiO films for device applications. *Thin Solid Films* **2020**, *707*, 138097. [CrossRef]
18. Sarkar, K.; Debnath, A.; Deb, K.; Bera, A.; Saha, B. Effect of NiO incorporation in charge transport of polyaniline: Improved polymer based thermoelectric generator. *Energy* **2019**, *177*, 203–210. [CrossRef]
19. Nakate, U.T.; Ahmad, R.; Patil, P.; Yu, Y.; Hahn, Y.-B. Ultra thin NiO nanosheets for high performance hydrogen gas sensor device. *Appl. Surf. Sci.* **2020**, *506*, 144971. [CrossRef]
20. Gagaoudakis, E.; Michail, G.; Kampylafka, V.; Tsagaraki, K.; Aperathitis, E.; Moschovis, K.; Binas, V.; Kiriakidis, G. Room Temperature p-Type NiO Nanostructure Thin Film Sensor for Hydrogen and Methane Detection. *Sens. Lett.* **2017**, *15*, 663–667. [CrossRef]
21. Zhang, S.; Li, Y.; Sun, G.; Zhang, B.; Wang, Y.; Cao, J.; Zhang, Z. Enhanced methane sensing properties of porous NiO nanaosheets by decorating with SnO_2. *Sens. Actuators B Chem.* **2019**, *288*, 373–382. [CrossRef]
22. Urso, M.; Leonardi, S.G.; Neri, G.; Petralia, S.; Conoci, S.; Priolo, F.; Mirabella, S. Room temperature detection and modelling of sub-ppm NO_2 by low-cost nanoporous NiO film. *Sens. Actuators B Chem.* **2020**, *305*, 127481. [CrossRef]
23. Su, C.; Zhang, L.; Han, Y.; Ren, C.; Chen, X.; Hu, J.; Zeng, M.; Hu, N.; Su, Y.; Zhou, Z.; et al. Controllable synthesis of crescent-shaped porous NiO nanoplates for conductometric ethanol gas sensors. *Sens. Actuators B Chem.* **2019**, *296*, 126642. [CrossRef]
24. Demin, V.S.; Krasovskii, A.N.; Lyudchik, A.M.; Pokatashkin, V.I.; Grigorishin, I.L.; Kudanovich, O.N. Measurement of ozone over a wide range of concentrations using semiconductor NiO gas sensors. *Meas. Tech.* **2008**, *51*, 1038–1044. [CrossRef]
25. Nandy, S.; Maiti, U.N.; Ghosh, C.K.; Chattopadhyay, K.K. Enhanced p-type conductivity and band gap narrowing in heavily Al doped NiO thin films deposited by RF magnetron sputtering. *J. Phys. Condens. Matter* **2009**, *21*, 115804. [CrossRef]
26. Wang, S.; Huang, D.; Xu, S.; Jiang, W.; Wang, T.; Hu, J.; Hu, N.; Su, Y.; Zhang, Y.; Yang, Z. Two-dimensional NiO nanosheets with enhanced room temperature NO_2 sensing performance via Al doping. *Phys. Chem. Chem. Phys.* **2017**, *19*, 19043–19049. [CrossRef] [PubMed]
27. Kampitakis, V.; Gagaoudakis, E.; Zappa, D.; Comini, E.; Aperathitis, E.; Kostopoulos, A.; Kiriakidis, G.; Binas, V. Highly sensitive and selective NO_2 chemical sensors based on Al doped NiO thin films. *Mater. Sci. Semicond. Process.* **2020**, *115*, 105149. [CrossRef]
28. Wang, C.; Cui, X.; Liu, J.; Zhou, X.; Cheng, X.; Sun, P.; Hu, X.; Li, X.; Zheng, J.; Lu, G. Design of Superior Ethanol Gas Sensor Based on Al-Doped NiO Nanorod-Flowers. *ACS Sens.* **2016**, *1*, 131–136. [CrossRef]
29. Abdul-Hussein, Y.M.; Ali, H.J.; Latif, L.A.; Abdulsattar, M.A.; Fadhel, H.M. Preparation of Al-Doped NiO Thin Films by Spray Pyrolysis Technique for CO Gas Sensing. *J. Adv. Pharm. Educ. Res.* **2019**, *9*, 1–6.
30. Gagaoudakis, E.; Michail, G.; Katerinopoulou, D.; Moschovis, K.; Iliopoulos, E.; Kiriakidis, G.; Binas, V.; Aperathitis, E. Transparent p-type NiO:Al thin films as room temperature hydrogen and methane gas sensors. *Mater. Sci. Semicond. Process.* **2020**, *109*, 104922. [CrossRef]
31. Siddique, M.N.; Ahmed, A.; Tripathi, P. Electric transport and enhanced dielectric permittivity in pure and Al doped NiO nanostructures. *J. Alloys Compd.* **2018**, *735*, 516–529. [CrossRef]
32. Liu, L.; Li, T.; Yi, Z.; Chi, F.; Lin, Z.; Zhang, X.; Xu, K. Conductometric ozone sensor based on mesoporous ultrafine Co_3O_4 nanobricks. *Sens. Actuators B Chem.* **2019**, *297*, 126815. [CrossRef]
33. Mokoena, T.P.; Swart, H.C.; Motaung, D.E. A review on recent progress of p-type nickel oxide based gas sensors: Future perspectives. *J. Alloys Compd.* **2019**, *805*, 267–294. [CrossRef]
34. Sun, Q.; Wu, Z.; Cao, Y.; Guo, J.; Long, M.; Duan, H.; Jia, D. Chemiresistive sensor arrays based on noncovalently functionalized multi-walled carbon nanotubes for ozone detection. *Sens. Actuators B Chem.* **2019**, *297*, 126689. [CrossRef]
35. Onofre, Y.J.; Catto, A.C.; Bernardini, S.; Fiorido, T.; Aguir, K.; Longo, E.; Mastelaro, V.R.; da Silva, L.F.; de Godoy, M.P. Highly selective ozone gas sensor based on nanocrystalline $Zn_{0.95}Co_{0.05}O$ thin film obtained via spray pyrolysis technique. *Appl. Surf. Sci.* **2019**, *478*, 347–354. [CrossRef]
36. Wu, C.-H.; Jiang, G.-J.; Chang, K.-W.; Deng, Z.-Y.; Li, Y.-N.; Chen, K.-L.; Jeng, C.-C. Analysis of the Sensing Properties of a Highly Stable and Reproducible Ozone Gas Sensor Based on Amorphous In-Ga-Zn-O Thin Film. *Sensors* **2018**, *18*, 163. [CrossRef] [PubMed]
37. Catto, A.C.; Da Silva, L.F.; Ribeiro, C.; Bernardini, S.; Aguir, K.; Longo, E.; Mastelaro, V.R. An easy method of preparing ozone gas sensors based on ZnO nanorods. *RSC Adv.* **2015**, *5*, 19528–19533. [CrossRef]
38. Joshi, N.; Da Silva, L.F.; Jadhav, H.; M'Peko, J.-C.; Torres, B.B.M.; Aguir, K.; Mastelaro, V.R.; Oliveira, O.N. One-step approach for preparing ozone gas sensors based on hierarchical $NiCo_2O_4$ structures. *RSC Adv.* **2016**, *6*, 92655–92662. [CrossRef]

Article

Diurnal, Temporal and Spatial Variations of Main Air Pollutants Before and during Emergency Lockdown in the City of Novi Sad (Serbia)

Miloš Davidović [1,*], Sonja Dmitrašinović [2], Maja Jovanović [1], Jelena Radonić [2] and Milena Jovašević-Stojanović [1]

1. VINČA Institute of Nuclear Sciences—National Institute of th Republic of Serbia, University of Belgrade, 11000 Belgrade, Serbia; majaj@vin.bg.ac.rs (M.J.); mjovst@vin.bg.ac.rs (M.J.-S.)
2. Department of Environmental Engineering and Occupational Safety and Health, Faculty of Technical Sciences, University of Novi Sad, 21000 Novi Sad, Serbia; dmitrasinovic@uns.ac.rs (S.D.); jelenaradonic@uns.ac.rs (J.R.)
* Correspondence: davidovic@vin.bg.ac.rs

Citation: Davidović, M.; Dmitrašinović, S.; Jovanović, M.; Radonić, J.; Jovašević-Stojanović, M. Diurnal, Temporal and Spatial Variations of Main Air Pollutants Before and during Emergency Lockdown in the City of Novi Sad (Serbia). *Appl. Sci.* **2021**, *11*, 1212. https://doi.org/10.3390/app11031212

Academic Editor: Thomas Maggos
Received: 13 December 2020
Accepted: 18 January 2021
Published: 28 January 2021

Publisher's Note: MDPI stays neutral with regard to jurisdictional claims in published maps and institutional affiliations.

Copyright: © 2021 by the authors. Licensee MDPI, Basel, Switzerland. This article is an open access article distributed under the terms and conditions of the Creative Commons Attribution (CC BY) license (https://creativecommons.org/licenses/by/4.0/).

Abstract: Changes in air pollution in the region of the city of Novi Sad due to the COVID-19 induced state of emergency were evaluated while using data from permanently operating air quality monitoring stations belonging to the national, regional, and local networks, as well as ad hoc deployed low-cost particulate matter (PM) sensors. The low-cost sensors were collocated with reference gravimetric pumps. The starting idea for this research was to determine if and to what extent a massive change of anthropogenic activities introduced by lockdown could be observed in main air pollutants levels. An analysis of the data showed that fine and coarse particulate matter, as well as SO_2 levels, did not change noticeably, compared to the pre-lockdown period. Isolated larger peaks in PM pollution were traced back to the Aralkum Desert episode. The reduced movement of vehicles and reduced industrial and construction activities during the lockdown in Novi Sad led to a reduction and a more uniform profile of the $PM_{2.5}$ levels during the period between morning and afternoon air pollution peak, approximately during typical working hours. Daily profiles of NO_2, NO, and NO_X during the state of emergency proved lower levels during most hours of the day, due to restrictions on vehicular movement. CO during the state of the emergency mainly exhibited a lower level during night. Pollutants having transportation-dominated source profiles exhibited a decrease in level, while pollutants with domestic heating source profiles mostly exhibited a constant level. Considering local sources in Novi Sad, slight to moderate air quality improvement was observed after the lockdown as compared with days before. Furthermore, PM low-cost sensors' usefulness in air quality assessment was confirmed, as they increase spatial resolution, but it is necessary to calibrate them at the deployment location.

Keywords: PM and gaseous air pollutants; air pollution monitoring; low-cost PM sensors; sensor calibration; emergency lockdown

1. Introduction

The appearance of the highly contagious coronavirus [1] (COVID-19) at the end of 2019 in Wuhan, China, and many deaths all over the globe, forced the world's governments to adopt different levels of interventions and emergency measures, due to the virus' easy human transmission [1,2]. The emergency measures included travel restrictions and lockdowns. During the implementation of these measures worldwide, many countries reported air pollution reduction, which could result from reduced transport and other anthropogenic activities in some countries [3]. Besides transport sector restrictions, the industrial and manufacturing sectors are heavily affected by the pandemic [4,5], and their reduced or stopped activities could also be reflected in decrease of air pollution.

The lockdown interventions led to a reduction in population-weighted $PM_{2.5}$ of 14.5 µg/m^3 across China (−29.7%) and 2.2 µg/m^3 across Europe (−17.1%), with mean reduction in $PM_{2.5}$ concentrations during the lockdown period of around 20% in Europe when comparing the lockdown period in 2020 to previous years [6,7]. A study led by French National Institute for Industrial Environment and Risks (INERIS) [8], which analyzed short-term influence of COVID lockdowns on PM_{10} and nitrogen dioxide (NO_2), determined significant decreases in NO_2 and nitrogen oxides throughout Europe. Regarding PM_{10} concentration, the situation is less uniform and European cities selected in the study [8] experienced a range of different outcomes depending on the city region. At Iberian Peninsula, reductions in PM_{10} were observed. In Western Europe, PM_{10} increase (or only a slight decrease) after lockdown was observed due to dominating effect of emission advections from anthropogenic sources (e.g., agricultural, industrial). At Scandinavian Peninsula, only a small decrease of PM_{10} levels was identified due to considerable effect of road dust due to use of studded tires, road dusting and salting, i.e., high road dust emissions, which caused enhancement of PM_{10} effectively masking the lockdown reduction effect. At Central and Eastern Europe mixed decrease/increase of the average PM_{10} concentrations was identified as the effect of lockdown emission reductions was disturbed by a large natural dust episode on 26–29 March. In addition to PM, the main gaseous air pollutants NO_2, sulfur dioxide (SO_2), carbon monoxide (CO), and ozone (O_3) that are usually collected at monitoring sites were analyzed in different areas all over Europe [8]. From these preliminary studies and reports [7,8], it seems evident that nitrogen dioxide (NO_2), nitrogen oxide (NO) and nitrogen oxides (NO_X) noticeably decrease while O_3 increases.

This study is devoted to short-term air quality changes in the city of Novi Sad, Serbia (Lat/Long: 45.26714°, 19.83355°), focused on the period immediately before and during the COVID-19 lockdown. In order to create data sets, online and off-line data from national [9], regional [10], and local monitoring networks [11] were combined with newly acquired data from an ongoing measurement campaigns utilizing PM low-cost sensors (LCS). The aim of this study was to investigate the impacts of lockdown and local meteorology on the level of selected main air pollutants [12]. The analysis was focused on the period of approximately three months, 1 February–30 April, six weeks before and six weeks during lockdown, with additional attention focused to period of about one week immediately before and during the COVID-19 lockdown, for which high-resolution LCS data were available. Selected air pollution data was supplemented (underpinned) with available local meteorological parameters in order to better understand and explain the interactions between pollution and meteorology. Depending on the temporal resolution of data provided by used monitoring tools, variations of main pollutants at the daily level were analyzed in cases when data with sufficient temporal resolution were available.

2. Materials and Methods

2.1. Study Area

Novi Sad represents an urban-industrial agglomeration and, it is the second largest city in Serbia. The city is situated at about 80 m above sea level and it experiences a regional climate from moderately continental to continental. The speed of all recorded winds is mostly between 2.2 and 3.1 m/s, and in terms of frequency, winds that prevail in Novi Sad are north, northeast, and northwest [13]. The main sources of outdoor air pollution in Serbia include the energy sector, the transport sector, waste dump sites, and industrial activities, while the specific sources of air pollution in Novi Sad include the petrochemical industry complex and increasing road traffic [14].

2.2. Air Pollution Monitoring at National, Regional and Local Networks at Novi Sad

The monitoring of outdoor ambient air quality is usually done via networks at the national and local levels. However, despite having a high quality of instrumentation, these kinds of networks are usually very sparse, may not monitor all of the needed parameters, and give little insight into personal exposure. In specific cases, when rapid deployment and increased temporal resolution may be of more interest, IoT (Internet of Things) enabled low-cost sensors may provide interesting complementary data.

In the city of Novi Sad, there are three main groups of monitoring stations: stations that belong to national networks, stations in the regional network, and stations in the local monitoring networks.

National Network, (website http://www.amskv.sepa.gov.rs/pregledstanica.php)

- NN1, Novi Sad-Rumenačka (air quality variables that are monitored: SO_2, NO_2, CO, PM_{10}, $PM_{2.5}$, wind direction, wind speed) type of station: traffic (Lat/Long: 45.26263°, 19.81902°)
- NN2, Novi Sad—Liman (air quality variables that are monitored: SO_2, O_3, NO_2, CO) type of station: background (Lat/Long: 45.23864°, 19.83570°)

Regional Network, (website http://www.amskv.sepa.gov.rs/pregledstanica.php)

- RN1, Novi Sad—Šangaj (air quality variables that are monitored: BTEX, H_2S, SO_2, t, RH, wind direction, wind speed), type of station: industrial, (Lat/Long: 45.27237°, 19.87333°)

Local network, (website https://environovisad.rs/air_points/mm3)

- LN1, Novi Sad—Intersection of Rumenačka and Bulevar Jaše Tomića, type of station: urban/traffic, (Lat/Long: 45.26348°, 19.81903°)
- LN2, Kać—Kralj Petar 1, type of station: suburban/traffic, (Lat/Long: 45.29980°, 19.93926°)
- LN3, Novi Sad- Sunčani kej 41, type of station: urban/background, (Lat/Long: 45.24000°, 19.85139°)
- LN4, Sremska Kamenica, Kamenički park 1-14, type of station: suburban/background, (Lat/Long: 45.22931°, 19.84898°)

2.3. Ongoing Sampling Campaign with Low-Cost Sensors and Reference Pumps

An ongoing measuring campaign with low-cost sensors and reference gravimetric pumps in the city of Novi Sad started in February 2020 and was conducted during mid-February and March. Our monitoring campaign utilizing low-cost sensors was already underway in Novi Sad when Emergency State Measures were introduced on 16 March. The more, or less severe lockdown was used as a unique opportunity to conduct this additional experiment, in which results about the possible relation between massive change of general population daily habits and change in traffic and industrial emissions could be obtained. At selected sampling sites, besides low-cost sensors that measured $PM_{2.5}$ with a temporal resolution of several seconds, $PM_{2.5}$ was collected on Quartz fiber filters by reference gravimetric pumps set to sample air for 48 h. A note about sampling duration is in place here. Timing and duration of the sampling can be adapted for specific purpose in order to use the available resources most effectively. For example, within the ESCAPE project sampling schedule (for purposes of Europe-wide LUR modeling) was two-week sampling, with timers which were set to sample for 15 min every 2 h so that effectively a 42-h sample was collected over 14 days [15]. This compromise must encompass frequency of accessing device in the field conditions on one hand and quality and quantity of samples necessary for calibration of low-cost sensor on the other hand. In this way "best of both worlds" are obtained: the low-cost sensors provide high temporal resolution, while use of reference gravimetric pumps minimizes the errors. Our sampling schedule produces approximately 8 samples over period of 16 days. Additional details are given in Appendix A. Particulate matter low-cost sensor suite (LCS), *ekoNET* was made by Dunavnet [16]. LCS platform was equipped with a PMS7003 particulate matter sensor. Plantower PMS7003 has counting efficiency of 50% at 0.3 μm and 98% at 0.5 μm, and maximum consistency error ±10% at 100~500 μg/m^3 and ±10 μg/m^3 at 0~100 μg/m^3 range [17,18]. Reference gravimetric pump that was collocated with each LCS was Leckel Model LVS3 [19], with standard $PM_{2.5}$ inlet and 2.3 lpm flowrate.

In the period before the Emergency State Measures, the sampling campaign was performed at 4 measuring sites (MS), namely: MS1 (Hajduk Veljkova 4), MS2 (Rumenački put 20), MS3 (Reljkovićeva 2), MS5 (Račkog 78). All four localities can be characterized as traffic sites since they are situated in the vicinity of traffic intersections and streets with a high or medium volume of traffic (Figure 1). After lockdown, the campaign was continued during the following 8–10 days at 3 sites (MS1, MS2, MS3) and repeated at MS5 where sampling was performed previously during February. The instruments were situated within 1m of each other, and no additional inlets were used in order to minimize particle losses. At measuring sites, the instrumentation was at about 2–3 m above the ground. The sampling campaign was carried out at all 4 measuring sites continuously, 7 days before the state of emergency (abbreviation BEMS in further text) and 8 days during the state of emergency (abbreviation EMS in further text) that was declared due to COVID-19 in Serbia.

Figure 1. Map with location of measuring sites (MS1, MS2, MS3, MS5) and monitoring stations at national (NN), regional (RN) and local level (LN).

Measuring site 1 (MS1, Lat/Long: 45.24968°, 19.82467°) was located in a narrow city center, at a distance of 50 m from the intersection of Futoški put, Hajduk Veljkova, and Cara Dušana Street. Possible sources of $PM_{2.5}$ in addition to traffic may include residential heating, small boilers and stoves for individual households in the surrounding area. Types of vehicles passing through the intersection are light vehicles and public transportation (buses). A small heating plant is placed within close proximity (about 1 km). Measuring site 2 (MS2, Lat/Long: 45.27573°, 19.80066°) was located in the Industrial Zone South, at the very border of the residential area in the Rumenačka 20 street. Measuring instruments were located at the height of about 2.5 m next to a very busy roundabout within the courtyard of the Veterinary Institute in Novi Sad. The distance between the roundabout and measuring site was about 30 m. The traffic at MS2 consists mainly of light and heavy trucks and buses since the roundabout represents the main way that leads to the highway in that city area. The rest of the Industrial zone consists of a couple of industrial plants whose production capacities are very low, so the main source of suspended particles at this location is traffic. The third measuring site, (MS3, Lat/Long: 45.2494°, 19.87729°), was situated in Petrovaradin, associated municipality of Novi Sad. Instruments were placed at the height of 3 m, near the intersection of Reljkovićeva and Preradovićeva Street, which is one of the busiest intersections in Petrovaradin municipality. The distance between the measuring site and Reljkovićeva Street was 5 m, and the distance to the intersection was 60 m. Reljkovićeva Street is the main street with heavy traffic that passes through Petrovaradin from Novi Sad. Traffic mainly consists of trucks and cars and, to a lesser extent, of bus traffic. Sources of $PM_{2.5}$ at the measuring site beside traffic are residential heating. The fourth site (MS5,

Lat/Long: 45.23406°, 19.88453°), was also situated in Petrovaradin, as was MS3. Račkog Street serves as a transportation hub and is in the vicinity of the intersections of regional roads. It is the busiest street in Petrovaradin, with high intensity of both heavy vehicle and light vehicle traffic. The majority of heavy transportation from Reljkovićeva Street passes through this street. The measuring devices were set at the height of 3 m, at a distance of 12 m from the street. At this site, particulate matter can be emitted mainly by traffic and by residential heating boilers. Considering MS positions, possible sources of $PM_{2.5}$ in the ambient air at all four localities include traffic, along with households and facilities in the vicinity that are using natural gas and other fossil fuels as a heating source.

3. Results and Discussion

As an initial analysis step, we have plotted particulate matter concentration at available stations that belong to local and national monitoring networks before and after entering the state of emergency (Figure 2) in the period from 1 February to 30 April. Local and regional monitoring stations publicly report daily average concentrations, while data collected at national monitoring station are available with 1 h resolution. A large peak in both $PM_{2.5}$ (Figure 2b) and PM_{10} (Figure 2a) concentration is clearly visible on 26, 27 and 28 March, and is similar in magnitude for all monitoring stations. For these several days, the daily average levels of PM_{10} and $PM_{2.5}$ were up to 250 and 100 µg/m^3, respectively. Extremely high concentrations of particulate matter were recorded at automatic monitoring stations in the whole of Serbia and surrounding countries. The level of PM_{10} at the higher temporal resolution (e.g., 1 min) in some cities in Serbia, including Novi Sad, reached up to 600 µg/m^3. No similar peaks exist for PM_{10} for other days, while for $PM_{2.5}$ similar magnitude peaks do exist, albeit only locally (i.e., for only one of the observed monitoring stations). This kind of pattern indicates some isolated regional event that happened in the period on 26–28 March. Figure A2 shows the back trajectory that was calculated using Hybrid Single-Particle Lagrangian Integrated Trajectory (HYSPLIT) transport and dispersion model. The air mass back trajectory was calculated at 270 m above the ground, on 27 March, when the maximal concentration of PM_{10} was measured at Novi Sad's national and local network sites. Based on the back-trajectory analysis, the presence of a possible source of non-local air pollution is evident. This source of dust can be associated via back trajectory tracing to the Aralkum Desert that is located at the Kazakhstan and Uzbekistan border. Back trajectory for surrounding days is also given in Appendix B.

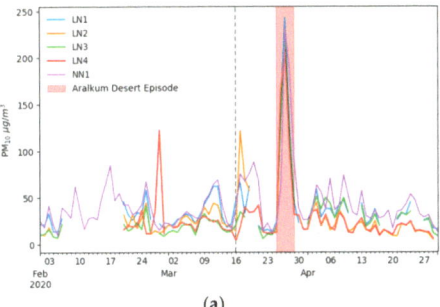 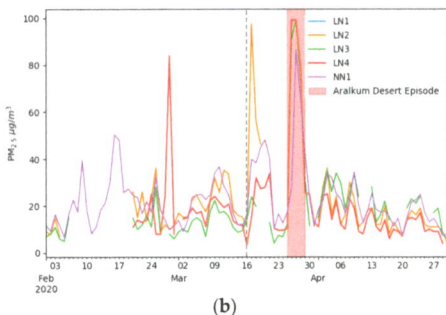

(a) (b)

Figure 2. Daily concentrations at sites belonging to national and local monitoring networks before and after entering the state of emergency (dashed gray line) (a) PM_{10} (b) $PM_{2.5}$.

Tables 1 and 2 show descriptive statistics for fine and coarse particulate matter data collected at stations that belong to the national and local monitoring network in the area of Novi Sad for the period from 1 February to 30 April, split into periods before and after lockdown measures introduced due to COVID-19 pandemic. Tables also include descriptive statistic when 3-day Aralkum Desert episode is excluded from data. Looking into data from

EMS and EMS* columns in Tables 1 and 2, it becomes evident that natural distant sources may be very dominant in total particulate matter pollution and thus effectively mask local changes originating from emergency state, strongly swaying mean and maximum values of PM pollution. When comparing BEMS (before emergency state) and EMS* (during emergency state) columns in Table 1, it can be seen that the median slightly increased for LN1 and NN1 (both stations are urban traffic), strongly increased for LN3 (urban background), and decreased for LN2 and LN4 (LN2 and LN4 are suburban traffic and suburban background stations respectively). When comparing BEMS and EMS* columns in Table 2, it can be seen that the median decreased for LN2, LN4 and NN1 (suburban traffic, suburban background and urban traffic) and, again, strongly increased for LN3 (urban background). While it is expected that traffic sites experienced decrease in $PM_{2.5}$ pollution, a strong increase in both PM_{10} and $PM_{2.5}$ for LN3 is somewhat surprising and may indicate presence of source which emits both fine and coarse particles that was active during EMS. However, LN3 had lowest concentration of all stations in BEMS period. The often-nonlinear relationships between changes in emissions and changes in concentrations may also explain why lower air pollution may not occur at all locations.

Table 1. Descriptive statistics for PM_{10} [µg/m^3] data collected via the national and local network for air pollution monitoring in Novi Sad for the period of BEMS and EMS campaign in Novi Sad. EMS* denotes EMS data with Aralkum Desert episode excluded from data.

	LN1			LN2			LN3			LN4			NN1		
Period	BEMS	EMS	EMS*	BEMS	EMS	EMS*	BEMS	EMS	EMS*	BEMS	EMS	EMS*	BEMS	EMS	EMS*
Mean	30.97	43.97	32.59	23.77	/	24.97	17.72	39.69	28.88	25.21	29.49	20.38	36.83	49.58	41.89
St.dev.	13.24	43.56	13.84	10.31	/	22.30	7.62	40.17	12.49	21.85	37.64	9.83	16.89	39.95	20.30
Median	28.50	33.00	30.50	21.10	/	17.00	16.05	31.00	28.50	20.40	19.00	17.00	33.35	36.50	36.30
MIN	14.00	13.00	13.00	8.70	/	5.00	7.30	6.00	6.00	11.70	4.00	4.00	9.70	13.20	13.20
MAX	62.00	243.00	66.00	44.60	/	122.00	40.90	219.00	51.00	122.60	209.00	42.00	85.10	231.60	88.50

Table 2. Descriptive statistics for $PM_{2.5}$ [µg/m^3] data collected via the national and local network for air pollution monitoring in Novi Sad for the period of BEMS and EMS campaign in Novi Sad. EMS* denotes EMS data with Aralkum Desert episode excluded from data.

	LN2			LN3			LN4			NN1		
Period	BEMS	EMS	EMS*	BEMS	EMS	EMS*	BEMS	EMS	EMS*	BEMS	EMS	EMS*
Mean	19.08	/	19.95	12.09	25.22	19.44	18.71	19.95	14.77	20.95	23.95	21.42
St.dev.	8.26	/	17.52	5.19	21.25	8.51	14.85	21.06	7.63	10.08	15.44	10.65
Median	17.00	/	14.00	10.95	20.35	19.00	14.80	14.00	12.40	18.30	18.50	17.60
MIN	7.00	/	4.00	5.00	4.10	4.10	8.00	3.20	3.20	6.80	7.30	7.30
MAX	36.00	/	97.60	28.00	99.00	35.00	84.00	99.00	33.60	50.50	86.40	48.20

In the subsequent section, diurnal variation will be able to reveal more details about changes and differences in air pollution levels since we will use additional data obtained from high-resolution low-cost sensors.

Weather conditions can also noticeably contribute to the changes that are seen in pollutant concentrations, and in some cases changes in meteorology can lead to increased air pollution. At the time of this study, meteorological data with 1 h temporal resolution originating from national network station NN1 were available. Meteorological conditions in the interaction with the landscape's physical features are key factors that influence the rate of change, movement, and dispersal of gaseous and particulate matter pollution in the air. Sampling campaign in both BEMS and EMS period was conducted during the heating season (lasting from 15 October until 15 April) in the Republic of Serbia. It is well known that particulate matter concentration may correlate with temperature and relative humidity, and that often it strongly negatively correlates with wind speed [20]. This means that wind speed and temperature are two key factors affecting $PM_{2.5}$ and PM_{10} concentration distributions emitted from local sources. These two factors will be discussed next.

Before and during the emergency state lowest $PM_{2.5}$ concentrations were observed when the wind was coming from the southeast direction and to a lesser extent northeast direction (Figure 3 upper row), while most of the highest PM$_{2.5}$ concentrations were observed for low wind speeds within the 1.5 m/s, (Figure 3 upper and middle row).

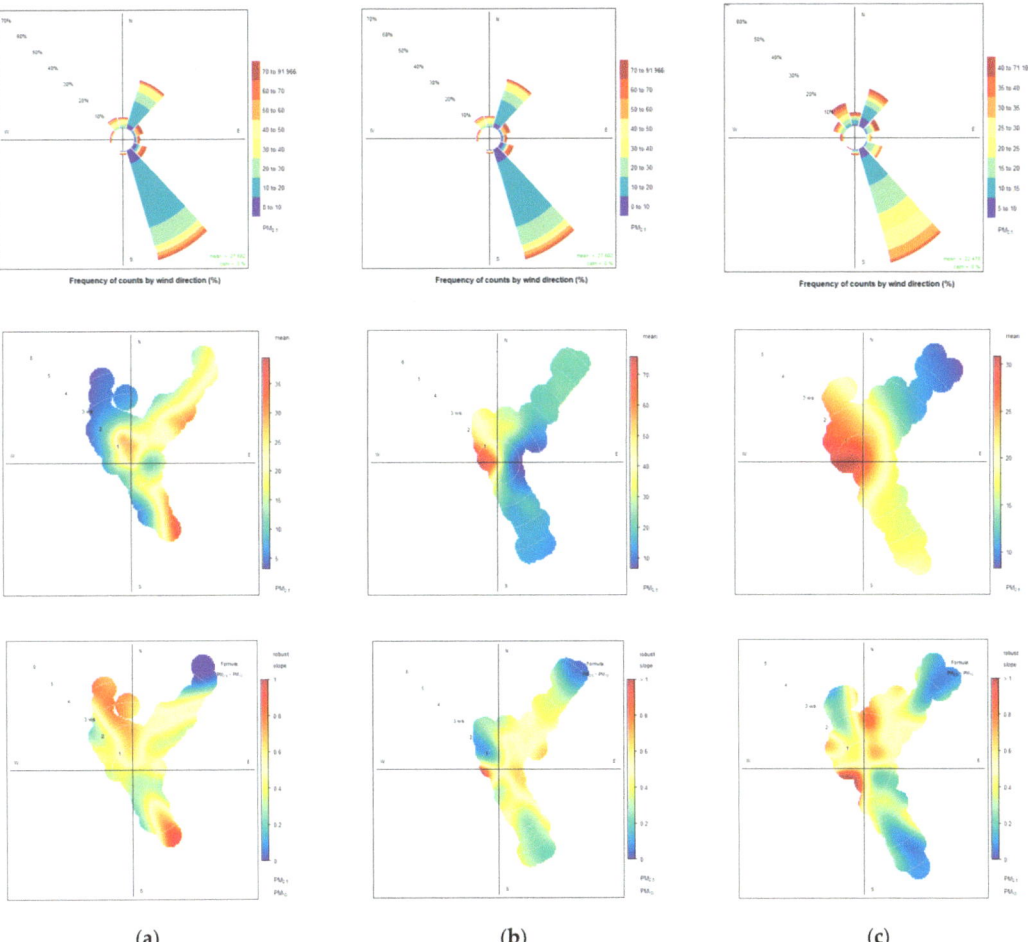

Figure 3. PM$_{2.5}$ [µg/m^3] counts by wind direction (upper row), PM$_{2.5}$ [µg/m^3] pollution rose (middle row) and PM$_{2.5}$~PM$_{10}$ robust slope (lower row) (**a**) before emergency state 1–15 March (**b**) during emergency state 16–31 March, Aralkum episode excluded (**c**) during emergency state 1–15 April.

Having in mind the position and type (traffic) of Rumenačka monitoring station, this indicates the presence of strong local sources of air pollution, mainly related to traffic and residential heating, whose influence is minimized mainly by favorable meteorological conditions. These conditions may come in form of stronger winds, for example over 5 m/s, as can be seen from low PM$_{2.5}$ concentrations in Figure 3b,c middle row. This is also confirmed by lower row in Figure 3, depicting PM$_{2.5}$~PM$_{10}$ robust slope. Different types of sources have different size distribution signature, for example combustion sources emit a large proportion of PM$_{2.5}$, while on the other hand material used during construction, geological matter, polen and similar have larger coarse fraction. As can be seen from Figure 3b,c, within wind speeds smaller than 1 m/s there is an indication of strong local combustion processes, indicated by high PM$_{2.5}$~PM$_{10}$ ratio. Stronger wind noticeably change local air pollution composition and reduce PM$_{2.5}$ concentration, thus making PM$_{2.5}$~PM$_{10}$ ratio very small.

Figure 4 shows the mean values and 95% confidence interval for the normalized pressure, temperature and PM$_{2.5}$ measured at station Rumenačka, before and during emergency state. All of the air quality variables are min-max normalized taking into account the whole period 1 March to 15 April. Min-max values for the pressure, temperature and PM$_{2.5}$ concentrations are 989.0–1024.0 mbar, −2.06–28.4 °C, and 0.0–91.97 µg/m^3, respectively.

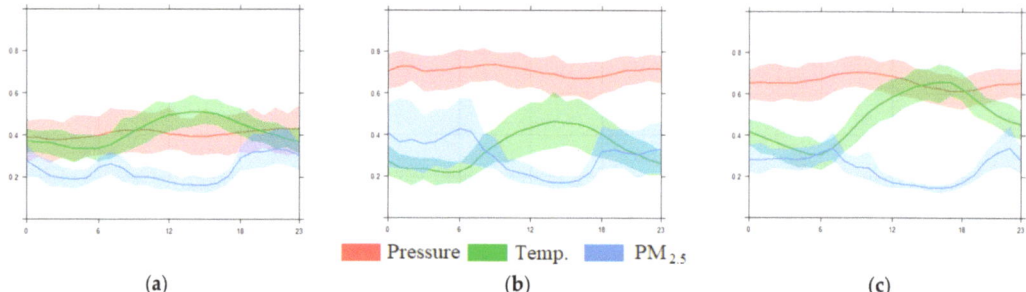

Figure 4. Normalized pressure, temperature and PM$_{2.5}$ concentrations in different hours of the day (min-max normalization for the whole period) (**a**) before emergency state 1–15 March (**b**) during emergency state 16–31 March, Aralkum episode excluded (**c**) during emergency state 1–15 April.

The period during the emergency state had a lower mean temperature in 0–6 h period of the day, as evident from the plots. This is also accompanied by increased PM$_{2.5}$ concentrations in the 0–6 h period of the day and higher pressure throughout the day compared to the period before the emergency state. In the city of Novi Sad, residential heating is realized either as individual boilers or via district heating facilities, which are a substantial contributor to PM concentrations. Residential heating had to be used more since citizens have been instructed to stay at home as much as possible [21]. This is evident in Figure 4b, where lower temperatures correspond to higher PM$_{2.5}$ pollution. On the other hand, as evident from Figure 4c, higher temperatures, at least in part, can be associated with reduction in PM$_{2.5}$ pollution.

3.1. Descriptive Statistic of PM$_{2.5}$ Levels before and at the Beginning of Emergency Measure State Measured with Low-Cost Sensors

In the current and following section, we will try to gain a bit more insight into periods before and during COVID lockdown by utilizing higher temporal resolution of low-cost sensors. Table 3 shows descriptive statistics for data collected via low-cost sensors at 4 different locations in Novi Sad. For each measuring spot, two columns are given, first referring to the period before the emergency state, and second referring to the period after the emergency state. These periods are identical for measuring sites MS1, MS2 and MS3, going from 8 to 15 March 2020 for the BEMS period, and from 16 to 24 March 2020 for the EMS period. From MS5 site, BEMS period data was collected somewhat earlier from 6 to 23 February 2020, while the EMS period mostly coincides with the EMS period for the remaining sites, 18 to 26 March 2020.

Table 3. Descriptive statistics for data collected via low-cost sensors, at 4 different locations in Novi Sad.

	MS1		MS2		MS3		MS5	
Campaign	BEM	EM	BEM	EM	BEM	EM	BEM	EM
Mean	24.05	25.85	26.79	31.02	26.27	28.72	23.50	21.31
St.dev.	17.70	16.71	20.94	23.54	16.14	16.86	15.50	6.38
Median	21.46	22.13	22.54	23.61	26.61	24.02	20.01	21.11
MIN	2.56	6.37	1.90	4.47	6.50	8.28	7.99	11.49
MAX	74.21	74.26	78.84	88.54	66.34	75.64	61.83	39.14
25th perc.	8.73	13.24	7.97	12.15	11.03	16.34	12.59	17.08
75th perc.	34.52	35.26	40.81	47.50	36.07	38.12	30.13	24.70

The median of $PM_{2.5}$ concentration was similar in the BEMS and EMS period for all sites, a slight decrease in the median was observed only for MS3. Maximum values of $PM_{2.5}$ (98th percentile) increased for MS2 and MS3 sites, remained the same for MS1, and decreased for MS5 site. Another interesting aspect is the standard deviation of the pollution, which may signal change in pollution sources and their distribution: it remained the same at MS1 and MS3 sites but showed a noticeable decrease at MS5 site. More insight into $PM_{2.5}$ pollution and change that occurred during EMS at these sites is possible via time series and 24 h boxplots shown in further text.

3.2. Changes in Diurnal Variation of $PM_{2.5}$ at Sampling Site MS1, MS2, MS3 and MS5 before and after Measures of the State Emergency

The reduced movement of vehicles as well as reduced industrial activities during lockdown in Novi Sad, leading to the reduction of the exhaust emissions, were a probable cause for a reduction (as seen from reduced interquartile range of hourly boxplots in Figure 5) and a more uniform profile of the $PM_{2.5}$ levels approximately during typical working hours (8 AM–5 PM).

According to recently published studies, similar findings were identified in several cities in various parts of the world [22]. In the first month of lockdown, period 15 March–15 April 2020, the official heating session was still ongoing. For the most part, mean daily temperatures in the week before and in the first weeks of lockdown, during which the sampling campaign was conducted, were below 10 °C. Residential heating sources caused higher concentration in the morning and during the night at almost all sampling sites and can be clearly recognized from the boxplot representing 1 h variations in the week before the EMS and period at the EMS beginning (Figure 5).

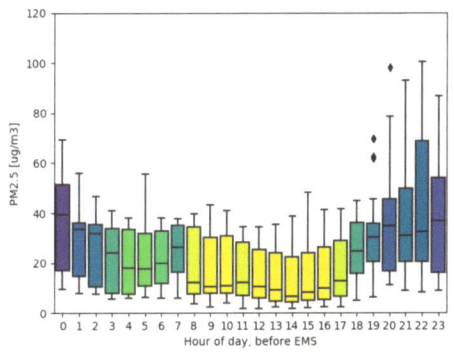
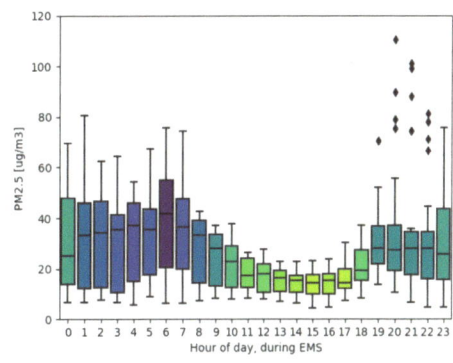

(a)

Figure 5. *Cont.*

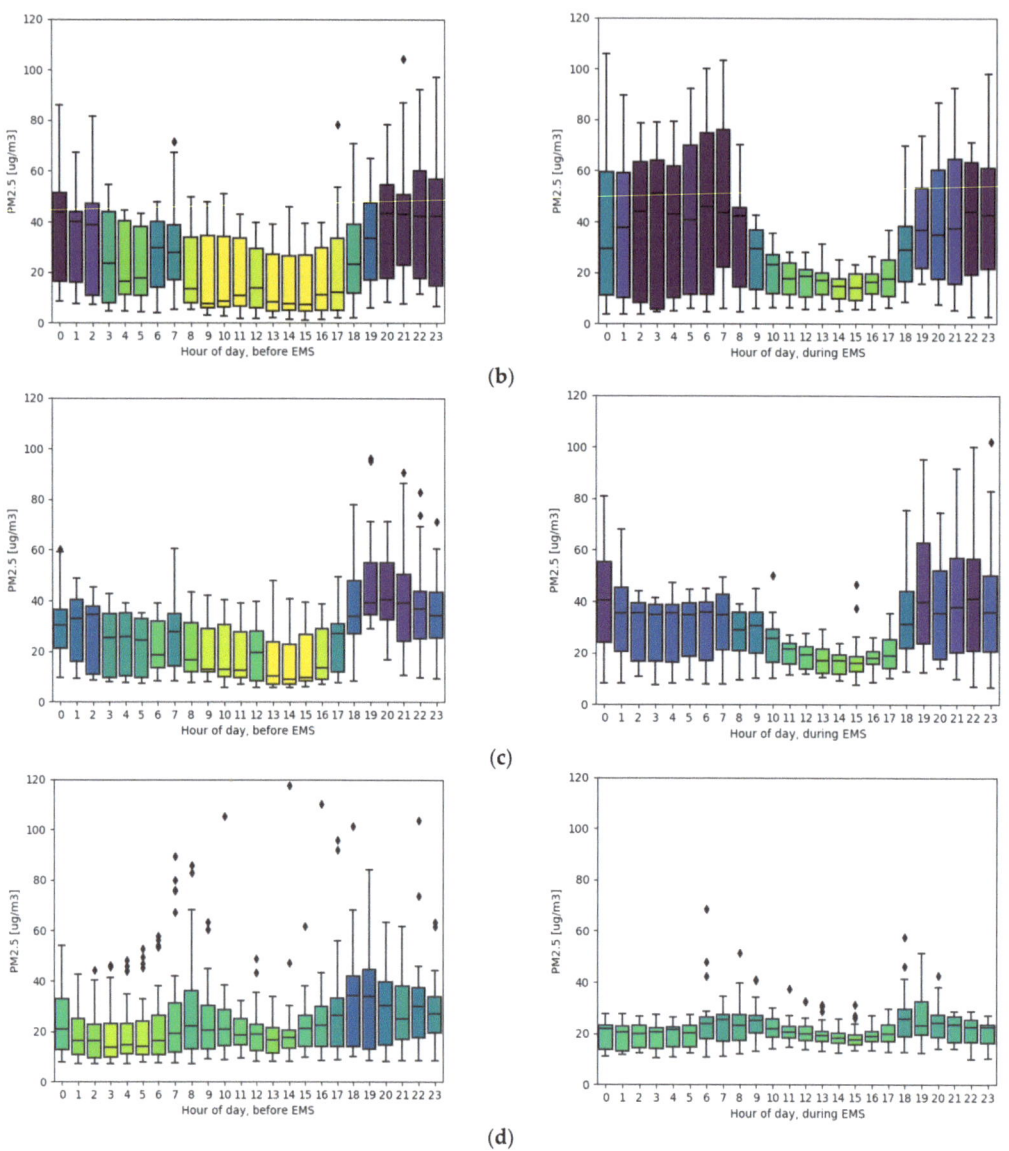

Figure 5. Daily profile of $PM_{2.5}$ measured before measures of the state of emergency (**left**) and during measures of the state of emergency (**right**) at (**a**) MS1 (**b**) MS2 (**c**) MS3 (**d**) MS5.

3.3. Changes in Diurnal Variation of Gaseous Pollutants and $PM_{2.5}$ at NN1 before and after Measures of the State Emergency

For the identification of changes of main air pollutants levels, it was necessary to create diurnal variation whenever it was possible, i.e., data have been collected using monitors with high temporal resolution. Figure 6 depicts diurnal variations of measured main gaseous pollutants at site that belongs to the national monitoring network: Rumenačka SEPA, NN1 before and after measures of the state emergency. At NN1 the following was observed:

- Daily profiles of SO_2, the main pollutant, depicted in the form of a boxplot (Figure 6a), show that, on average, there was no noticeable change a week before and 10 days during the emergency state. For most hours of day, the SO_2 concentration was below 20 µg/m^3, with occasional "outliers" in the range from 35–40 µg/m^3 in the morning and afternoon pollution peaks.
- Daily profiles for NO_x show a clear change in daily patterns (Figure 6b). Because boxplot color corresponds to the value of median, it is evident that the median of these main pollutants was reduced during the first 10 days of the emergency state during most hours of the day. NOx boxplots show clear morning and afternoon pollution peaks, both before and during the emergency state, but the median of these peaks is reduced during the emergency state. The median was around 80–95 µg/m^3 before emergency state from 18–19 h, and an evident reduction to the median of about 15–25 µg/m^3 is visible during the emergency state. Additionally, during the most active hours of the day (period from 9 h to 17 h), a noticeable drop in the level of air pollution was observed, indicated by lower median and smaller interquartile range (median around 50 µg/m^3 before emergency state during this time of day, and about 25 µg/m^3 during emergency state in this time of day).
- Observing CO daily profiles following conclusions can be reached. During the active hours of the day, in the period from 9 h to 17 h, the level of air pollution stayed similar in, going around 0.8 mg/m^3 with a relatively narrow interquartile range (Figure 6c). During morning hours, the situation is similar, with a slight increase during the emergency state, with median still being in the range from 0.8 to 1 mg/m^3, but with larger interquartile ranges during the emergency state. On the other hand, from 18 h to 23 h, there is a slight reduction in the CO level during the emergency state. Since there are no noticeable changes in the daily profiles before and during the emergency state for CO levels, it can be argued that pollutants with domestic heating source profiles mostly exhibited constant levels.
- An interesting increase in $PM_{2.5}$ levels in the early morning and in the nighttime at different monitoring sites over the area of the city of Novi Sad, even during lockdown, has also been noticed (Figure 6d). However, during the most active hours of the day (period from 9 h to 17 h) a noticeable drop in the level of air pollution was evident, as indicated by a lower median and smaller interquartile range.

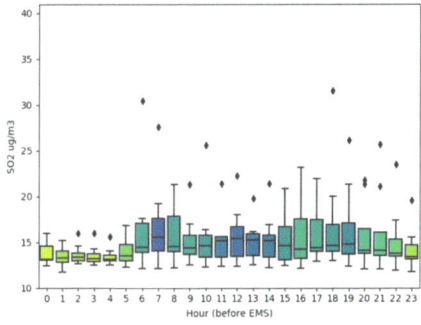 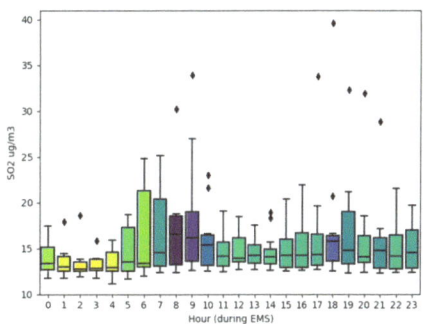

(a)

Figure 6. Cont.

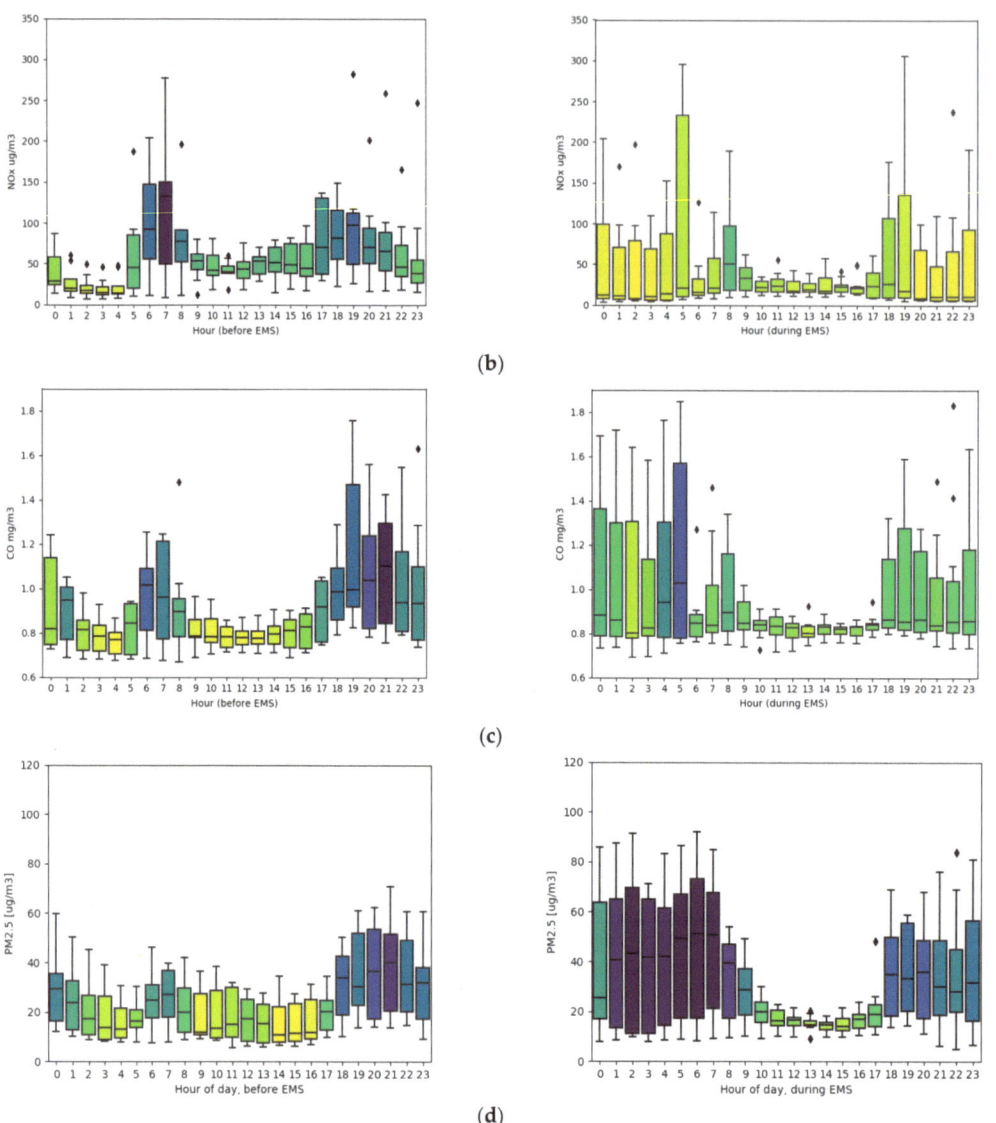

Figure 6. Rumenačka SEPA, NN1 station, hourly boxplots before (**left**) and during emergency state (**right**) (a) SO$_2$ (b) NO$_x$ (c) CO (d) PM$_{2.5}$.

Equivalent analysis was also done for Liman SEPA NN2 station and is shown in Appendix D. Similar and comparable results were observed all over the EU in various studies focusing on COVID-19 influences [23,24]. Results for levels of NO$_2$, PM$_{10}$ and PM$_{2.5}$ over the EU member countries have been presented by EEA [25]. Data show, as in the city of Novi Sad, that NO$_2$, a pollutant mainly emitted by road transport, has decreased in many European cities where lockdown measures were applied. Changes in levels of PM$_{2.5}$ may also be expected due to lockdown measures, however, a consistent reduction cannot yet be seen across European cities, as is the case in Novi Sad. This is likely due to the fact that the main sources of this pollutant are more varied, including, at the European level, the combustion of fuel for the heating of residential, commercial and institutional buildings, industrial activities and road traffic.

4. Conclusions

In this paper, we have presented a study targeting particulate matter pollution in the city of Novi Sad that was conducted during February–April of 2020. On 15 March 2020, when the COVID-19 lockdown started and emergency state measures were declared, a sampling campaign of collecting data with gravimetric pumps and IoT low-cost sensors has been already in progress. New circumstances were promptly integrated into the campaign's planning, as they opened possibilities for examining the influence of broader societal changes on air pollution. Additional campaign efforts, which now also included examining COVID-19 lockdown influence, were carried out and data were collected both before and at the beginning of COVID-19 lockdown in Novi Sad.

Looking at descriptive aggregate statistical data, we have not identified a noticeable reduction in $PM_{2.5}$ concentration levels at a daily level. The measuring campaign and data collection were finished at the last sampling site (MS5) on March 26th. No significant non-local sources were influential up until that point. On 27 and 28 March, when the concentrations of $PM_{2.5}$ and PM_{10} were extremely high in the whole of Serbia and surrounding countries, Aralkum desert was pointed out as the external source of high increase in PM. This source of dust was confirmed via back trajectory tracing to Aralkum Desert. The trends and levels of fine and coarse particulate matter in Novi Sad and other cities in Serbia before and after lockdown were under multiple factors similar to those that were observed in cities in the region, e.g., Budapest and Sofia, as shown in [8].

In the city of Novi Sad, air pollution mainly comes from traffic and residential heating, where citizens of Novi Sad use a variety of fuels for heating of individual houses: gas, biomass, and fossil fuels. More detailed analysis showed that the morning peak and higher concentration during the night period remained present in the daily profiles, even after the COVID-19 lockdown and in that period, there were also no noticeable changes in the daily profile at all sampling sites. Morning peak and higher concentrations during the night and variability characteristics for that period stayed the same at all four low-cost sampling sites and part of the daily profile for that period was similar before and after emergency state measures were declared. For the period corresponding to working hours, between 8 AM and 5 PM, when people are the most active and traffic is usually the most frequent, the PM concentrations and their variability were lower during COVID-19 lockdown than during the week before COVID-19 lockdown. This can be attributed to the fact that majority of citizens stayed at home, leading to reduced traffic and associated emissions.

These analyses prove that lowering anthropogenic activities intensity contributes to changes in the diurnal pattern of fine particulate matter and some other main air pollutants. Natural events like long–range transport of air masses that may happen during lock–down, also strongly contribute to the increased air pollutants concentrations. These influences may sway the summary statistics and illustrates the need for higher temporal resolution monitoring. The perspective for further analyses is in additional analyses of chemical content of $PM_{2.5}$ from collected filters.

Author Contributions: Conceptualization, M.D. and M.J.-S.; Data curation, M.D., S.D., M.J. and M.J.-S.; Funding acquisition, M.D. and M.J.-S.; Investigation, M.D., S.D., M.J. and M.J.-S.; Methodology, M.D. and M.J.-S.; Project administration, M.D. and M.J.-S.; Resources, M.J.-S.; Supervision, M.D., J.R. and M.J.-S.; Validation, M.D., M.J. and M.J.-S.; Visualization, M.D. and S.D.; Writing—Original draft, M.D., S.D. and M.J.-S.; Writing—Review & editing, M.D., S.D., M.J., J.R. and M.J.-S. All authors have read and agreed to the published version of the manuscript.

Funding: The research was funded by the Ministry of Education, Science and Technological Development of the Republic of Serbia and H2020 project VIDIS (GA 952433).

Institutional Review Board Statement: Not applicable.

Informed Consent Statement: Not applicable.

Data Availability Statement: Data is contained within the article. Additional data referenced but not presented in this study could be made available on request from the corresponding author.

Acknowledgments: We thank the United Nations Development Programme (UNDP) Serbia, for support of our research efforts, especially during COVID-19 pandemic.

Conflicts of Interest: The authors declare no conflict of interest.

Appendix A

Field Calibration of $PM_{2.5}$ Low-Cost Device Performed via Collocation with Reference Pump

As our extensive descriptive data analysis has shown, low-cost $PM_{2.5}$ devices are suitable for giving insights into air pollution trends, which can be confirmed via strong correlation between low-cost device readings and data obtained from the national network. However, despite this and despite the fact that the low-cost devices have gone through initial calibration, in order to avoid mismatch between calibration location and deployment location, low-cost sensors were additionally calibrated by collocating them with the reference gravimetric pumps at their deployment location. Data calibration is also essential to mitigate unfounded risk perception from low-cost sensors [26,27]. Calibration curves are shown in Figure A1, along with the Pearson's correlation coefficient and calibration equations.

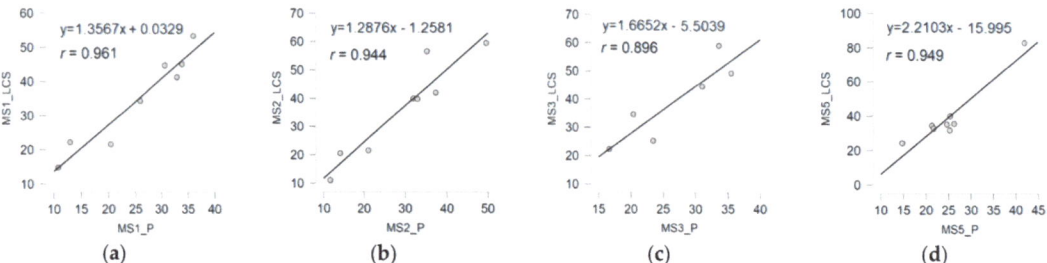

Figure A1. $PM_{2.5}$ correlation: (**a**) MS1 site reference pump vs. low-cost device (**b**) MS2 site reference pump vs. low-cost device (**c**) MS3 site reference pump vs. low-cost device (**d**) MS5 site reference pump vs. low-cost device.

Only the data points that were obtained from unscathed filters were used for calibration purposes. Despite the sensors having gone through initial calibration by the manufacturer, calibration coefficients obtained via collocation at the deployment location are modestly (figure panels (a), (b), (c), or significantly (d) different from 1.0). It can be thus concluded that the additional round calibration at the deployment location is a desirable and necessary step to obtain useful results from low-cost sensors. All results and data reported in this study, originating from low-cost sensors, are calibrated using deployment location calibration.

Appendix B

HYSPLIT Analysis of Sand Episode Originating from Aralkum Desert

During the episode [28], air mass trajectory came from the northeast, passed through Romania, Ukraine, Russia, Kazakhstan and Uzbekistan (Figure A2b). Finally, extremely high concentrations of PM_{10} were probably, in large part, due to the dust coming from the Aralkum Desert located at the Kazakhstan and Uzbekistan border.

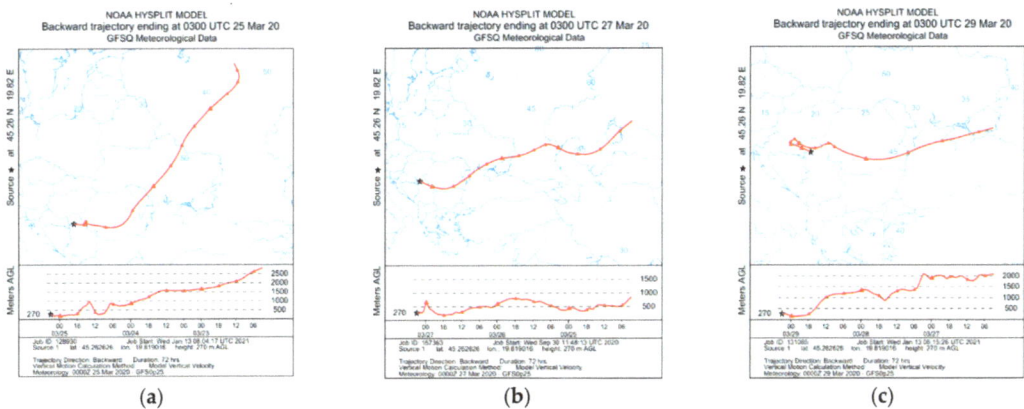

Figure A2. The air mass back trajectory for the city of Novi Sad on (**a**) 25 of March (**b**) 27 of March (**c**) on 29 of March.

In the period before and after PM pollution episodes, back trajectories showed less capability for long-range transport as evident from Figure A2a,c (note the differences in zoom level for same 72 h trajectory duration).

Appendix C

Correlation of PM$_{2.5}$ Concentration in the City of Novi Sad between Spatially Distributed Measuring Sites

In order to better understand spatial variability of PM pollution in the city of Novi Sad, data for PM$_{2.5}$ fine particulate matter at additional sampling sites, which was measured simultaneously using low-cost sensors and reference pumps set to 48 h averaging, were correlated with the NN1 SEPA station.

Pearson's correlation coefficient between 48 h concentrations measured at the site that belongs to the national network of air quality monitoring NN1 collected with equivalent monitor GRIMM Aerosol EDM 180 [29] and concentrations measured using reference gravimetric pumps LVS3 Sven Leckel [19] at MS1, MS2, MS3 and MS5 site vary between 0.85 and 0.99 (Figure A3).

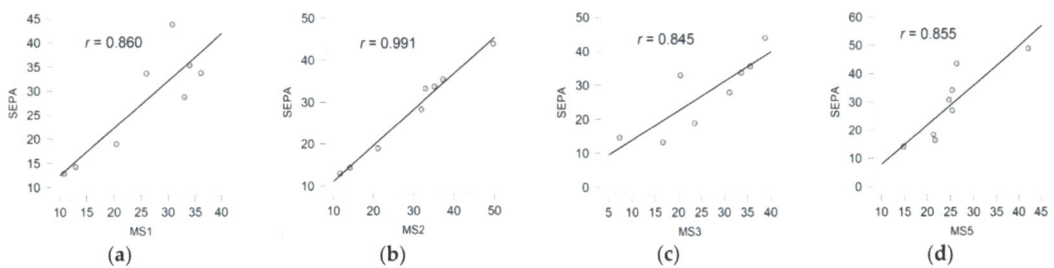

Figure A3. PM2.5 correlation: (**a**) MS1 site reference pump vs. SEPA equivalent instrument (**b**) MS2 site reference pump vs. SEPA equivalent instrument (**c**) MS3 site reference pump vs. SEPA equivalent instrument (**d**) MS5 site reference pump vs. SEPA equivalent instrument.

The highest Pearson's correlation coefficient was identified, not surprisingly, between sites situated along the same street, Rumenačka (Figure A3b), at the distance of about 2.7 km. High correlation despite the relatively large distance between the sites indicates that sources of pollution covary, most probably due to traffic consisting of mainly light and heavy trucks and buses, since the roundabout represents the main way that leads to the highway. Pearson's correlation coefficient was also high for the other three sites, approximately 0.86 for the MS1 site and 0.85 for the other two sites located on the other side of the Danube river at Petrovaradin, a location where there are currently no monitoring sites from national, regional or local monitoring network.

Pearson's correlation coefficient between 1 h concentrations measured at the site that belongs to the national network of air quality monitoring NN1 and concentrations measured using low-cost sensors at MS1, MS2, MS3 and MS5 shown in Figure A4 vary between 0.61 and 0.97. The highest Pearson's correlation coefficient can be noticed between the sites at the same street, Rumenačka (Figure A4b), at a distance of about 2.7 km, r = 0.97. Pearson's coefficient correlation of 0.89 is identified between NN1 measured with equivalent instrument GRIMM Aerosol EDM 180 and Dunavnet *ekoNET* device equipped with PMS7003 low-cost sensor located at MS1 (Figure A4a) and MS3 (Figure A4c).

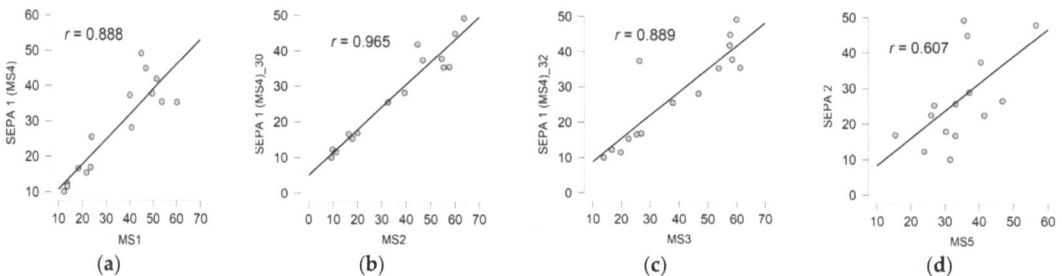

Figure A4. PM2.5 correlation: (**a**) MS1 low-cost sensor vs. SEPA equivalent instrument (**b**) MS2 low-cost sensor vs. SEPA equivalent instrument (**c**) MS3 low-cost sensor vs. SEPA equivalent instrument (**d**) MS5 low-cost sensor vs. SEPA equivalent instrument (at different location).

A Pearson's correlation coefficient that is lowest is identified between the equivalent device at NN1 and low-cost device at MS5 (Figure A4d). The possible reason for the lower coefficient correlation between NN1 and MS5 is probably due to differences in particulate matter levels and probably due to divergent content as well as local meteorological conditions at two locations. This illustrates the need for higher spatial resolution of PM monitoring efforts and, to some extent, locations where there is a need to deploy additional sensors (locations with lower correlation to reference stations).

Appendix D

Changes in Diurnal Variation of Gaseous Pollutants at NN2 before and after Measures of the State Emergency

In Figure A5, diurnal variation of the main gaseous pollutants at Liman SEPA, NN2 automatic monitoring station, before and after the state emergency measures, are shown:

- Daily profiles of SO_2, the main pollutant, depicted in the form of a boxplot (Figure A5a), show that, on average, there is a slight change a week before and 10 days during the emergency state. In the week before, for most hours of day, SO_2 concentration was approximately 5 µg/m³, while in the first 10 days of the emergency state, the median is a bit higher between 7 and 8 µg/m³ from midnight to morning hours and about 10 µg/m³ later on during the working hours and evening.
- Daily profiles for O_3, the main pollutant, follow diurnal cycle with maximal levels from 10 to 16 h with a median of about 70 µg/m³ in the week before and 80 µg/m³ in the first 10 days of the emergency state (Figure A5b). During morning and evening hours, median values are also 10 µg/m³ lower in the week before (about 50 µg/m³) than in the first 10 days of the emergency state (60 about µg/m³). Interquartile ranges are larger during the emergency state in periods of day with maximal levels of O_3.
- Daily profiles for NO, NO_2 and NO_x at Liman NN2 site, although lower in general, show a clear change in daily patterns similar to Rumenačka NN1 site. NOx boxplots

(Figure A5c) show clear morning and afternoon pollution peaks, both before and during the emergency state, but the median of these peaks is reduced during the emergency state (median around 30–35 µg/m³ before emergency state from 18 to 19 h, and an evident reduction of the median of about 15 µg/m³ during the emergency state). In addition, during the most active hours of the day (period from 9 to 17 h), a significant drop in the level of air pollution can be observed, indicated by a lower median (median around 15–30 µg/m³ before the emergency state during this time of day, and about 10–15 µg/m³ during emergency state during this time of day).

- Observing CO daily profiles at Liman (Figure A5d), NN2 site, the following conclusions can be reached. During the active hours of the day, from 9 h to 17 h level of CO stayed similar, going between 0.4 and 0.5 mg/m³, with a narrower interquartile range during the emergency state. During morning hours, the median level is similar, but with larger interquartile ranges before the emergency state. Similar as at the Rumenačka site, from 18 to 23 h, there is a slight reduction in CO level during the emergency state. Daily profiles before and during the emergency state for CO levels are connected with domestic heating sources that stay the same and traffic, leading to a slight reduction.

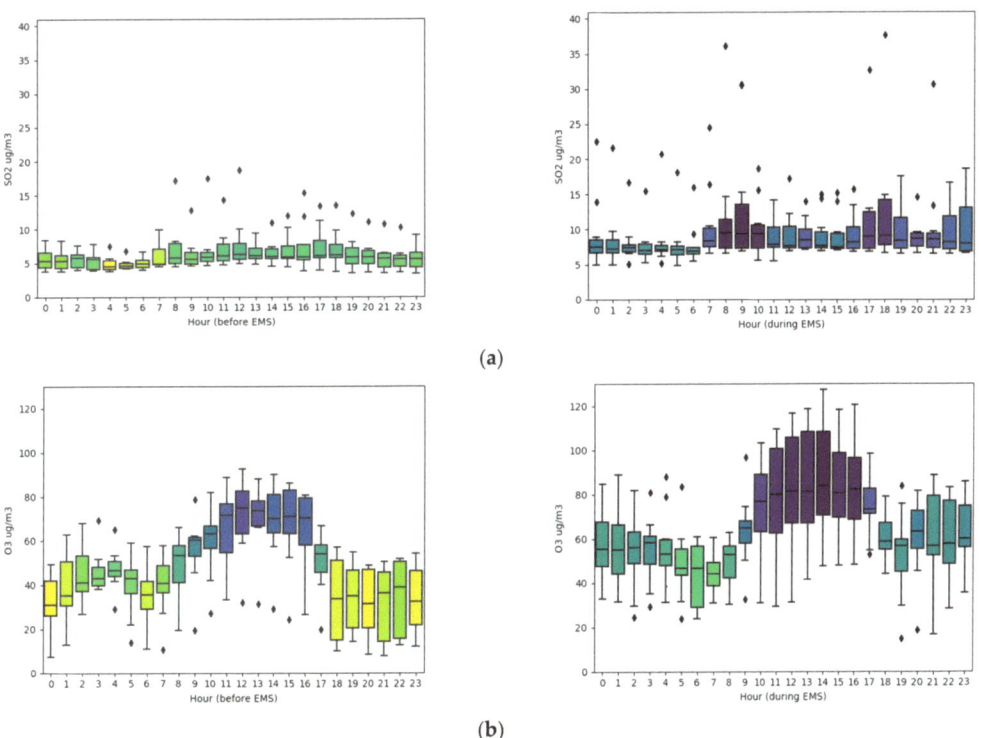

Figure A5. *Cont.*

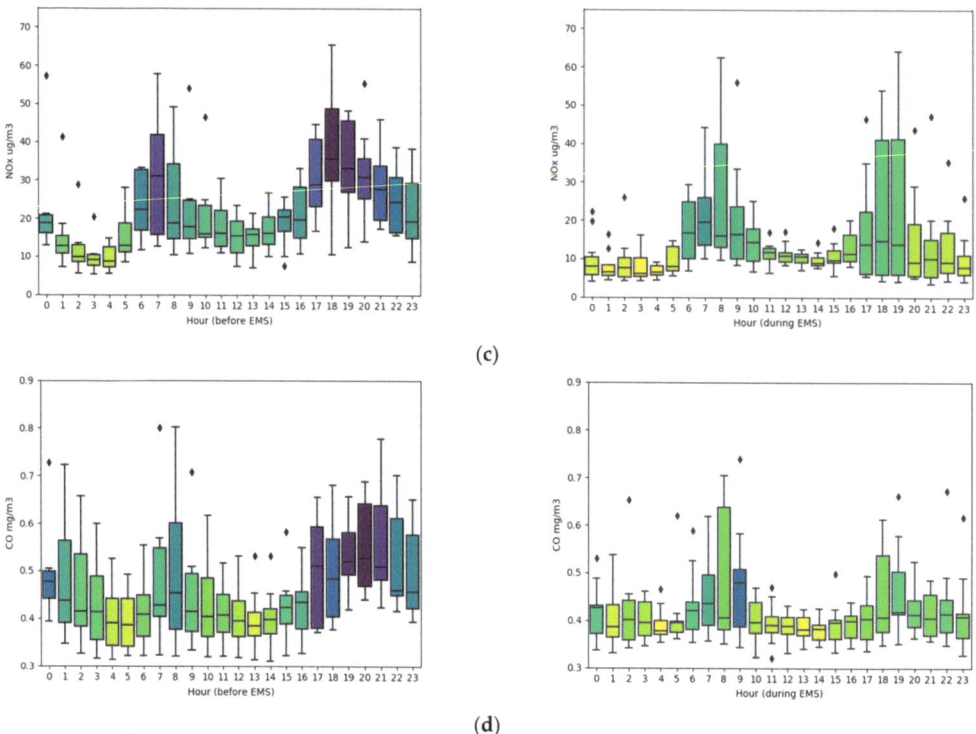

Figure A5. Liman SEPA, NN2, station, hourly boxplots before (**left**) and during emergency state (**right**) (**a**) SO_2 (**b**) O_3 (**c**) NO_x (**d**) CO.

References

1. Shrestha, A.M.; Shrestha, U.B.; Sharma, R.; Bhattarai, S.; Tran, H.N.T.; Rupakheti, M. Lockdown Caused by COVID-19 Pandemic Reduces Air Pollution in Cities Worldwide. 2020. Available online: https://eartharxiv.org/repository/view/304/ (accessed on 28 January 2021). [CrossRef]
2. Sharma, S.; Zhang, M.; Anshika; Gao, J.; Zhang, H.; Kota, S.H. Effect of restricted emissions during COVID-19 on air quality in India. *Sci. Total Environ.* **2020**, *728*, 138878. [CrossRef] [PubMed]
3. Muhammad, S.; Long, X.; Salman, M. COVID-19 pandemic and environmental pollution: A blessing in disguise? *Sci. Total Environ.* **2020**, *728*, 138820. [CrossRef] [PubMed]
4. Dutheil, F.; Baker, J.S.; Navel, V. COVID-19 as a factor influencing air pollution? *Environ. Pollut.* **2020**, *263*, 2019–2021. [CrossRef] [PubMed]
5. Li, L.; Li, Q.; Huang, L.; Wang, Q.; Zhu, A.; Xu, J.; Liu, Z.; Li, H.; Shi, L.; Li, R. Air quality changes during the COVID-19 lockdown over the Yangtze River Delta Region: An insight into the impact of human activity pattern changes on air pollution variation. *Sci. Total Environ.* **2020**, 139282. [CrossRef] [PubMed]
6. Giani, P.; Castruccio, S.; Anav, A.; Howard, D.; Hu, W.; Crippa, P. Short-term and long-term health impacts of air pollution reductions from COVID-19 lockdowns in China and Europe: A modelling study. *Lancet Planet. Health* **2020**, *4*, e474–e482. [CrossRef]
7. European Environment Agency. *Air Quality in Europe—2020 Report*; European Environment Agency: Copenhagen, Denmark, 2020.
8. INERIS. COVID Impact on Air Quality in Europe: A Preliminary Regional Model Analysis. Available online: https://policy.atmosphere.copernicus.eu/reports/CAMS71_COVID_20200626_v1.3.pdf (accessed on 1 December 2020).
9. Serbian Environmental Protection Agency. Combined Review of Automatic Air Quality Monitoring in the Republic of Serbia. Available online: http://www.amskv.sepa.gov.rs/ (accessed on 12 February 2020).
10. Provincial Secretariat for Urban Planning and Environmental Protection. Available online: http://www.ekourbapv.vojvodina.gov.rs (accessed on 12 February 2020).
11. Monitoring the Air Quality of the City of Novi Sad. Available online: https://environovisad.rs/vazduh (accessed on 5 December 2020).

12. Cleaner Air for All. Available online: https://ec.europa.eu/environment/air/cleaner_air/ (accessed on 5 December 2020).
13. Petrović, J. Environmental Aspects of Thermal Power Plants Utilization in Novi Sad—Air Pollution. Ph.D. Thesis, University Educons, Faculty of Environmental Protection, Sremska Kamenica, Serbia, April 2017.
14. WHO Regional Office for Europe. Health Impact of Ambient Air Pollution in Serbia, A Call to Action. Available online: https://serbia.un.org/en/22141-health-impact-ambient-air-pollution-serbia-call-action (accessed on 5 December 2020).
15. Eeftens, M.; Tsai, M.-Y.; Ampe, C.; Anwander, B.; Beelen, R.; Bellander, T.; Cesaroni, G.; Cirach, M.; Cyrys, J.; de Hoogh, K. Spatial variation of $PM_{2.5}$, PM_{10}, $PM_{2.5}$ absorbance and PMcoarse concentrations between and within 20 European study areas and the relationship with NO2–results of the ESCAPE project. *Atmos. Environ.* **2012**, *62*, 303–317. [CrossRef]
16. DunavNet. Environmental Monitoring. Available online: https://dunavnet.eu/solutions/environmental-monitoring/ (accessed on 5 December 2020).
17. Plantower. Digital Universal Particle Concentration Sensor, PMS7003 Series Data Manual. Plantower: 2016. Available online: https://download.kamami.com/p564008-p564008-PMS7003%20series%20data%20manua_English_V2.5.pdf (accessed on 5 December 2020).
18. Levy Zamora, M.; Xiong, F.; Gentner, D.; Kerkez, B.; Kohrman-Glaser, J.; Koehler, K. Field and laboratory evaluations of the low-cost plantower particulate matter sensor. *Environ. Sci. Technol.* **2018**, *53*, 838–849. [CrossRef] [PubMed]
19. Sven Leckel. LVS3 Poduct Page. Available online: https://www.leckel.de/ (accessed on 1 December 2020).
20. Wang, J.; Ogawa, S. Effects of meteorological conditions on PM2. 5 concentrations in Nagasaki, Japan. *Int. J. Environ. Res. Public Health* **2015**, *12*, 9089–9101. [CrossRef] [PubMed]
21. Menut, L.; Bessagnet, B.; Siour, G.; Mailler, S.; Pennel, R.; Cholakian, A. Impact of lockdown measures to combat Covid-19 on air quality over western Europe. *Sci. Total Environ.* **2020**, *741*, 140426. [CrossRef] [PubMed]
22. Singh, V.; Singh, S.; Biswal, A.; Kesarkar, A.P.; Mor, S.; Ravindra, K. Diurnal and temporal changes in air pollution during COVID-19 strict lockdown over different regions of India. *Environ. Pollut.* **2020**, *266*, 115368. [CrossRef] [PubMed]
23. Donzelli, G.; Cioni, L.; Cancellieri, M.; Llopis Morales, A.; Morales Suárez-Varela, M.M. The Effect of the Covid-19 Lockdown on Air Quality in Three Italian Medium-Sized Cities. *Atmosphere* **2020**, *11*, 1118. [CrossRef]
24. Fu, F.; Purvis-Roberts, K.L.; Williams, B. Impact of the COVID-19 Pandemic Lockdown on Air Pollution in 20 Major Cities around the World. *Atmosphere* **2020**, *11*, 1189. [CrossRef]
25. European Environment Agency. Air Quality and COVID19. Available online: https://www.eea.europa.eu/themes/air/air-quality-and-covid19 (accessed on 1 December 2020).
26. Lee, C.-H.; Wang, Y.-B.; Yu, H.-L. An efficient spatiotemporal data calibration approach for the low-cost PM2. 5 sensing network: A case study in Taiwan. *Environ. Int.* **2019**, *130*, 104838. [CrossRef] [PubMed]
27. Trilles, S.; Vicente, A.B.; Juan, P.; Ramos, F.; Meseguer, S.; Serra, L. Reliability Validation of a Low-Cost Particulate Matter IoT Sensor in Indoor and Outdoor Environments Using a Reference Sampler. *Sustainability* **2019**, *11*, 7220. [CrossRef]
28. EUMETSAT. Aralkum Desert Dust Pollutes Air in South-East European Countries on March 27th. Available online: https://www.eumetsat.int/aralkum-desert-dust-pollutes-air-south-east-europe (accessed on 5 December 2020).
29. Grimm Aerosol. MODEL EDM180 Product Page. Available online: https://www.grimm-aerosol.com/products-en/environmental-dust-monitoring/approved-pm-monitor/edm180-the-proven/ (accessed on 1 December 2020).

Article

Emissions reduction of Greenhouse Gases, Ozone Precursors, Aerosols and Acidifying Gases from Road Transportation during the COVID-19 lockdown in Colombia

Yiniva Camargo-Caicedo [1,*], Laura C. Mantilla-Romo [1] and Tomás R. Bolaño-Ortiz [2,3,4,*]

1. Environmental Systems Modeling Research Group (GIMSA), University of Magdalena, Santa Marta 470001, Colombia; lauramantillacr@unimagdalena.edu.co
2. Mendoza Regional Faculty-National Technological University (FRM-UTN), Mendoza M5500, Argentina
3. National Scientific and Technical Research Council (CONICET), Mendoza M5500, Argentina
4. Centre for Environmental Technologies (CETAM), Universidad Técnica Federico Santa María, Valparaíso 46383, Chile
* Correspondence: ycamargo@unimagdalena.edu.co (Y.C.-C.); tomas.bolano@frm.utn.edu.ar (T.R.B.-O.)

Citation: Camargo-Caicedo, Y.; Mantilla-Romo, L.C.; Bolaño-Ortiz, T.R. Emissions reduction of Greenhouse Gases, Ozone Precursors, Aerosols and Acidifying Gases from Road Transportation during the COVID-19 lockdown in Colombia. *Appl. Sci.* **2021**, *11*, 1458. https://doi.org/10.3390/app11041458

Academic Editor: Thomas Maggos

Received: 27 December 2020
Accepted: 26 January 2021
Published: 5 February 2021

Publisher's Note: MDPI stays neutral with regard to jurisdictional claims in published maps and institutional affiliations.

Copyright: © 2021 by the authors. Licensee MDPI, Basel, Switzerland. This article is an open access article distributed under the terms and conditions of the Creative Commons Attribution (CC BY) license (https://creativecommons.org/licenses/by/4.0/).

Abstract: The aim of this work was to analyze the changes in the emissions from the transport sector during the COVID-19 lockdown in Colombia. We compared estimated emissions from road transportation of four groups of pollutants, namely, greenhouse gases (CO_2, CH_4, N_2O), ozone precursor gases (CO, NMVOC, NOx), aerosols (BC, $PM_{2.5}$, PM_{10}), and acidifying gases (NH_3, SO_2), during the first half of 2020 with values obtained in the same period of 2018. The estimate of emissions from road transportation was determined using a standardized methodology consistent with the 2006 Intergovernmental Panel on Climate Change (IPCC) Guidelines for National Greenhouse Gas Inventories and the European Environment Agency/European Monitoring and Evaluation Program. We found a substantial reduction in GHG emissions for CH_4, N_2O, and CO_2 by 17%, 21%, and 28%, respectively. The ozone precursors CO and NMVOC presented a decrease of 21% and 22%, respectively, while NOx emissions were reduced up to 15% for the study period. In addition, BC decreased 15%, and there was a reduction of 17% for both PM_{10} and $PM_{2.5}$ emissions. Finally, acidifying gases presented negative variations of 19% for SO_2 and 23% for NH_3 emissions. Furthermore, these results were consistent with the Ozone Monitoring Instrument (OMI) satellite observations and measurements at air quality stations. Our results suggest that the largest decreases were due to the reduction in the burning of gasoline and diesel oil from the transport sector during the COVID-19 lockdown. These results can serve decision makers in adopting strategies to improve air quality related to the analyzed sector.

Keywords: COVID-19; lockdown; acidifying gases; aerosols; greenhouse gases; ozone precursors; road transportation; Colombia

1. Introduction

COVID-19 emerged on 30 December 2019 [1] and was declared a global pandemic by the World Health Organization on 11 March 2020 [2]. The outbreak of the virus started in Wuhan, the capital of Hubei Province, China, and in a few weeks, it had spread to dozens of other countries in Asia [3]. Since then, the SARS-CoV-2 virus has spread in Africa, America, Asia, Europe and Oceania [4]. It led to most countries adopting isolation measures to stop its spread and avoid the collapse of health systems [5]. The first case in Colombia was confirmed by the National Health Institute on 6 March 2020. The Ministry of Health and Social Protection declared a public health emergency in the country on 12 March 2020, and a few weeks later, the Ministry of Interior ordered preventive lockdown and containment measures starting on 25 March 2020, whereby many human activities in the educational, cultural, transportation, and industrial manufacturing sectors were constrained. Consequently, educational institutes and non-essential factories remained closed, public events were cancelled, and work at home was implemented, to prevent the further spread of the COVID-19 pandemic.

The anthropogenic changes caused by the lockdown led to a decline in industrial production and energy consumption, up to 30% in some countries [6,7]. Energy demand has been altered drastically worldwide, and due to forced confinement, many international borders were closed and populations were isolated in their homes [8]. This led to a change in some consumption patterns for energy, e.g., those related to the transport sector, because of a reduction in mobility. These restrictions on economic activity during the pandemic have reduced NO_2 emissions in China, Europe and the United States during COVID-19 [9].

Mobility has also been one of the things most affected by the COVID-19 restrictions. The changes in patterns of mobility indicate a reduction in vehicular traffic; as a consequence, a decrease in emissions associated with this sector is to be expected, given that the greenhouse gas (GHG) emissions from road and aviation transportation make up 72% and 11% of all GHG emissions, respectively [10]. Consequently, containment measures implemented in various countries have shown changes in the air quality [11–14]. The use of fossil fuels by road vehicles is the main source of four groups of pollutants, including GHG [10], including carbon dioxide (CO_2), methane (CH_4), and nitrous oxide (N_2O); ozone precursor gases, such as carbon monoxide (CO) [15], non-methane volatile organic compounds (NMVOC) [16], and nitrogen oxides (NOx) [17,18]; aerosols, including black carbon (BC) [10] and particulate matter ($PM_{2.5}$, PM_{10}) [15]; acidifying gases, such as ammonia (NH_3) [19] and sulfur dioxide (SO_2) [20]. GHG emissions, such as CO_2, are mainly produced by power generation and road transport. Other GHG emissions, such as CH_4, are generated by fermentation processes, fossil fuel extraction and use, landfills and waste. In addition, N_2O is produced from soil emissions [21]. Ozone precursor gases, such as CO, are emitted by incomplete fuel combustion of road transport as well as industrial processes [22]. NMVOCs are important air pollutants because of their contributions of secondary compounds (aerosols and ozone), generated from gasoline combustion [16,23,24]. The emissions of NOx (NOx = NO + NO_2) mainly include biomass burning and fuel combustion (e.g., power plant combustion, industrial emissions and transportation emissions) [25]. Aerosol emissions are contributed mostly as by-products of combustion from thermal power stations, vehicle engines and factories [26], with on-road vehicles being the source of fine particulate matter ($PM_{2.5}$) [27]. In addition, one of the main anthropogenic emissions sources of BC is the incomplete combustion of fossil fuels (especially diesel) in vehicles [10]. Acidifying gases are emitted by the combustion of biomass and fossil fuels as well as by industrial activity [19,20]. NH_3 emissions related to road traffic are due to use of catalytic NOx reduction systems on light and heavy-duty vehicles [19], whose devices use an injection of urea or ammonia [28]. Recent studies showed that the containment measures to minimize the spread of SARS-CoV-2 have resulted in reductions of 15% to 40% in industrial sectors and temporarily reduced China's CO_2 emissions by 25%. The European Public Health Alliance (EPHA) states that, in Italy, the urban NO_2 pollution comes mainly from traffic, especially diesel vehicles, which are also a major source of particulate matter; the COVID-19 pandemic has resulted in a remarkable drop in these pollutants. France also showed a drop in NOx emissions as a result of the reduction in economic activities and transportation. During the spread of the COVID-19 pandemic in New York, traffic levels were estimated to be down 35% compared with the previous year; significant decreases in the emissions of CO and CO_2 were registered, with a 5–10% reduction in CO_2 [26].

Some studies have examined the effects of the COVID-19 lockdown on urban mobility [18,29–31]. The data show that mobility has dropped around the world as the spread of the virus has increased; public transportation systems were the most affected due to users refusing to use them in order to avoid social contact, and therefore the risk of contagion [32]. Other studies have shown an improvement in air quality in some Colombian cities due to mobility restrictions during the COVID-19 lockdown [33,34]. However, these studies did not look at the changes in atmospheric emissions associated with the observed air quality changes. Google, in its COVID-19 Community Mobility Reports for Colombia (https://www.google.com/covid19/mobility/), reports that, in April 2020, the country saw the biggest reduction in visits to retail and recreation places (77%), transport stations (77%), parks (67%), grocery stores and pharmacies (59%), and workplaces (58%), while the trend of mobility in residential areas increased by 28%. At the beginning of May, the opening of some economic sectors caused an increase in mobility in relation to the previous month, especially in workplaces (17%), grocery stores and pharmacies (13%), retail and recreation places (10%), and transport stations (9%).

Therefore, the aim of this study is to analyze the changes in the emissions associated with road transportation during the COVID-19 lockdown in Colombia, comparing these emissions with values obtained in the same period of 2018 for four groups of pollutants, namely, GHGs (CH_4, CO_2, N_2O), ozone precursor gases (CO, NMVOC, NOx), aerosols (BC, PM_{10}, $PM_{2.5}$), and acidifying gases (NH_3, SO_2). The results can serve decision makers in the development of strategies to improve air quality related to the road transport sector in Colombia. This article is ordered as follows. Section 2 describes the methodology applied to estimate emissions in Colombia and details the changes in air quality observed by Bogotá's air quality network and from the OMI satellite. Section 4 details the results of the emissions changes and improvements in air quality in Colombia due to its COVID-19 pandemic lockdown, while Section 5 discusses the results and provides further analysis in light of updated literature. Finally, Section 5 reports the main conclusions and perspectives.

2. Materials and Methods

2.1. Study Area

Colombia occupies a total surface of 1,140,000 km^2 in the northern part of South America (Figure 1). It has a population of approximately 49.5 million inhabitants, distributed into 32 departments and one capital district, Bogotá D.C., with a population of 7.8 million [35]. The gross domestic product (GDP) was 323.80 billion USD (at current prices), with a per capita income of 7842 USD (GDP/capita) in 2019, according to the World Bank data and its trading economics projections [36]. The country's vehicle fleet reached 15.6 million units in 2020 [37], with a fuel consumption during the first half of the year equivalent to 2.5 million m^3 diesel oil, 2.6 million m^3 gasoline and 600,000 m^3 compressed natural gas (CNG). According to the last Colombia GHG national inventory presented to the Intergovernmental Panel on Climate Change (IPCC) [38], from a sectorial point of view, annual Carbon Dioxide Equivalent (CO_2 eq.) emissions (for the year 2012) correspond to 158.6 Tg to agriculture, 78.0 Tg to energy, 13.3 Tg to waste, and 8.9 Tg to industry. While the transport sub-sector emitted 28.2 Tg, contributing 36% of energy sector emissions and 11% of the total emissions of the country.

2.2. Emission Estimation

We studied emissions from road transportation in four groups of pollutants that affect climate change, air quality and health, namely, GHGs (CH_4, CO_2 and N_2O), ozone precursor gases (CO, NMVOC and NOx), aerosols (BC, PM_{10} and $PM_{2.5}$), and acidifying gases (NH_3 and SO_2).

Several studies have been conducted to estimate the emissions from road transportation based on fuel consumption [17,39,40]. To estimate these emissions, we selected a standardized methodology consistent with the 2006 IPCC Guidelines for National Greenhouse Gas Inventories [41] and the method from the EEA/EMEP Emission Inventory Guidebook 2019. [42]. Thus, we used tier 1 methods that use activity data derived from available statistical information (energy statistics, production statistics, traffic counts, population size, etc.). In addition, tier 1 emission factors were chosen to represent "typical" or "averaged" process conditions; they tended to be independent of technology. Furthermore, we used an additional level of detail (tier 2) for the calculation of SO_2 emissions, since Colombian fuel emission factors were used [43]. This is consistent with previous studies that showed that this methodology was adequate to estimate inventories at the national level, when detailed information by city was not available [20,44–47]. Overall, the method was based on estimating emissions through a linear relationship between activity data and emission factors (Table A1). The calculation was made using Equation (1), as follows:

$$E(p) = \sum_{p,f,v} \left(Fuel_{f,v} * Ef_{p,f,v} \right) \quad (1)$$

where $E(p)$ is the total emission for species or pollutant p, $Fuel\,(f,v)$ is the fuel sold (diesel, gasoline and CNG) for type of vehicle v, $Ef(p,f,v)$ is the emission factor for pollutant species p, for type of fuel f and vehicle v.

Figure 1. Location of Colombia in South America. Study area covers the entire territory of Colombia.

Therefore, the emission estimate for each polluting species was calculated using Equation (1) with the following data:

Fuel: We used the Statistical Bulletin by Ministry of Mines and Energy [48], which includes activities such as monthly sales of fuels for the first half of 2018. We used the Liquid Fuel Information System (SICOM) [49], which includes monthly sales to fuel retail distributors for the first half of 2020. In addition, we used the Mercantile Exchange

Colombia [50] (as shown in Table A2), for data on the consumption of CNG (Figure A1). Furthermore, considering that consumption was only focused on the transport sector, fuel distribution data was obtained from automotive service stations, assuming 96% of the distribution of the total of this category (retail distributors) was the total of the fuel supply, according to SICOM data [49]. For fuel consumption by vehicle type, the consumption distribution percentages (Table 1) of the indicative action plan for energy efficiency [51] were selected, calculating consumption by vehicle category.

Table 1. Fuel consumption by vehicle category [51].

Fuel	Consumption (%)				
	Cars	Cargo	Public Transport	Motorcycles	Others
Gasoline	77	-	-	22	1
Diesel oil	18	53	26	-	3
CNG	91	7	2	-	-

Number of vehicles: Census of number of vehicles by type (vehicle category) from the Single National Traffic Registry of Colombia [37].

Emission factors: The emission factors considered were those established by the EMEP/EEE Joint Inventory Guide to Air Pollutant Emissions database [42], for vehicle type and pollutant (GHGs, ozone precursors, aerosols and acidifying gases). The SO_2 emission factor and the power calorific value by type of fuel was obtained from the 2016 UPME FECOC calculator (Colombian fuel emission factors) in energy units (Kg Tj^{-1}) [43], except the CNG power calorific value was taken from the PROMIGAS technical notes [52]. These values were assumed for all types of vehicle under study. Additionally, CO_{2eq} from the main GHGs (CH_4 and N_2O) was estimated. CO_{2eq} emissions with a 100-year horizon global warming potential (GWP100: CH_4 = 28 and N_2O = 298) have been considered through the IPCC's suggestion in the 5th Assessment Report (AR5) [53]. We analyzed the monthly variations in emissions from January to June 2018 and 2020. The emissions reduction during the COVID-19 pandemic lockdown in Colombia was calculated based on the year 2018.

2.3. Emissions Reduction vs. Air Quality Improvement

We analyzed the improvements in air quality to relate them to the emission reductions analyzed during the quarantine period. We used data from five air quality traffic stations in Bogotá (Carvajal-Sevillana, Estación Móvil, Fontibón, Las Ferias, Minambiente) available in the Bogotá Air Quality Monitoring Database (BAQMD) [54]. These data were used to assess the air quality concentration of CO, SO_2, NO_2 and O_3; the equipment used by BAQMD is specified in Table A3. For each station, data from April, May and June of 2018 were used to calculate the mean concentrations of each pollutant for each month. Similarly, data from April, May and June of 2020 were used to calculate mean levels of each pollutant during the lockdown. It is worth clarifying that BAQMD reports pollutant concentrations under standard conditions (1 atm and 25 °C). Thus, this allowed us to perform a comparison with concentrations during the same period of a base year (2018). This base-year comparison was also performed to control for meteorological conditions. We used tropospheric NO_2 data for April to June 2018 and 2020, retrieved from the ozone monitoring instrument (OMI), a visual and ultraviolet spectrometer aboard the NASA Aura spacecraft [55]. This information enabled the emissions analysis and estimation associated with road transportation in the four groups of pollutants previously cited. In addition, the average NO_2 retrieved from OMI data was estimated for the period of April to June 2018 and 2020 to evaluate the NO_2 level variation during the pandemic lockdown in Colombia [56].

3. Results

Figure 2 shows the monthly emissions of analyzed pollutants for the compared periods. In the first half of 2020, the emissions of the four groups of pollutants associated with road transportation decreased starting in March compared with those estimated for the base year 2018. In late March 2020, the national government adopted vehicle restrictions, so April showed a higher reduction of GHG emissions for CO_2, CH_4 and N_2O in percentages equivalent to 58%, 40% and 71%, respectively. CO_2 reduction was the most representative due its contribution of 97.62% of the total emissions from the transport sector, specifically the burning of fossil fuels by road transportation (lightweight and cargo vehicles) [38]. Later, GHG emissions increased in May by 24%, 16% and 27% for CO_2, CH_4 and N_2O, respectively, owing the reactivation of some economic sectors. Restrictions began to be relaxed, allowing the opening of some activities that were restricted during the confinement. As a result, GHG emissions in June continued to increase, though they remained lower than those of 2018.

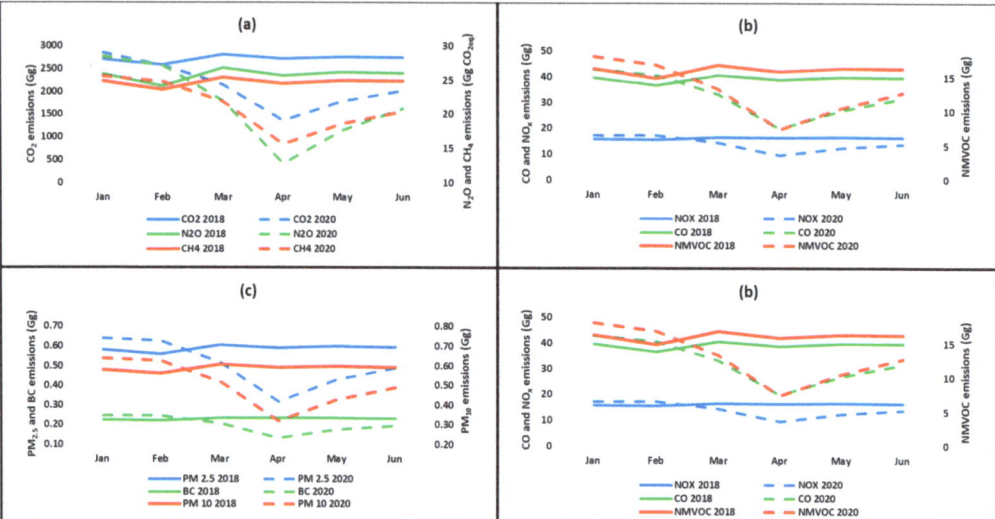

Figure 2. Estimated total emissions (Gg) of the four groups of pollutants that affect climate change, air quality and health: (**a**) GHGs (CH_4, CO_2 and N_2O); (**b**) ozone precursors (CO, NMCOV and NOx); (**c**) aerosols (BC, PM_{10} and $PM_{2.5}$); (**d**) acidifying gases (NH_3 and SO_2) for January to June of 2018 and 2020.

As shown in Figure 3, all estimated pollutants showed reductions between January and June 2020 due to the pandemic lockdown in Colombia. Negative variations in GHG emissions were 28%, 17% and 20% for CO_2, N_2O and CH_4, respectively. While the ozone precursor group showed a reduction of up to 21% and 22% for CO and NMVOC, respectively. The emissions of these pollutants were mostly the result of burning gasoline and diesel oil, which represent 90% of the total emissions. In addition, the NOx emissions variation was −15% for the study period, with 50% of the total emissions by this pollutant attributed to the burning of diesel oil.

Aerosol emissions of PM_{10} and $PM_{2.5}$ each showed a negative emissions variation of 17%, which was associated mostly with the fuel consumption by cargo vehicles and public transport [57]. BC emissions showed a decrease of 15%, and acidifying gases also displayed reductions. SO_2 emissions showed a negative variation of 19%, while NH_3 emissions were reduced by 23% of its. These emissions reductions were mainly produced by the reduction in consumption of gasoline and diesel oil. In general, pollutant groups that registered the most reduction in emissions variations were GHGs (−22%) and acidifying gases (−21%), while CO_2 presented the greatest reduction among all pollutants analyzed.

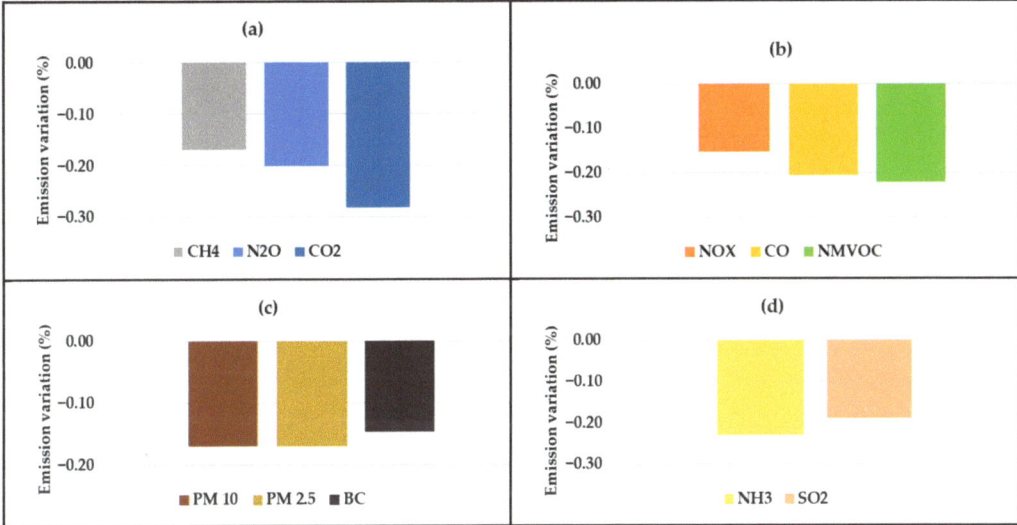

Figure 3. Emissions variations of the four groups of pollutants in the study: (**a**) GHGs; (**b**) ozone precursors; (**c**) aerosols; (**d**) acidifying gases, during the January to June 2020 in relation to the same period of 2018.

Figure 4 shows the variations in CO_2 emissions in Colombia. Territorial divisions that showed the greatest reduction in CO_2 emissions were Bogotá D.C. (−4168 Gg CO_2), Magdalena (−1381 Gg CO_2), Bolívar (−308 Gg CO_2), Atlántico (−118 Gg CO_2), and Caquetá (−30 Gg CO_2). These contrast with positive emission variations in departments such as Valle del Cauca (295 Gg CO_2), Cundinamarca (278 Gg CO_2), Norte de Santander (248 Gg CO_2), Antioquia (232 Gg CO_2), and Cesar (209 Gg CO_2), during the study period.

Colombian administrative divisions that showed the greatest reduction in CO_2 (Bogota, Magdalena, Bolivar, Atlántico and Caquetá) make up 44.5% of the national population. The circulation of people was reduced to avoid contagion by COVID-19. Thus, these territories registered (between March and June 2020) a decrease of 6005 Gg of CO_2 compared to the same period in 2018. While Valle del Cauca, Cundinamarca, Norte de Santander, Antioquia, and Cesar departments reported a total increase of 1262 Gg CO_2. Overall, the net reduction in Colombia was approximately of 4743 Gg CO_2 (Table A4).

Considering the significant emission reduction of CO_2 in Bogotá D.C., associated with road transportation and its population density, we also analyzed data from five air quality traffic stations in Bogotá. In addition, we evaluated the concentrations of CO, SO_2, NO_2 and O_3 during the lockdown period ranging from April to June 2020 and compared these to the same period in 2018. We observed significant air quality improvements through a decrease in CO, SO_2 and NO_2 in areas influenced by vehicular traffic. Drastic reductions in CO (up to −60.85%), SO_2 (up to −73.23%), and NO_2 (up to −60.60%) concentrations were observed in the urban area during the lockdown, as shown in Table 2. By contrast, an increase of up to 106.32% (in May) in ozone concentrations was observed in urban areas of Bogotá.

Figure 5 shows NO_2 concentration reductions visualized by satellite measurement of background tropospheric data available from OMI. The levels of NO_2 over Colombia decreased substantially in the Central Region during the lockdown (April to June 2020) compared to the same period in 2018. Nevertheless, the north region showed an increase in the levels of NO_2 over Atlántico, Bolívar, Cesar, La Guajira and Magdalena departments.

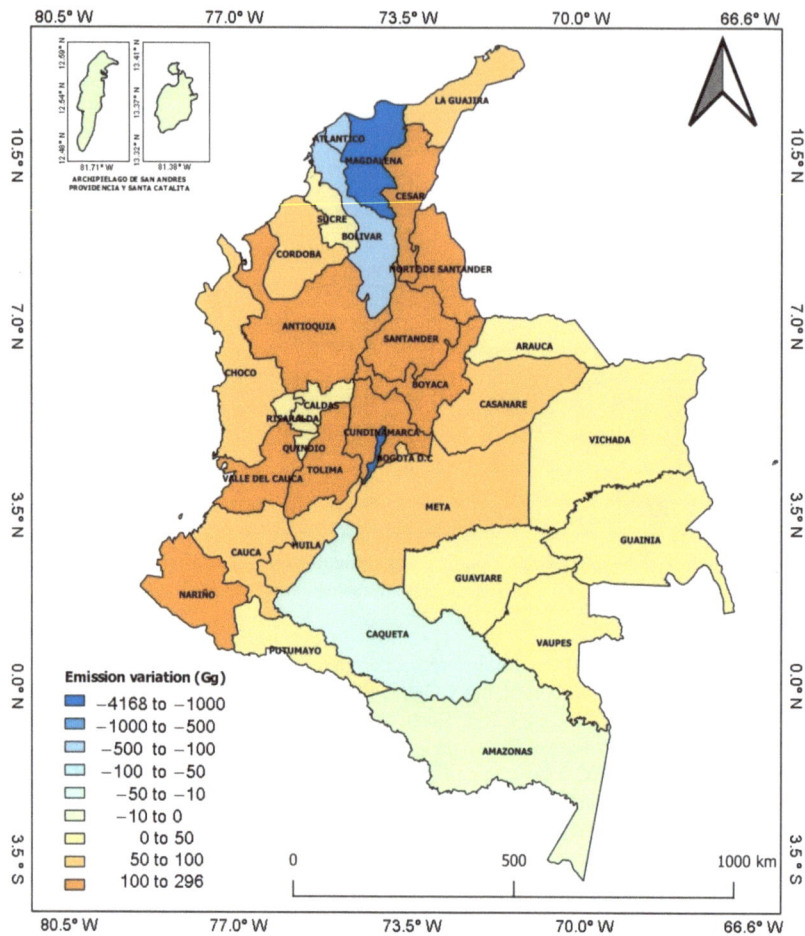

Figure 4. Spatial distribution of CO_2 emissions variation through internal political and territorial divisions (departments). Warm and cold colors indicate an increase and decrease, respectively, in emissions between the March and June 2018 and 2020.

Table 2. Mean concentration and standard deviation of CO, SO_2, NO_2 and O_3 in Bogotá during the lockdown (April to June 2020) compared to the same period in 2018 [54].

Air Pollutant	Mean Concentration 2018 ($\mu g.m^{-3}$)			Mean Concentration 2020 ($\mu g.m^{-3}$)			Variation of Mean Concentrations (%) from 2018 to 2020		
	Apr	May	Jun	Apr	May	Jun	Apr	May	Jun
CO	1408.75 ± 363.64	1269.63 ± 304.62	1074.67 ± 396.98	551.55 ± 276.74	787.13 ± 301.38	920.46 ± 422.75	−60.85	−38.18	−13.8
SO_2	3.90 ± 2.00	3.41 ± 1.26	4.50 ± 1.50	2.83 ± 1.19	3.30 ± 2.06	4.13 ± 2.15	−27.39	−16.57	−8.22
NO_2	56.10 ± 12.20	44.70 ± 7.82	46.60 ± 7.46	22.10 ± 9.75	27.80 ± 14.46	29.30 ± 12.59	−60.6	−37.81	−37.12
O_3	14.66 ± 6.88	10.28 ± 3.95	13.06 ± 6.75	37.53 ± 13.61	21.21 ± 7.07	18.16 ± 7.43	60.92	106.32	27.66

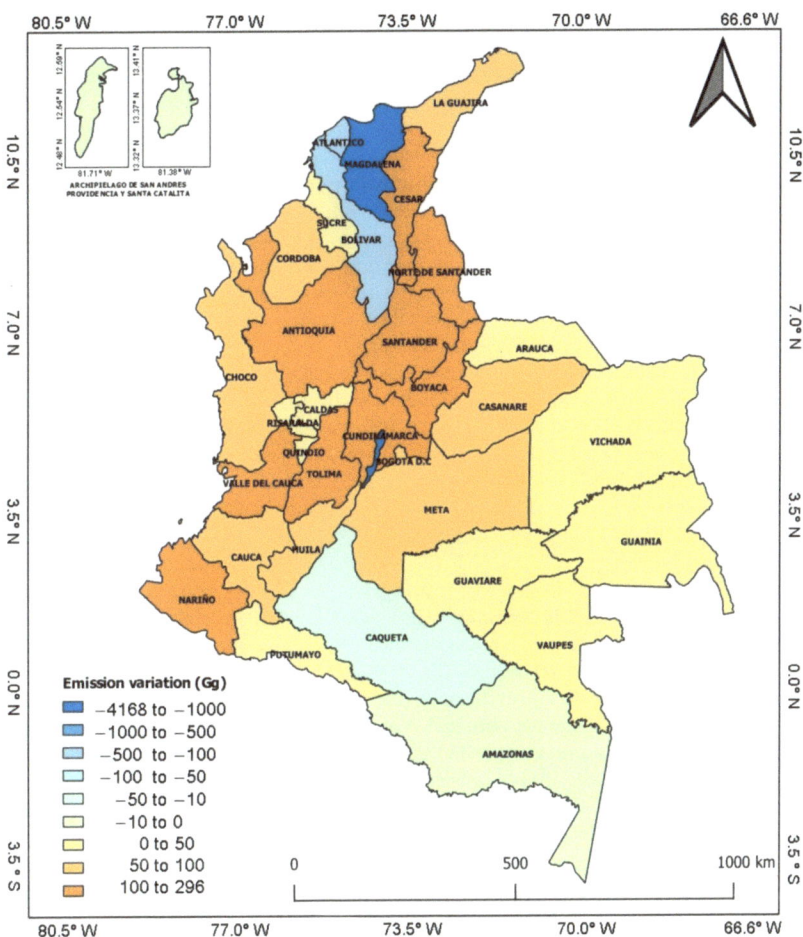

Figure 5. Spatial distribution of mean levels of tropospheric NO_2 through internal political and territorial divisions (departments) between April and June 2018 and 2020. Source: Time averaged map of NO_2 tropospheric column (30% cloud screened) daily 0.25 deg. (OMI OMNO2dv003) $1/cm^2$.

4. Discussion

The Colombian government's restrictions to curb the spread of the COVID-19 pandemic have had a significant impact in several sectors of its economy due to the cessation of some activities [58]. Our results showed reductions for the four groups of pollutants analyzed. In particular, a total of 6010 Gg were eliminated, mainly in seven territorial subdivisions of Colombia where close to 50% of the national population live [59]. One of the positive impacts identified is the emissions reduction from decreased road transport. This is registered by recent studies on air quality improvements carried out in Sao Paulo (Brazil), which reported high reductions of air pollutant concentrations during its partial lockdown due to the decrease in vehicular traffic in analyzed areas [60]. In Barcelona (Spain), the most significant reductions were estimated for pollutants related to traffic emissions [61]. Emissions in China caused by road transport have been affected by the lockdown, generating a reduction of the pollutants associated with this sector [62]. Therefore, the lockdown significantly reduced the air pollution (air pollutants and warming gases) in most cities across the world [26].

Emissions of the four pollutant groups selected in this study depend on the consumption of fossil fuels, which during the lockdown decreased in accordance with the lower vehicle traffic in Colombia. According to Colombian government reports, diesel and gasoline consumption experienced a drop of 50% and 65%, respectively, since mid-March when the lockdown began [63]. Furthermore, the Mercantile Exchange Colombia did not register increases in the consumption data for CNG [50]. This led to a reduction in the estimated emissions of the four pollutant groups studied, with the most variation in GHGs, specifically CO_2. It is consistent with recent studies that affirm the first sector with the greatest reduction in global emissions of CO_2 during isolation was transportation [64].

The emissions reduction of ozone precursor gases (CO, NMVOC, NOx) registered in this study is consistent with the highest reduction of CO and NO_2 that occurred in China due the lockdown measures taken to control the COVID-19 pandemic, which dramatically reduced the number of vehicles on the road, and consequently led to an improvement in air quality due most likely to reduced emissions from some sectors (such as the transportation linked to the NO_2 emissions). This occurred chiefly in those provinces with large fleet vehicular and secondary industries, which suggests that the reduced emissions from the transportation and industrial sectors caused a decrease in concentrations of these gases [18]. In addition, it was reported that NO_2 emissions were reduced by up to 60% in the city of Santander (Spain) [32]. Other studies found a 20–30% reduction in emissions of NO_2 in China, Spain, France, Italy, and the USA due to the lockdown [9] and a drastic reduction of NO (up to -77.3%), NO_2 (up to -54.3%), and CO (up to -64.8%) in Sao Paulo (Brazil). In the case of NO, one recent study demonstrated that heavy-duty diesel trucks are the major sources of this pollutant [65]. While the NMVOC emissions reduction was -22% in this study, other research has shown a $PM_{2.5}$ emissions reduction of -17% [23].

Aerosol reductions (BC, PM_{10}, $PM_{2.5}$) in our study were consistent with recent studies. Chinese researchers carried out an analysis of $PM_{2.5}$ data in cities such as Beijing, Shanghai, Guangzhou, and Wuhan during COVID-19 and found a pronounced reduction in air pollution attributed to the reduction of emissions in transportation and industrial sectors [18]. As well, it was observed over the major cities of India, such as Delhi, Mumbai, Hyderabad, Kolkata and Chennai, that a decline in $PM_{2.5}$ during the lockdown period registered a significant improvement in air quality, which provides important information to the cities' administration about the implementation of regulations [14]. Other studies conducted during the lockdown suggested the main sources of atmospheric particulate matter PM_{10} and $PM_{2.5}$ (include fossil fuel combustion, motor vehicle exhaust emissions, industrial production, secondary particulate matter generation, among others) experienced a significant reduction up to -48.9% in three of China's provinces [66]. The decline in $PM_{2.5}$ emissions due to the lockdown to control the spread of SARS-CoV-2 in New York, Los Angeles, Zaragoza, Rome, Dubai, Delhi, Mumbai, Beijing and Shanghai reflected the positive changes that contributed to improve air quality [67]. BC emissions reduction can be attributed to on-road diesel sources [68], so the mitigation of transportation-related BC emissions decreased the global emissions significantly [69].

In this study, acidifying gases (NH_3, SO_2) also showed a significant emissions reduction, up to -23% for NH_3. Other studies found that decreasing emissions were identified in Kannur district, India (-16%), due to a complete shutdown of traffic and industrial activities [70], as NH_3 emissions come mainly from heavy-duty diesel vehicles [65]. In addition, SO_2 emissions registered a decrease (-19%), which was identified in China as a decrease attributed to lower emissions from traffic and coal combustion [62]. Kannur, India, reported decreased emissions (-62%), and a diurnal variation most pronounced during peak traffic hours was absent during the lockdown owing to the roads being deserted [70].

Figure 4 shows NO_2 emissions increased in the northern Colombian region due to events of long-range pollution transport, like regional biomass burning beginning at the end of March, during the lockdown, according to recent studies [33,34]; the air quality improvement shown in this period was partially annulled by the impact of these events.

Despite an emissions reduction in the four pollutant groups selected, an increase of ozone concentration was observed in urban areas of Bogotá. This result was consistent with recent studies, in which Sao Paulo (Brazil) urban areas, highly influenced by road transportation, had an increase of approximately 30% in ozone emissions [60]. The increase of ozone concentration is related to nitrogen monoxide decreases, which may cause a reduction in ozone consumption during the photochemical reactions [61,71]. Moreover, VOCs are often the limiting precursors for O_3 production in urban areas [23,31]. O_3 levels increased up to 57%, probably due to lower titration of O_3 by NO (titration, $NO + O_3 = NO_2 + O_2$), and the decrease of NOx added to the increase of solar irradiation and temperatures in this period of the year [61]; ozone levels are a major concern in tropical cities, where the temperature and insolation favor the atmospheric processes leading to O_3 formation [31]. In this sense, recent studies also showed that reductions in $PM_{2.5}$ during the COVID-19 pandemic favored the formation of O_3 due to a reduction in NO_x levels due to reduced transport and an increase in solar radiation [31,72]. On the other hand, the increase in ozone seems to be associated with the decrease in $PM_{2.5}$, because the sinking of hydroperoxy radicals is slowed down, and therefore, ozone production accelerates [73].

Therefore, these results showed that a reduction in the transport sector contributed to lower emissions of the four pollutant groups (GHGs, ozone precursors, aerosols and acidifying gases), but was not able to cut down ozone concentrations, which leads us to consider other strategies aimed at reducing emissions and the reactivity in the troposphere, such as fuel composition and the control of vehicular emission systems. However, these results indicate that today, more than ever, we must take measures that are focused on individual behavioral changes.

Previous studies recommend high-impact actions for emissions savings >0.8 Mg CO_2 eq per year for countries, with potential contribution to systemic change and substantial reduction in annual emissions, such as living without vehicles (2.4 Mg CO_2 eq saved per year) and opting for more efficient vehicles or switching to electric cars (1.19 Mg CO_2 eq saved per year) [74]. Using the cleanest available technology (electric cars) results in significant reductions. Despite the fact that these actions can be effective, the dependence of people on the use of conventional cars is increasingly noticeable, and it is evidenced by the vehicle fleet records in Colombia. Therefore, governments should consider the adoption of incentives to use fewer polluting vehicles [75]. Also, Wynes et al. [74] show significant emissions reductions through moderate-impact actions (emissions savings 0.2–0.8 Mg CO_2 eq per year), such as replacing gasoline-burning vehicles with hybrid cars, and even the use of public transportation, which reduces emissions by 26–76% [76], as well as biking and walking. In Colombia, incentive measures should encourage the use of CNG or hybrid vehicles, as natural gas represents the lowest emissions compared to the other fuels under study.

The changes in air pollution during the COVID-19 lockdown can provide insight into the achievability of air quality improvement when there are significant restrictions in emissions related to the sectors with the greatest impact, thus giving regulators better ability to control air pollution [13]. However, it is likely that most of the changes observed in 2020 in terms of emissions are temporary, since no structural changes are reflected in the economic or transport systems [8]. Moreover, several studies have shown that poor air quality is related to increases in infections and mortality due to COVID-19 [77–80]. This would indicate that a reduction in emissions and improvements in air quality could also reduce the rate of infection and mortality due to COVID-19 [47,81–85]. Thus, it would be expected that prevention measures (such as social distancing and lockdowns, among others) are actually more profitable than a cure [78,86,87].

5. Conclusions

The effect of restricted human activities due to the COVID-19 pandemic in Colombia since mid-March of 2020 was studied by analyzing emissions variations of eleven criteria pollutants, comparing the first half of 2020 with values obtained in the same period of 2018.

In general, the air quality improved during the COVID-19 lockdown, and it was apparently caused by reductions in emissions of some human activities, such as in the transportation sector. Lifting the lockdown and the normalization of activities in the productive sectors may reverse the reduction of global air pollution and even increase air pollution levels if researchers, decision makers, productive sectors, and governments do not articulate efforts to maintain the economy with minimum emissions. COVID-19 has allowed us to analyze the positive impacts of the measures adopted during the lockdown, specifically those that have generated reductions in pollution emissions with evident consequences for the air quality. Thus, it is important to identify the impact of low, moderate and high actions on reducing emissions, with emphasis in the agricultural and energy sectors, and especially the contributions of the transport sub-sector. The circumstances under which we have lived, and the measures adopted during the pandemic, taking in consideration changes for improving environmental conditions, can be the subject of dialogue at the next conference of the United Nations for Climate Change, COP26. Additionally, future work may use more detailed methodologies, such as tier 3 [42,88], to achieve high-resolution spatial inventories in Colombia.

Author Contributions: Conceptualization, Y.C.-C. and T.R.B.-O.; data curation, Y.C.-C., L.C.M.-R. and T.R.B.-O.; formal analysis, Y.C.-C., L.C.M.-R. and T.R.B.-O.; investigation, Y.C.-C., L.C.M.-R. and T.R.B.-O.; methodology, Y.C.-C., L.C.M.-R. and T.R.B.-O.; writing—original draft preparation, Y.C.-C., L.C.M.-R. and T.R.B.-O.; project administration, Y.C.-C.; software, L.C.M.-R.; supervision, Y.C.-C. and T.R.B.-O.; visualization, Y.C.-C., L.C.M.-R. and T.R.B.-O.; writing—original draft, Y.C.-C., L.C.M.-R. and T.R.B.-O.; funding acquisition, Y.C.-C. All authors have read and agreed to the published version of the manuscript.

Funding: The work of T.R.B.-O. was funded in part by the Consejo Nacional de Investigaciones Científicas y Técnicas (CONICET) (National Council for Scientific and Technical Investigations), the National Agency of Scientific and Technological Promotion (ANPCyT) (Agencia Nacional de Promoción Científica y Tecnológica) under project PICT 2016 1115, and Universidad Tecnológica Nacional (UTN) (National Technological University) in Argentina. T.R.B.-O. also thanks GORE-FNDR-V Región in Valparaíso and Programa de Asignación Rápida de Recursos para Proyectos de Investigación Sobre el Coronavirus (COVID−19) (project CO-VID0581-2020), ANID, Ministerio de Ciencia, Tecnología, Conocimiento e Innovación, both in Chile. Y.C.-C. and L.M.R. thank the grants and fund research (AA N° 014-2015, AA N° 031-2016, Res N° 0388-2017, Res N° 0709-2017, Res N° 0119-2018, Res N° 0305-2018 and Res N°0347-2020) from Universidad del Magdalena (Santa Marta, Colombia).

Institutional Review Board Statement: Not applicable.

Informed Consent Statement: Not applicable.

Data Availability Statement: The data presented in this study are available on request from the corresponding authors.

Acknowledgments: We would like to thank Viverlys Diaz-Gutierrez for her help to graph Figure 4. We also would like to thank NASA's Aura Mission scientific teams, and their associated personnel, for the production of the data used in this research effort.

Conflicts of Interest: The authors declare no conflict of interest.

Appendix A

Table A1. Emission factors by vehicle classification and fuel type.

Vehicle Type	Fuel	CO_2 * (Kg/m^3)	CH_4 ** (kg/TJ)	N_2O ** (Kg/m^3)	CO * (Kg/m^3)	NMVOC * (Kg/m^3)	NO_x * (Kg/m^3)	BC * (Kg/m^3)	$PM_{2.5}$ * (Kg/m^3)	PM_{10} * (Kg/m^3)	NH_3 * (Kg/m^3)	SO_2 + (kg/TJ)
Personal cars	Gasoline	2329	25.00	0.15	61.74	7.39	6.42	3×10^{-3}	0.02	0.02	0.81	3.57
	Gas Oil	2678	3.90	0.07	2.81	0.59	10.95	0.53	0.93	0.93	0.05	2.91
	CNG	1972	92.00	0.06	60.90	9.81	10.93	-	-	-	0.06	-
Light commercial vehicles	Gasoline	2329	25.00	0.14	111.94	10.72	9.72	7×10^{-4}	0.01	0.01	0.49	3.57
	Gas Oil	2678	3.90	0.05	6.25	1.30	12.60	0.71	1.28	1.28	0.03	2.91
	CNG	1972	92.00	-	4.10	0.14	9.35	-	0.01	0.01	-	-
Heavy duty vehicles	Gas Oil	2678	3.90	0.04	6.41	1.62	28.20	0.42	0.79	0.79	0.01	2.91
	CNG	1972	92.00	-	4.10	0.19	9.35	-	0.01	0.01	-	-
Motorcycles	Gasoline	2329	25.00	0.04	39.54	96.58	4.88	0.18	1.62	1.62	0.04	3.57

* EMEP/EEA air pollutant emission inventory guidebook 2019 [42]. ** 2006 IPCC Guidelines for National Greenhouse Inventories [41].
+ 2016 UPME FECOC calculator (Colombian fuel emission factors) [43].

Table A2. Fuel sales (m^3) by department for March to June 2018 and 2020.

Departments	Gasoline		Diesel	
	2018	2020	2018	2020
Amazonas	1167	1808	1167	491
Antioquia	154,614	195,364	154,614	206,015
Arauca	2456	13,591	2456	11,024
San Andrés y Providencia	1844	2397	1844	621
Atlántico	73,855	46,499	73,855	53,498
Bogotá D.C.	1,112,981	184,188	892,924	144,130
Bolívar	118,597	45,119	124,428	73,072
Boyacá	25,132	39,383	21,619	53,587
Caldas	17,596	24,280	11,606	19,639
Caquetá	8680	14,167	22,570	6439
Casanare	7751	15,883	12,365	37,034
Cauca	18,299	41,413	11,317	22,147
Cesar	30,871	59,565	59,162	112,311
Choco	6606	19,768	5091	15,555
Córdoba	22,813	37,646	16,226	36,267
Cundinamarca	74,089	88,491	74,260	165,606
Guainía	956	3576	670	696
Guaviare	2257	4270	1314	2293
Huila	21,200	34,236	16,067	26,438
La Guajira	3781	31,629	10,080	16,893
Magdalena	268,517	22,459	320,757	18,756
Meta	19,924	34,320	22,967	47,797
Nariño	33,581	72,826	21,191	46,554
Norte de Santander	14,744	68,938	21,411	67,054
Putumayo	7304	17,838	6064	7558
Quindío	11,741	16,622	7083	13,289
Risaralda	20,316	29,554	14,635	22,211
Santander	45,700	63,675	42,618	65,638
Sucre	11,014	18,367	6122	14,021
Tolima	28,855	38,160	26,876	60,583
Valle del Cauca	106,590	143,532	84,913	163,125
Vaupés	183	458	244	93
Vichada	572	2877	636	1887

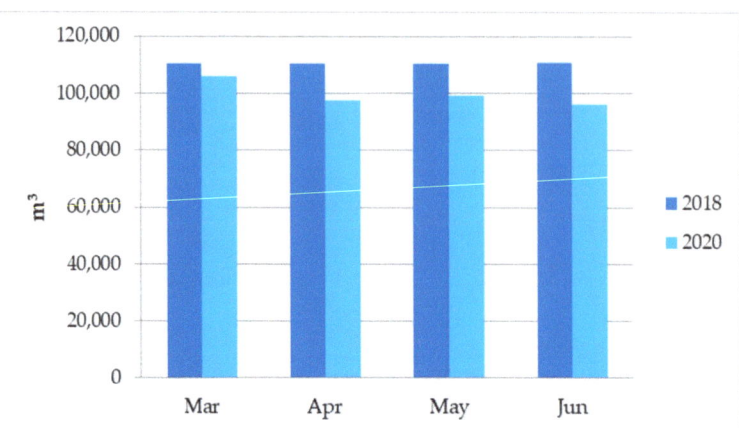

Figure A1. Fuel consumption from March to June 2018 and 2020. GNG sales data was only available at the national level.

Table A3. Equipment used by BAQMD to monitor air quality in Bogota city [54]. Note that only the equipment that measures the parameters (pollutants) used in this comparison is shown.

Pollutants	Measurement Principle Used	Equipment
CO	Infrared absorption spectrophotometry	CO Thermo Scientific 48i
SO_2	Ultraviolet pulsed fluorescence	SO_2 Thermo Scientific 43i
NO_2	Chemiluminescence	NOx Ecotech 9841
O_3	Absorption spectrophotometry in the ultraviolet	O_3 Ecotech 9841

Table A4. CO_2 emissions (Gg) by departments from March to June 2018 and 2020.

Departments	2018	2020
Amazonas	5.84	5.53
Antioquia	774.15	1006.71
Arauca	12.30	61.18
San Andrés y Providencia	9.23	7.25
Atlántico	369.79	251.56
Bogotá D.C.	4983.45	814.97
Bolívar	609.43	300.76
Boyacá	116.43	235.23
Caldas	72.06	109.14
Caquetá	80.65	50.24
Casanare	51.17	136.16
Cauca	72.93	155.77
Cesar	230.33	439.49
Choco	29.02	87.70
Córdoba	96.59	184.80
Cundinamarca	371.42	649.57
Guainía	4.02	10.19
Guaviare	8.78	16.09
Huila	92.40	150.54
La Guajira	35.80	118.91
Magdalena	1484.36	102.54
Meta	107.91	207.93
Nariño	134.96	294.31
Norte de Santander	91.67	340.13

Table A4. *Cont.*

Departments	2018	2020
Putumayo	33.25	61.79
Quindío	46.31	74.30
Risaralda	86.51	128.31
Santander	220.57	324.08
Sucre	42.05	80.33
Tolima	139.18	251.11
Valle del Cauca	475.65	771.13
Vaupés	1.08	1.31
Vichada	3.04	11.75

References

1. Aylward, B.; Liang, W. Report of the WHO-China Joint Mission on Coronavirus Disease 2019 (COVID-19). *WHO-China Jt. Mission Coronavirus Dis.* **2020**, *2019*, 16–24.
2. World Health Organization (WHO) Coronavirus disease 2019 Situation Report 51 11th March 2020. *World Health Organ.* **2020**, *2019*, 2633. [CrossRef]
3. Wang, C.; Horby, P.W.; Hayden, F.G.; Gao, G.F. A novel coronavirus outbreak of global health concern. *Lancet* **2020**, *395*, 470–473. [CrossRef]
4. JHU Coronavirus Resources Center. Available online: https://coronavirus.jhu.edu/map.html (accessed on 25 April 2020).
5. Wilder-Smith, A.; Freedman, D.O. Isolation, quarantine, social distancing and community containment: Pivotal role for old-style public health measures in the novel coronavirus (2019-nCoV) outbreak. *J. Travel Med.* **2020**, *27*, taaa020. [CrossRef] [PubMed]
6. McKibbin, W.; Fernando, R. The Global Macroeconomic Impacts of COVID-19. *Brook. Inst.* **2020**, 1–43. [CrossRef]
7. May, B. World Economic Prospects Monthly. *Econ. Outlook* **2020**, *44*, 1–33. [CrossRef]
8. Le Quéré, C.; Jackson, R.B.; Jones, M.W.; Smith, A.J.P.; Abernethy, S.; Andrew, R.M.; De-Gol, A.J.; Willis, D.R.; Shan, Y.; Canadell, J.G.; et al. Temporary reduction in daily global CO2 emissions during the COVID-19 forced confinement. *Nat. Clim. Chang.* **2020**, *10*, 647–654. [CrossRef]
9. Muhammad, S.; Long, X.; Salman, M. COVID-19 pandemic and environmental pollution: A blessing in disguise? *Sci. Total Environ.* **2020**, *728*, 138820. [CrossRef]
10. IPCC Proposed outline of the special report in 2018 on the impacts of global warming of 1.5 °C above pre-industrial levels and related global greenhouse gas emission pathways, in the context of strengthening the global response to the threat of climate cha. *Ipcc Sr15* **2018**, *2*, 17–20.
11. Dantas, G.; Siciliano, B.; França, B.B.; da Silva, C.M.; Arbilla, G. The impact of COVID-19 partial lockdown on the air quality of the city of Rio de Janeiro, Brazil. *Sci. Total Environ.* **2020**. [CrossRef]
12. O'Reilly, K.M.; Auzenbergs, M.; Jafari, Y.; Liu, Y.; Flasche, S.; Lowe, R. Effective transmission across the globe: The role of climate in COVID-19 mitigation strategies. *Lancet Planet. Health* **2020**, *4*, e172. [CrossRef]
13. Sharma, S.; Zhang, M.; Anshika; Gao, J.; Zhang, H.; Kota, S.H. Effect of restricted emissions during COVID-19 on air quality in India. *Sci. Total Environ.* **2020**. [CrossRef] [PubMed]
14. Singh, R.P.; Chauhan, A. Impact of lockdown on air quality in India during COVID-19 pandemic. *Air Qual. Atmos. Health* **2020**, *13*, 921–928. [CrossRef] [PubMed]
15. Pant, P.; Harrison, R.M. Estimation of the contribution of road traffic emissions to particulate matter concentrations from field measurements: A review. *Atmos. Environ.* **2013**, *77*, 78–97. [CrossRef]
16. Ly, B.T.; Kajii, Y.; Nguyen, T.Y.L.; Shoji, K.; Van, D.A.; Do, T.N.N.; Nghiem, T.D.; Sakamoto, Y. Characteristics of roadside volatile organic compounds in an urban area dominated by gasoline vehicles, a case study in Hanoi. *Chemosphere* **2020**, *254*, 126749. [CrossRef]
17. Fameli, K.M.; Kotrikla, A.M.; Psanis, C.; Biskos, G.; Polydoropoulou, A. Estimation of the emissions by transport in two port cities of the northeastern Mediterranean, Greece. *Environ. Pollut.* **2020**, *257*, 113598. [CrossRef]
18. Wang, Y.; Yuan, Y.; Wang, Q.; Liu, C.; Zhi, Q.; Cao, J. Changes in air quality related to the control of coronavirus in China: Implications for traffic and industrial emissions. *Sci. Total Environ.* **2020**, *731*, 139133. [CrossRef]
19. Tournadre, B.; Chelin, P.; Ray, M.; Cuesta, J.; Kutzner, R.D.; Landsheere, X.; Fortems-Cheiney, A.; Flaud, J.-M.; Hase, F.; Blumenstock, T.; et al. Atmospheric ammonia (NH$_3$) over the Paris megacity: 9 years of total column observations from ground-based infrared remote sensing. *Atmos. Meas. Tech.* **2020**, *13*, 3923–3937. [CrossRef]
20. Puliafito, S.E.; Bolaño-Ortiz, T.; Berná, L.; Pascual Flores, R. High resolution inventory of atmospheric emissions from livestock production, agriculture, and biomass burning sectors of Argentina. *Atmos. Environ.* **2020**, *223*, 117248. [CrossRef]
21. Janssens-maenhout, G. Supplement of EDGAR v4.3.2 Global Atlas of the three major greenhouse gas emissions for the period 1970–2012. *Earth Syst. Sci. Data* **2019**, *11*, 959–1002. [CrossRef]

22. Janssens-Maenhout, G.; Crippa, M.; Guizzardi, D.; Dentener, F.; Muntean, M.; Pouliot, G.; Keating, T.; Zhang, Q.; Kurokawa, J.; Wankmüller, R.; et al. HTAP-v2.2: A mosaic of regional and global emission grid maps for 2008 and 2010 to study hemispheric transport of air pollution. *Atmos. Chem. Phys.* **2015**, *15*, 11411–11432. [CrossRef]
23. Huang, X.; Zhang, Y.; Wang, Y.; Ou, Y.; Chen, D.; Pei, C.; Huang, Z.; Zhang, Z.; Liu, T.; Luo, S.; et al. Evaluating the effectiveness of multiple emission control measures on reducing volatile organic compounds in ambient air based on observational data: A case study during the 2010 Guangzhou Asian Games. *Sci. Total Environ.* **2020**, *723*, 138171. [CrossRef] [PubMed]
24. Franco, J.F.; Pacheco, J.; Belalcázar, L.C.; Behrentz, E. Characterization and source identification of VOC species in Bogotá, Colombia. *Atmósfera* **2015**, *28*, 1–11. [CrossRef]
25. Xue, R.; Wang, S.; Li, D.; Zou, Z.; Chan, K.L.; Valks, P.; Saiz-Lopez, A.; Zhou, B. Spatio-temporal variations in NO_2 and SO_2 over Shanghai and Chongming Eco-Island measured by Ozone Monitoring Instrument (OMI) during 2008–2017. *J. Clean. Prod.* **2020**, *258*, 120563. [CrossRef]
26. Anjum, N. Good in The Worst: COVID-19 Restrictions and Ease in Global Air Pollution. *Preprints* **2020**. [CrossRef]
27. Kheirbek, I.; Haney, J.; Douglas, S.; Ito, K.; Matte, T. The contribution of motor vehicle emissions to ambient fine particulate matter public health impacts in New York City: A health burden assessment. *Environ. Health A Glob. Access Sci. Source* **2016**, *15*, 89. [CrossRef]
28. Chang, Y.; Zou, Z.; Deng, C.; Huang, K.; Collett, J.L.; Lin, J.; Zhuang, G. The importance of vehicle emissions as a source of atmospheric ammonia in the megacity of Shanghai. *Atmos. Chem. Phys.* **2016**, *16*, 3577–3594. [CrossRef]
29. Huang, Y.; Zhou, J.L.; Yu, Y.; Mok, W.; Lee, C.F.C. Uncertainty in the Impact of the COVID-19 Pandemic on Air Quality in Hong Kong, China. *Atmosphere* **2020**, *11*, 914. [CrossRef]
30. Musselwhite, C.; Avineri, E.; Susilo, Y. Editorial JTH 16 –The Coronavirus Disease COVID-19 and implications for transport and health. *J. Transp. Health* **2020**, *16*, 4–7. [CrossRef]
31. Siciliano, B.; Dantas, G.; Cleyton, M.; Arbilla, G. Increased ozone levels during the COVID-19 lockdown: Analysis for the city of Rio de Janeiro, Brazil. *Sci. Total Environ.* **2020**, *737*, 139765. [CrossRef]
32. Aloi, A.; Alonso, B.; Benavente, J.; Cordera, R.; Echániz, E.; González, F.; Ladisa, C.; Lezama-Romanelli, R.; López-Parra, Á.; Mazzei, V.; et al. Effects of the COVID-19 lockdown on urban mobility: Empirical evidence from the city of Santander (Spain). *Sustainability* **2020**, *12*, 3870. [CrossRef]
33. Mendez-Espinosa, J.F.; Rojas, N.Y.; Vargas, J.; Pachón, J.E.; Belalcazar, L.C.; Ramírez, O. Air quality variations in Northern South America during the COVID-19 lockdown. *Sci. Total Environ.* **2020**, *749*, 141621. [CrossRef] [PubMed]
34. Arregocés, H.A.; Rojano, R.; Restrepo, G. Impact of lockdown on particulate matter concentrations in Colombia during the COVID-19 pandemic. *Sci. Total Environ.* **2020**, 142874. [CrossRef] [PubMed]
35. DANE Departamento Administrativo Nacional de Estadística (DANE). Available online: https://www.dane.gov.co/index.php/estadisticas-por-tema/demografia-y-poblacion/proyecciones-de-poblacion (accessed on 18 January 2021).
36. Economics, T. Colombia—Economic Forecasts—2020–2022 Outlook. Available online: https://tradingeconomics.com/colombia/forecast (accessed on 30 July 2020).
37. Runt Runt—Registro Único Nacional de Tránsito. Available online: https://www.runt.com.co/runt-en-cifras/parque-automotor (accessed on 1 July 2020).
38. IDEAM. *Inventario Nacional y Departamental de Gases Efecto Invernadero. Tercera Comunicación Nacional de Cambio Climático*; IDEAM: Bogota, Colombia, 2016; ISBN 9789588971254.
39. Hao, H.; Wang, H.; Ouyang, M. Fuel conservation and GHG (Greenhouse gas) emissions mitigation scenarios for China's passenger vehicle fleet. *Energy* **2011**, *36*, 6520–6528. [CrossRef]
40. Bebkiewicz, K.; Chłopek, Z.; Lasocki, J.; Szczepański, K.; Zimakowska-Laskowska, M. Analysis of emission of greenhouse gases from road transport in Poland between 1990 and 2017. *Atmosphere* **2020**, *11*, 387. [CrossRef]
41. Eggleston, H.S.; Miwa, K.; Srivastava, N.; Tanabe, K. *IPCC 2006 IPCC Guidelines for National Greenhouse Inventories—A primer, Prepared by the National Greenhouse Gas Inventories Programme*; Iges: Hayama, Japan, 2006; ISBN 4-88788-032-4.
42. EMEP. *EEA Report No 13/2019*; EMEP: Geneva, Switzerland, 2019; ISBN 978-92-9480-098-5.
43. FECOC Unidad de Planeación Minero Energética—UPME. Available online: http://www.upme.gov.co/Calculadora_Emisiones/aplicacion/calculadora.html (accessed on 26 July 2020).
44. Puliafito, S.E.; Allende, D.G.; Castesana, P.S.; Ruggeri, M.F. High-resolution atmospheric emission inventory of the argentine energy sector. Comparison with edgar global emission database. *Heliyon* **2017**, *3*, 489. [CrossRef] [PubMed]
45. Puliafito, S.E.; Allende, D.; Pinto, S.; Castesana, P. High resolution inventory of GHG emissions of the road transport sector in Argentina. *Atmos. Environ.* **2015**, *101*, 303–311. [CrossRef]
46. Puliafito, S.E.; Bolaño-Ortiz, T.R.; Berná Peña, L.L.; Pascual-Flores, R.M. Dataset supporting the estimation and analysis of high spatial resolution inventories of atmospheric emissions from several sectors in Argentina. *Data Br.* **2020**, *29*, 105281. [CrossRef]
47. Bolaño-Ortiz, T.R.; Puliafito, S.E.; Berná-Peña, L.L.; Pascual-Flores, R.M.; Urquiza, J.; Camargo-Caicedo, Y. Atmospheric Emission Changes and Their Economic Impacts during the COVID-19 Pandemic Lockdown in Argentina. *Sustainability* **2020**, *12*, 8661.
48. Ministry of Mines and Energy Information System of Liquid Fuels of Colombia-SICOM. Available online: https://www.sicom.gov.co/index.php/boletin-estadistico (accessed on 18 January 2021).
49. Ministry of Mines and Energy Information System of Liquid Fuels of Colombia-SICOM. Available online: https://www.sicom.gov.co/sicom/identificacionAction.do?method=pRedirectHttps (accessed on 1 July 2020).

50. GMGNC Manager of the Natural Gas Market in Colombia-GMGNC. Available online: https://www.bmcbec.com.co/informes/informes-mensuales (accessed on 20 November 2020).
51. UPME Plan de Acción Indicativo de Eficiencia Energética 2017–2022; Bogotá, 2016; pp.1–157. Available online: https://www1.upme.gov.co/DemandaEnergetica/MarcoNormatividad/PAI_PROURE_2017-2022.pdf (accessed on 20 July 2020).
52. PROMIGAS Promigas S.A. E.S.P. Available online: http://www.promigas.com/Es/Noticias/Paginas/Revista-Magasin/Edicion-27/Pagina-69.aspx (accessed on 23 July 2020).
53. Myhre, G.; Shindell, D.; Bréon, F.-M.F.-M.; Collins, W.; Fuglestvedt, J.; Huang, J.; Koch, D.; Lamarque, J.-F.J.-F.; Lee, D.; Mendoza, B.; et al. Anthropogenic and Natural Radiative Forcing: Supplementary Material. *Clim. Chang.* **2013**. [CrossRef]
54. Secretary of Environment Bogotá, Bogotá Air Quality Monitoring Network—RMCAB. Available online: http://201.245.192.252:81/home/map (accessed on 4 September 2020).
55. OMI National Aeronautics and Space Administration. Available online: https://www.nasa.gov/mission_pages/aura/spacecraft/omi.html (accessed on 24 July 2020).
56. Acker, J.G.; Leptoukh Eos, G. Trans. AGU. In Proceedings of the Online Analysis Enhances Use of NASA Earth Science Data; AGU: Washington, DC, USA, 2007; pp. 14–17.
57. De Souza, C.D.R.; Silva, S.D.; Da Silva, M.A.V.; de D'Agosto, M.A.; Barboza, A.P. Inventory of conventional air pollutants emissions from road transportation for the state of Rio de Janeiro. *Energy Policy* **2013**, *53*, 125–135. [CrossRef]
58. Undp Undp in Latin America and the Caribbean. Available online: https://www.latinamerica.undp.org/content/rblac/en/home/library/crisis_prevention_and_recovery/social-and-economic-impact-of-covid-19-and-policy-options-in-arg.html (accessed on 29 July 2020).
59. DANE Statistics by Demography and Population in Colombia. Available online: https://www.dane.gov.co/index.php/estadisticas-por-tema/demografia-y-poblacion (accessed on 19 January 2021).
60. Nakada, L.Y.K.; Urban, R.C. COVID-19 pandemic: Impacts on the air quality during the partial lockdown in São Paulo state, Brazil. *Sci. Total Environ.* **2020**, 139087. [CrossRef] [PubMed]
61. Tobías, A.; Carnerero, C.; Reche, C.; Massagué, J.; Via, M.; Minguillón, M.C.; Alastuey, A.; Querol, X. Changes in air quality during the lockdown in Barcelona (Spain) one month into the SARS-CoV-2 epidemic. *Sci. Total Environ.* **2020**, *726*, 138540. [CrossRef] [PubMed]
62. Silver, B.; He, X.; Arnold, S.R.; Spracklen, D.V. The impact of COVID-19 control measures on air quality in China. *Environ. Res. Lett.* **2020**, *15*. [CrossRef]
63. Dinero Dinero. Available online: https://www.dinero.com/pais/articulo/consumo-de-gasolina-en-colombia-en-junio-de-2020/290275 (accessed on 24 July 2020).
64. Liu, Z.; Ciais, P.; Deng, Z.; Lei, R.; Davis, S.J.; Feng, S.; Zheng, B.; Cui, D.; Dou, X.; Zhu, B.; et al. Near-real-time monitoring of global CO2 emissions reveals the effects of the COVID-19 pandemic. *Nat. Commun.* **2020**, *11*, 5172. [CrossRef]
65. He, L.; Zhang, S.; Hu, J.; Li, Z.; Zheng, X.; Cao, Y.; Xu, G.; Yan, M.; Wu, Y. On-road emission measurements of reactive nitrogen compounds from heavy-duty diesel trucks in China. *Environ. Pollut.* **2020**, *262*, 114280. [CrossRef]
66. Xu, K.; Cui, K.; Young, L.H.; Wang, Y.F.; Hsieh, Y.K.; Wan, S.; Zhang, J. Air quality index, indicatory air pollutants and impact of covid-19 event on the air quality near central china. *Aerosol Air Qual. Res.* **2020**, *20*, 1204–1221. [CrossRef]
67. Chauhan, A.; Singh, R.P. Decline in PM2.5 concentrations over major cities around the world associated with COVID-19. *Environ. Res.* **2020**, *187*, 109634. [CrossRef]
68. de Miranda, R.M.; Perez-Martinez, P.J.; de Fatima Andrade, M.; Ribeiro, F.N.D. Relationship between black carbon (BC) and heavy traffic in São Paulo, Brazil. *Transp. Res. Part D Transp. Environ.* **2019**, *68*, 84–98. [CrossRef]
69. Chen, W.T.; Lee, Y.H.; Adams, P.J.; Nenes, A.; Seinfeld, J.H. Will black carbon mitigation dampen aerosol indirect forcing? *Geophys. Res. Lett.* **2010**, *37*, 1–5. [CrossRef]
70. Resmi, C.T.; Nishanth, T.; Satheesh Kumar, M.K.; Manoj, M.G.; Balachandramohan, M.; Valsaraj, K.T. Air quality improvement during triple-lockdown in the coastal city of Kannur, Kerala to combat Covid-19 transmission. *PeerJ* **2020**, *8*, e9642. [CrossRef] [PubMed]
71. De Fatima, M.; Kumar, P.; Dias, E.; Freitas, D.; Yuri, R.; Martins, J.; Martins, L.D.; Nogueira, T.; Perez-martinez, P.; Maura, R.; et al. ~ o Paulo: Evolution over the last 30 Air quality in the megacity of S a years and future perspectives. *Atmos. Environ.* **2017**, *159*, 66–82. [CrossRef]
72. Sicard, P.; De Marco, A.; Agathokleous, E.; Feng, Z.; Xu, X.; Paoletti, E.; Rodriguez, J.J.D.; Calatayud, V. Amplified ozone pollution in cities during the COVID-19 lockdown. *Sci. Total Environ.* **2020**, *735*, 139542. [CrossRef] [PubMed]
73. Li, K.; Jacob, D.J.; Liao, H.; Shen, L.; Zhang, Q.; Bates, K.H. Anthropogenic drivers of 2013–2017 trends in summer surface ozone in China. *Proc. Natl. Acad. Sci. USA* **2019**, *116*, 422–427. [CrossRef]
74. Wynes, S.; Nicholas, K.A. The climate mitigation gap: Education and government recommendations miss the most effective individual actions. *Environ. Res. Lett.* **2017**, *12*, 074024. [CrossRef]
75. Andrade-Castañeda, H.J.; Arteaga-Céspedes, C.C.; Segura-Madrigal, M.A. Emisión de gases de efecto invernadero por uso de combustibles fósiles en Ibagué, Tolima (Colombia). *Corpoica Cienc. y Tecnol. Agropecu.* **2017**, *18*, 103–112. [CrossRef]
76. Chester, M.; Pincetl, S.; Elizabeth, Z.; Eisenstein, W.; Matute, J. Infrastructure and automobile shifts: Positioning transit to reduce life-cycle environmental impacts for urban sustainability goals. *Environ. Res. Lett.* **2013**, *8*. [CrossRef]

77. Anderson, R.M.; Heesterbeek, H.; Klinkenberg, D.; Hollingsworth, T.D. How will country-based mitigation measures influence the course of the COVID-19 epidemic? *Lancet* **2020**, *395*, 931–934. [CrossRef]
78. Goscé, L.; Johansson, A. Analysing the link between public transport use and airborne transmission: Mobility and contagion in the London underground. *Environ. Health* **2018**, *17*, 84. [CrossRef]
79. Filippini, T.; Rothman, K.J.; Goffi, A.; Ferrari, F.; Maffeis, G.; Orsini, N.; Vinceti, M. Satellite-detected tropospheric nitrogen dioxide and spread of SARS-CoV-2 infection in Northern Italy. *Sci. Total Environ.* **2020**, *739*, 140278. [CrossRef]
80. Saha, B.; Debnath, A.; Saha, B. Analysis and finding the correlation of air quality parameters on the spread and deceased case of COVID-19 patients in India 2020. 2020, Research Square. Available online: https://assets.researchsquare.com/files/rs-34647/v1/d82c5de9-56a1-435b-a1e4-6ca206b8838f.pdf (accessed on 18 January 2021).
81. Bolaño-Ortiz, T.R.; Camargo-Caicedo, Y.; Puliafito, S.E.; Ruggeri, M.F.; Bolaño-Diaz, S.; Pascual-Flores, R.; Saturno, J.; Ibarra-Espinosa, S.; Mayol-Bracero, O.L.; Torres-Delgado, E.; et al. Spread of SARS-CoV-2 through Latin America and the Caribbean region: A look from its economic conditions, climate and air pollution indicators. *Environ. Res.* **2020**, *191*, 109938. [CrossRef] [PubMed]
82. Bolaño-Ortiz, T.R.; Pascual-Flores, R.M.; Puliafito, S.E.; Camargo-Caicedo, Y.; Berná-Peña, L.L.; Ruggeri, M.F.; Lopez-Noreña, A.I.; Tames, M.F.; Cereceda-Balic, F. Spread of COVID-19, Meteorological Conditions and Air Quality in the City of Buenos Aires, Argentina: Two Facets Observed during Its Pandemic Lockdown. *Atmosphere* **2020**, *11*, 1045. [CrossRef]
83. Bashir, M.F.; Bilal, B.M.A.; Komal, B. Correlation between environmental pollution indicators and COVID-19 pandemic: A brief study in Californian context. *Environ. Res.* **2020**, 109652. [CrossRef]
84. Travaglio, M.; Popovic, R.; Yu, Y.; Leal, N.; Martins, L.M. Links between air pollution and COVID-19 in England. *medRxiv* **2020**. [CrossRef]
85. Ogen, Y. Assessing nitrogen dioxide (NO_2) levels as a contributing factor to coronavirus (COVID-19) fatality. *Sci. Total Environ.* **2020**, *726*, 138605. [CrossRef]
86. Mankia, K.; Di Matteo, A.; Emery, P. Prevention and cure: The major unmet needs in the management of rheumatoid arthritis. *J. Autoimmun.* **2020**, *110*, 102399. [CrossRef]
87. Brianti, M.; Magnani, M.; Menegatti, M. Optimal choice of prevention and cure under uncertainty on disease effect and cure effectiveness. *Res. Econ.* **2018**, *72*, 327–342. [CrossRef]
88. EMEP EMEP/EEA air pollutant emission inventory guidebook—2016—European Environment Agency. *EEA Reports* **2016**. [CrossRef]

Article

Evaluating the Real-World NOx Emission from a China VI Heavy-Duty Diesel Vehicle

Peng Li * and Lin Lü

School of Energy and Power Engineering, Wuhan University of Technology, Wuhan 430063, China; lulinwhut@163.com
* Correspondence: leepeng@whut.edu.cn; Tel.: +86-134-7708-0396

Citation: Li, P.; Lü, L. Evaluating the Real-World NOx Emission from a China VI Heavy-Duty Diesel Vehicle. *Appl. Sci.* **2021**, *11*, 1335. https://doi.org/10.3390/app11031335

Academic Editor: Thomas Maggos
Received: 10 January 2021
Accepted: 28 January 2021
Published: 2 February 2021

Publisher's Note: MDPI stays neutral with regard to jurisdictional claims in published maps and institutional affiliations.

Copyright: © 2021 by the authors. Licensee MDPI, Basel, Switzerland. This article is an open access article distributed under the terms and conditions of the Creative Commons Attribution (CC BY) license (https://creativecommons.org/licenses/by/4.0/).

Abstract: The manufacturers of China VI heavy-duty vehicles were required to conduct in-service conformity (ISC) tests by using a portable emissions measurement system (PEMS). The moving averaging window (MAW) method was used to evaluate the NOx emission required by the China VI emission standard. This paper presented the results of four PEMS tests of a China VI (step B) N3 category vehicle. Our analyses revealed that the real NOx emission of the test route was much higher than the result evaluated by the MAW method. We also found the data produced during the urban section of a PEMS test was completely excluded from the evaluation based on the current required boundary conditions. Therefore, in order to ensure the objectivity of the evaluation, this paper proposed three different evaluation methods. Method 1 merely set the power threshold as 10% for valid MAWs; Method 2 reclassified the MAWs into "Urban MAWs", "Rural MAWs" and "Motorway MAWs" according to the vehicle speed. Method 3 reclassified the MAWs into "Hot MAWs" and "Cold MAWs" according to engine coolant temperature. The NOx emission evaluation results for Method 1 were not satisfactory, but those for Method 2 and Method 3 were close to the real NOx emission, the errors were all within ±10%.

Keywords: portable emissions measurement system; moving averaging window; real-world NOx emission; heavy-duty diesel vehicle; evaluation method

1. Introduction

On-road diesel vehicles produce a great amount of nitrogen oxides (NOx) worldwide, leading to deterioration of the environment and increasing health issues [1,2]. According to updated traffic emission inventory, considering nonlocal trucks, almost 80% of the total vehicular NOx emission in Beijing was emitted by diesel vehicles [2,3].

In the past few decades, increasingly stringent emission regulations had been adopted by many countries (e.g., Europe, US, Korea, Japan) [4,5]. China has also issued "Limits and measurement methods for emissions from diesel fueled heavy-duty vehicles (CHINA VI)" numbered "GB17691-2018" in June 2018.

In order to deal with the increasingly stringent emission standards, the manufacturers generally used selective catalytic reduction (SCR) technology for alleviating NOx emissions at the tailpipe combined with diesel particle filters (DPF) for particulate matter (PM) reduction, diesel oxidation catalysts (DOC) for the oxidation of incomplete combustion products and ammonia slip catalyst (ASC) for the oxidation of NH_3 [6,7].

What's more, in the DOC unity, in addition to oxidation of CO and unburned hydrocarbons, NO conversion to NO_2 takes place, thus increasing the quite low NO_2 concentration in the exhaust gas (about 5% to 10% of total NOx). This increase in NO_2 concentration speeds up the passive regeneration process of DPF and, thus, largely affects the decrease in back pressure, enhancing the operating performance and prolonging the life-time of the aftertreatment device [8–10].

Although regulated NOx emission limits had been progressively tightened, many researchers claimed that current diesel vehicles emitted far more NOx under real-world operating conditions than during laboratory certification testing [1,11,12]. Under this condition, the real driving emissions (RDE) test (also called PEMS test) protocols using a portable emissions measurement system (PEMS)—which is a compact equipment composed by a portable gas analyzer, a global positioning system (GPS) receiver, a data logging system and so on—had already been adopted by many countries (e.g., Europe, US, Korea) to check the on-road conformity of emissions [5,13–17]. In addition, manufacturers were also required to conduct in-service conformity (ISC) testing by using a portable emissions measurement system (PEMS) in China.

The PEMS test regulation establishes the requirements for route composition which must cover a wide range of real-world conditions by accounting with defined shares of urban, rural and motorway operation. Other parameters considered are trip duration, ranges of vehicle speed, cumulative work performed by the engine, etc. [18]. Ambient boundary conditions including altitude and ambient temperature for the PEMS test are also involved in the regulation [14].

However, even with all these requirements, repeatability of PEMS tests is hardly achievable, because the boundary conditions are unique for a PEMS test. For instance, it is scarcely possible to guarantee that the trip time and vehicle speed of two PEMS tests are exactly the same on the same test route, needless to say different test routes [14,19,20].

Therefore, PEMS tests still have some debatable points (e.g., trip composition, boundary conditions and data analysis methods) required further detailed study [14]. For instance, a study performed by Mendoza-Villafuerte et al. [21] revealed that up to 85% of the NOx emissions measured during the tests performed were not taken into consideration if the boundary conditions for data exclusion set in the current legislation were applied.

In order to overcome the repeatability issue of the PEMS tests, many data analysis methods (e.g., the vehicle specific power (VSP) method, the power binning (PB) method, not-to-exceed (NTE) method, moving averaging window (MAW) method, etc.) were introduced for processing the test data [22,23].

Varella et al. [23] tested three different methods (the MAW, PB and VSP), concluding that there were differences between all methods both for CO_2 and NOx emissions estimation due to the statistical and numerical treatment from each method. The current data analysis method regulated by the European Community (EC) and China is the moving average window (MAW) method.

This work aims to analyze the data produced during the four PEMS tests. Firstly, to analyze the NOx emission of each section (urban, rural and motorway); then, to calculate the MAW NOx emission under the required boundary conditions; finally, to explore proper methods to evaluate the real-world NOx emission based on MAW method.

2. Experiments and Materials

Detailed descriptions of the tested vehicle, test instrumentation and route, MAW method, boundary conditions for a valid PEMS test and data evaluation, judgement rule of pass-fail for emissions are provided in this section.

2.1. Tested Vehicle

A heavy-duty diesel vehicle (Figure 1a) which was the type approved to the China VI (step B) standard and registered in August 2019 was used to perform the on-road emissions measurement (PEMS test).

The tested vehicle which covered 2135.3 km at the beginning of Test 1 was equipped with the latest aftertreatment technologies comprised of a diesel oxidation catalyst (DOC) followed by a diesel particulate filter (DPF) in series with a selective catalytic reduction (SCR) catalyst and an ammonia slip catalyst (ASC) in sequence (Figure 1b,c). There are two on-board NOx sensors located at the DOC inlet for the engine output NOx measurement and ASC outlet for tailpipe NOx measurement, respectively (Figure 1c). The main characteristics of the tested vehicle are summarized in Table 1.

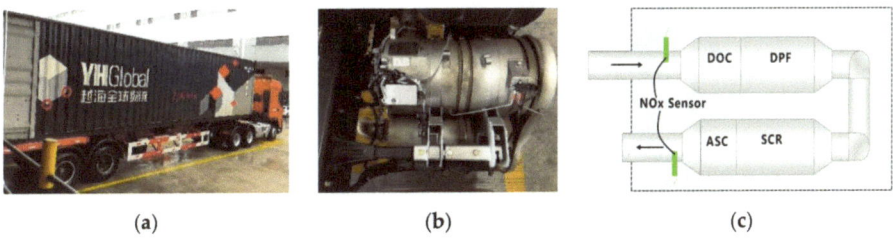

Figure 1. (a) Tested vehicle; (b) aftertreatment configuration (picture of real products); (c) NOx sensors location and aftertreatment configuration (schematic plot).

Table 1. Summary of vehicle, engine, aftertreatment, fuel and DEF specifications.

Type of Engine	XX13600-60
Type of Vehicle	Long–Haul
Year of production	2019
Engine rated power	441 kW
Reference Torque	3000 Nm
WHTC Cycle Work	38.72 kWh
Emission standard	China VI (step B)
Aftertreatment System	DOC + DPF + SCR + ASC
Gross vehicle weight	kg
Payload	50%
Category of vehicle	N3
Fuel	China VI Standard
DEF	Adblue (32.5%)

DEF: diesel exhaust fluid.

2.2. Portable Emissions Measurement System (PEMS)

AVL-M.O.V.E-PEMS (Figure 2) consists of tailpipe attachment, heated exhaust lines, exhaust flow meter (EFM), exhaust gas analyzers used to measure concentrations of gaseous emissions (including carbon monoxide (CO), carbon dioxide (CO_2), total hydrocarbon (THC), nitrogen monoxide (NO) and nitrogen dioxide (NO_2), etc.), PM module, PN module, a global positioning system (GPS) from which we can get vehicle speed, latitude, longitude and altitude, sensors for measuring ambient temperature and humidity, charger, system control, E-box, etc. NOx concentration is calculated by the sum of NO and NO_2 concentration. The electrical power needed for the PEMS operation (DC 22~28 V) is supplied by two external batteries.

The PEMS uses flame ionization detection (FID) for THC measurement, non-dispersive infrared (NDIR) for CO and CO_2 measurements, non-dispersive ultra-violet (NDUV) for NO, NO_2 measurement. The EFM uses a pitot tube based on Bernoulli's principle to calculate mass flow on the basis of airflow differential pressure measurement. The measurement principle and measurement range of gaseous emissions are shown in Table 2.

All emissions are measured on a wet basis, so that no corrections are required for the analysis. The PEMS is warmed up for at least 1.5 h, then zeroed and spanned with calibration gas before the test.

Figure 2. Installation of test instrument (AVL-M.O.V.E-PEMS).

Table 2. Measurement principle and measurement range of gaseous emissions for the PEMS used.

Measured Variable	Measurement Principle	Measurement Range
CO	NDIR	0~49,999 ppm
CO_2	NDIR	0~20 vol%
NO	NDUV	0~5000 ppm
NO_2	NDUV	0~2500 ppm
THC	FID	0~30,000 ppm

2.3. Test Route

The four PEMS tests were carried out in Suzhou, China along the same route. The test route shall always start with urban driving followed by rural and motorway driving specified in the regulation. We conducted the urban section of the route in the city, the rural and motorway sections on the beltway of Suzhou and a part of the China G2 expressway (Figure 3).

Figure 3. Topographic map of the PEMS test route.

For N3 category vehicles, the first short trip (referring to the driving process between the end of one idle speed and the beginning of the next idle speed) with vehicle speed exceeding 55 km/h is defined as the beginning of the rural section, and the first short trip with vehicle speed exceeding 75 km/h is defined as the beginning of the motorway section. The average vehicle speed of each section shall meet the following requirements: urban section ($\geq$15 to $\leq$30 km/h), rural section ($\geq$45 to $\leq$70 km/h), motorway Section (>70 km/h). The shares of operation shall be expressed as a percentage of the total trip duration, and the trip shall consist of approximately 20% urban, 25% rural and 55% motorway operation. Here, 'approximately' shall mean the target value ±5%.

The entire trip duration is decided by the cumulative work performed by the engine. All the tests considered for the analysis should perform between 4 and 7 times the amount of work performed over the WHTC cycle of the engine.

Moreover, the proportional cumulative positive altitude gain over the entire trip shall be less than 1200 m/100 km, the start and the end point shall not differ in their elevation above sea level by more than 100 m, etc. These requirements were all fulfilled since the test route was performed in a relatively flat area.

2.4. Moving Averaging Window (MAW) Method

The emissions shall be evaluated by the MAW method, based on the reference work (the amount of work performed over the WHTC cycle of the engine).

The principle of the calculation is as follows: the mass emissions are not calculated for the complete data set, but for sub-sets of the complete data set, the length of these sub-sets is determined by the work measured over the reference laboratory transient cycle (WHTC for "CHINA VI").

The moving average calculations are conducted with a time increment Δt equal to the data sampling period which was set as 500 ms in the four PEMS tests. The end point of the test is taken as the starting point of the first MAW shown in Figure 4.

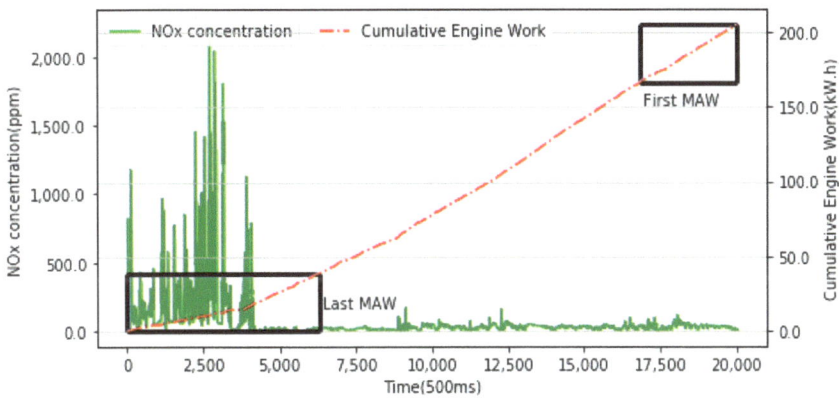

Figure 4. Definition of first and last moving averaging windows (MAWs).

The duration $t_{2,i} - t_{1,i}$ of the ith averaging window is determined by Equation (1):

$$W(t_{2,i}) - W(t_{1,i}) \geq W_{ref} \qquad (1)$$

where:
- $W(t_{j,i})$ is the cumulative engine work measured between the start and time $t_{j,i}$, kWh;
- W_{ref} is the amount of work produced over the WHTC, kWh;
- $t_{2,i}$ shall be selected by Equation (2):

$$W(t_{2,i} - \Delta t) - W(t_{1,i}) < W_{ref} \leq W(t_{2,i}) - W(t_{1,i}) \qquad (2)$$

where Δt is the data sampling period, equal to 1 s or less.

The brake-specific emissions EF_p (g/kWh) shall be calculated for each window and each pollutant by Equation (3):

$$EF_p = \frac{m}{W(t_{2,i}) - W(t_{1,i})} \quad (3)$$

where:
- m is the cumulative mass of the pollutant of the window, g/window;
- $W(t_{2,i}) - W(t_{1,i})$ is the cumulative engine work during the ith averaging window, kWh.

2.5. Boundary Conditions for Data Evaluation

The current PEMS procedure for heavy-duty vehicles is defined by a series of boundary conditions that prescribes the amount of data to be taken into consideration for the final analysis including the effectiveness of the test and pass-fail of the pollutants' emissions.

The main boundary conditions for a valid test are as follows:
- Ambient temperature: between -7 and 38 °C.
- Altitude: not more than 2400 m.
- Test route: as abovementioned in "2.3 Test Route" (including trip share and vehicle speed, etc.).
- Cold start: at the beginning of the PEMS test, the engine coolant temperature shall not exceed 30 °C unless the ambient temperature is higher than 30 °C, in this case, the engine coolant temperature shall not be 2 °C higher than the ambient temperature. The data used for emissions evaluation is recorded after the engine coolant temperature has reached 70 °C for the first time or after the coolant temperature is stabilized within ± 2 °C over a period of 5 min (whichever comes first but not later than 20 min after engine starts).
- Payload: 10 to 100% of the maximum vehicle payload.
- Cumulative work: 4~7 times the amount of work performed over the WHTC applicable to the engine used by the tested vehicle.
- Selection of valid windows: the valid windows are the windows whose average power exceeds the power threshold of 20% of the maximum engine power. The percentage of valid windows shall be equal or greater than 50%. If the percentage of valid windows is less than 50%, the data evaluation shall be repeated using lower power thresholds. The power threshold shall be reduced in steps of 1% until the percentage of valid windows is equal to or greater than 50%.
- Power Threshold: In any case, the power threshold shall not be lower than 10%, otherwise, the test shall be void.
- The pass-fail conditions for a valid test are as follows:
- The 90th cumulative percentile of the valid windows emissions shall be less than the limit required in the regulation (for China VI, NOx limit: 0.690 g/kWh; CO limit: 6 g/kWh; PN limit: 1.2×10^{12} #/kWh);
- NOx concentration is required to be less than or equivalent to 500 ppm for 95% of valid data points.

3. Results and Discussion

The data set used for emissions evaluation of the four PEMS tests were all recorded after the engine coolant temperature had reached 70 °C for the first time.

3.1. Overall Results

The results of the four PEMS tests conducted on the same route are shown in Table 3. As it can be seen, the four PEMS tests are all valid.

Table 3. Results of the four PEMS Tests of the tested vehicle.

	Test 1	Test 2	Test 3	Test 4
Altitude not more than 2400 m	Yes	Yes	Yes	Yes
Cold start	Yes	Yes	Yes	Yes
Average Ambient temperature (°C)	22.9	20.0	10.4	6.7
Average relative humidity (%)	96.3	69.6	66.2	65.9
Payload (%)	50	50	50	50
Urban first (urban-rural-motorway)	Yes	Yes	Yes	Yes
Urban share driving (%)	21.4	21.2	19.8	20.4
Rural share (%)	22.1	27.0	25.1	25.2
Motorway share (%)	56.5	51.7	55.0	54.4
Urban driving average speed (km/h)	25.3	24.4	21.6	23.5
Rural driving average speed (km/h)	55.4	60.9	60.9	57.7
Motorway driving average speed (km/h)	75.5	73.0	75.8	76.5
Odometer (km)	2135.3	10,557.4	47,443.5	82,258.3
Trip distance (km)	168.8	159.1	178.4	145.6
Trip duration (s)	9102	9117	10,009	8183
Total Work (kWh)	195.42	179.66	205.21	175.73
Cumulative work (*WHTC$_{Work}$)	5.047	4.640	5.300	4.539
Valid MAW Power Threshold (%)	20	17	20	20
Percentage of valid MAWs (%)	61.5	72.3	50.2	64.3
95th NOx concentration (ppm)	245.1	188.0	296.2	416.2
90th cumulative percentile of Valid MAWs NOx emission (g/kWh)	0.068	0.551	0.438	0.337
Test Valid or not	Valid	Valid	Valid	Valid

Test 1 was conducted on 1 September 2019; Test 2, on 8 October 2019; Test 3, on 26 November 2019; Test 4, on 14 January 2020. The payloads of the four PEMS tests were all 50% of the maximum vehicle load (Gross vehicle weight, 28,800 kg). All of the four PEMS tests started with a cold engine.

The power threshold which shall not be lower than 10% in any case used in the four PEMS tests was 20%, 17%, 20% and 20% respectively, under this circumstance, the percentage of valid MAWs of the four PEMS tests was 61.5%, 72.3%, 50.2%,64.3% respectively. The 90th cumulative percentile of valid MAWs NOx emissions (g/kWh) and 95th cumulative percentile of NOx concentration (ppm) of the four PEMS tests were all within the required limit under the boundary conditions described in Section 2.5 in this paper.

In addition, the odometer at the beginning of Test 1, regarded as customer acceptance testing, was 2135.3 km. The odometer at the beginning of Test 2, regarded as in-service conformity (ISC) testing, was 10,557.4 km. The vehicle odometer shall be at least 10,000 km when carrying out the in-service conformity (ISC) testing. As for Test 3 and Test 4, the odometer at the beginning of the test was chosen to meet the requirements of durability test of the in-service vehicle.

3.2. Section NOx Emission Analysis

Before the discussion, it had to be known that the engine output NOx concentration was from the NOx sensor located at the DOC inlet, and the tailpipe NOx concentration was calculated by the sum of NO and NO$_2$ concentration from gas analyzers of the AVL-M.O.V.E-PEMS. In this section, we mainly talked about NOx emission characteristics of each section, especially the urban section.

Figure 5a shows the cumulative mass of engine NOx emission of each section of the test route, and as it can be seen, the least amount of engine NOx emission was emitted during the urban section of the test route because of its shortest test duration. Figure 5b shows the contribution ratio of each section to the total mass of engine NOx emission. Specifically, the contribution ratio of the urban section was 12.39% in Test 1, 10.91% in Test 2, 9.04% in Test 3, 8.74% in Test 4. The contribution ratios of the motorway section were all more than 60% in the four PEMS tests. So, we may conclude that the contribution of each section to the total mass of engine NOx emission is positively correlated with the trip share.

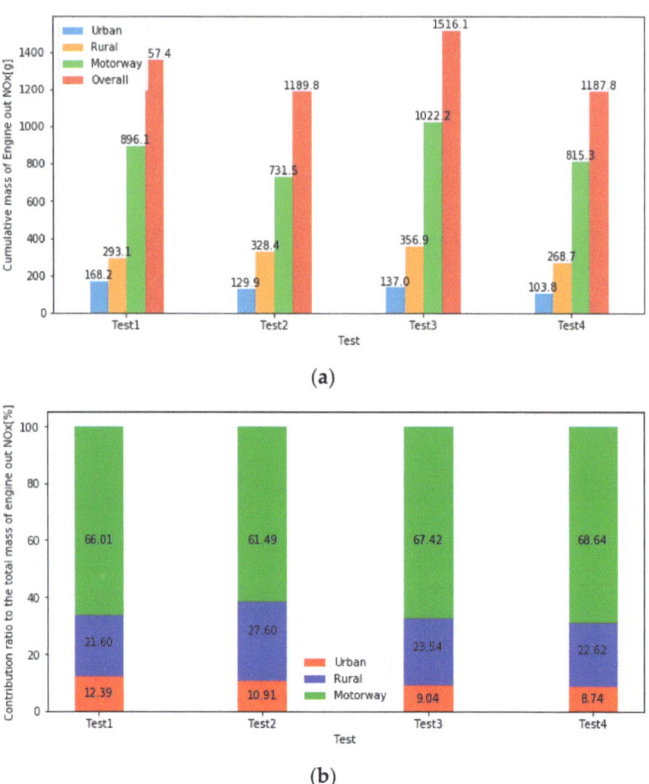

Figure 5. (a) Each section's cumulative mass of engine NOx emission; (b) each section's contribution ratio to the total mass of engine NOx emission.

Figure 6a shows the cumulative mass of tailpipe NOx of each section of the test route, as it can be seen, the greatest amount of tailpipe NOx emission was emitted during the urban section in spite of its lowest contribution to the total mass of the engine NOx emission and shortest test duration. The contribution ratio of urban section to the total mass of tailpipe NOx emission was as high as 69.10% in Test 1, 45.25% in Test 2, 55.52% in Test 3, 62.54% in Test 4 (Figure 6b). So, we may conclude that the tailpipe NOx emission of urban section is very terrible for the in-use N3 category heavy-duty vehicles.

Table 4 shows engine output of NOx and tailpipe NOx brake-specific emissions (BSNOx emission: g/kWh) in each section of the test route.

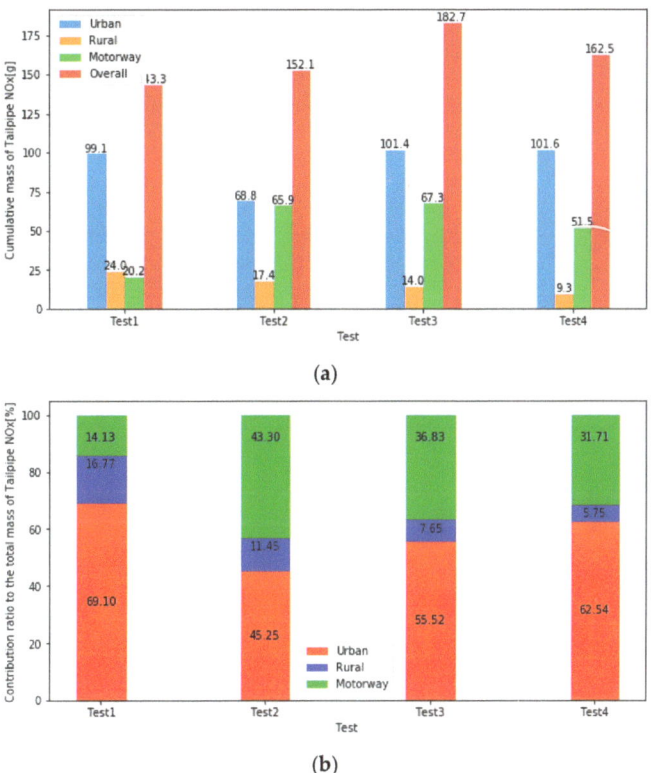

Figure 6. (a) Each section's cumulative mass of tailpipe NOx emission; (b) each section's contribution ratio to the total mass of tailpipe NOx emission.

Table 4. NOx brake-specific emissions of each section.

	Section	Test 1	Test 2	Test 3	Test 4
Engine out NOx emission [g/kWh]	Urban	8.162	6.647	8.105	7.114
	Rural	6.578	6.640	7.212	6.838
	Motorway	6.466	6.610	7.330	6.691
	Overall	6.665	6.622	7.363	6.759
Tailpipe NOx emission [g/kWh]	Urban	5.054	3.519	5.637	7.388
	Rural	0.395	0.370	0.344	0.271
	Motorway	0.059	0.502	0.413	0.317
	Overall	0.648	0.794	0.811	0.893

As it can be seen, there was a slight difference of the engine out BSNOx emissions between rural section and motorway section in all of the four PEMS tests. The engine output of BSNOx emission of urban section is a slightly higher than that of other sections in Test 1, Test 3 and Test 4, but, in Test 2, the engine out BSNOx emission of each section is almost the same. For the entire trip, the engine out BSNOx emissions of the four PEMS tests were 6.665, 6.622, 7.363 and 6.759 g/kWh, respectively.

The tailpipe BSNOx emissions of rural section and motorway section which were lower than the required limit of NOx emission (0.690 g/kWh) were also lower than that of the entire trip, but, the tailpipe BSNOx emission of urban section was significantly higher than that of other sections. For instance, the tailpipe BSNOx emission of urban section was 8.27 times greater than that of the entire trip in Test 4. For the entire trip, the tailpipe

BSNOx emissions of the four PEMS tests were 0.648 g/kWh, 0.794 g/kWh, 0.811 g/kWh, 0.893 g/kWh, respectively. So, we may conclude that the real-world NOx emissions may get worse as the increase of odometer.

Figure 7 shows the instantaneous emissions of the engine output of NOx, tailpipe NOx and the instantaneous vehicle speed of the entire trip in Test 3. As it can be seen, the tailpipe NOx concentration is almost close to the engine output of NOx concentration in the urban section since catalytic converter requires a certain temperature to work efficiently. A small urea solution injection was registered in the urban section because the temperature of catalytic converter was not high enough. The data points which conducted urea solution injection were only 6.3% of the data points recorded in the whole urban section in Test 3 (Figure 8).

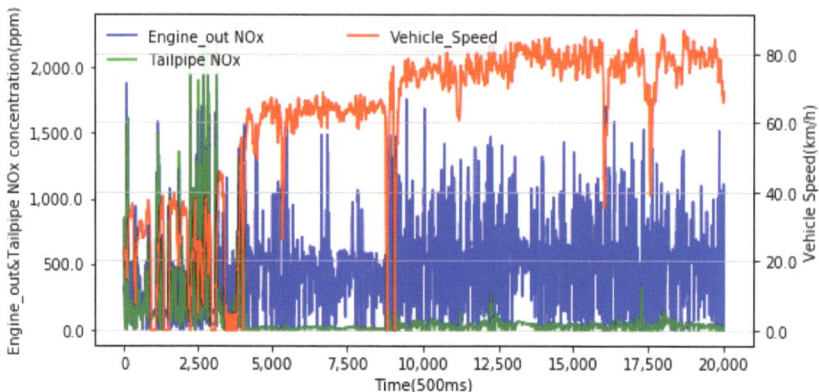

Figure 7. The instantaneous emissions of engine output of NOx, tailpipe NOx and vehicle speed of the entire trip in Test 3.

Figure 9a shows the instantaneous engine NOx emission and the vehicle acceleration profile of the entire trip in Test 3. As it can be seen, there was little abrupt acceleration during the rural or motorway section because the vehicle speed was relatively stable during these two sections. Most abrupt positive acceleration occurred during urban section. As shown in Figure 9b, NOx emission peaks were clearly linked to the vehicle acceleration peaks during the urban section, that means abrupt positive vehicle acceleration would lead to worse engine NOx emission. In fact, the urban section of the four PEMS test were mainly conducted on the city road with traffic jam, roundabouts and traffic light, driving under these circumstances may lead to more frequent "stop-go" events where abrupt positive vehicle acceleration would happen.

So, the lower temperature of SCR may lead to a smaller urea solution injection and more frequent "stop-go" events which may lead to higher engine NOx emission together would cause higher tailpipe NOx emission in the urban section.

We see clearly from the above discussion that the urban tailpipe NOx emission plays an important role in the real-world tailpipe NOx emission of N3 category heavy-duty diesel vehicle.

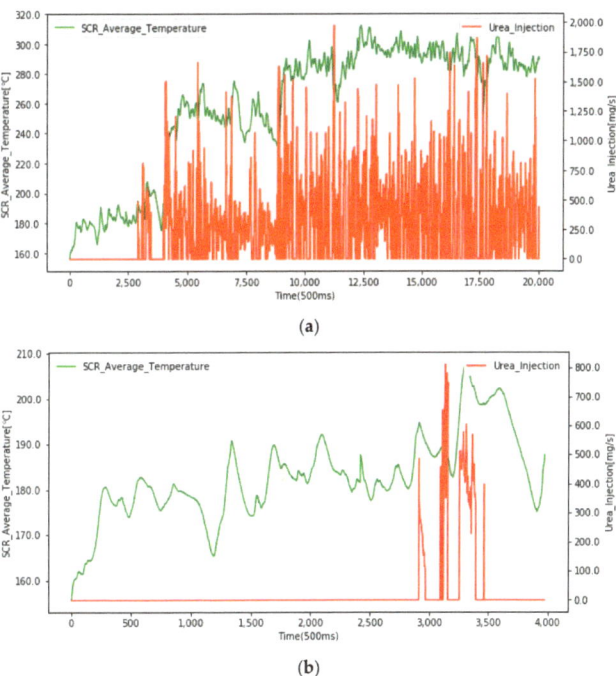

Figure 8. The average temperature of selective catalytic reduction (SCR) and the amount of urea solution injection in Test 3; (**a**) for entire trip; (**b**) for urban section.

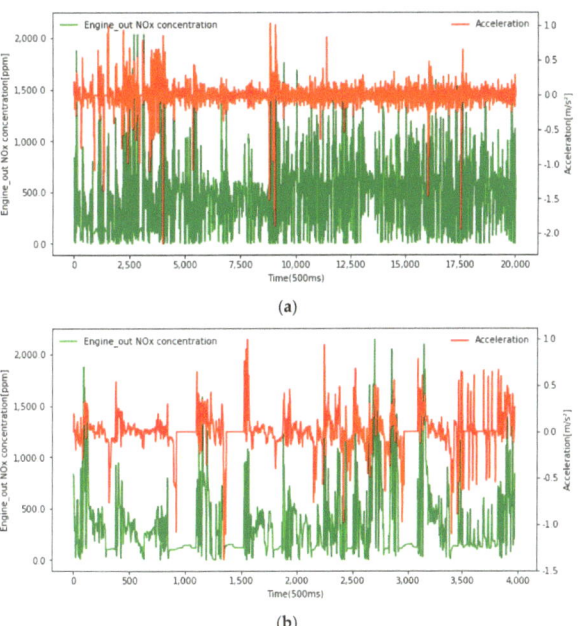

Figure 9. The instantaneous engine NOx emission and the vehicle acceleration in Test 3; (**a**) for entire trip; (**b**) for urban section.

3.3. MAW NOx Emission Analysis

In this section, the NOx emission we talk about refers to tailpipe NOx emission and all the basic data was from the AVL-M.O.V.E-PEMS. The boundary conditions meet the requirements described in Section 2.5 in this paper.

Table 5 shows the number of MAWs, number of valid MAWs, valid MAWs ratio, power threshold, 90th cumulative percentile of valid MAWs NOx emission (g/kWh), and the real NOx emission of the entire trip (g/kWh), and the difference between entire trip NOx emission and 90th cumulative percentile of valid MAWs NOx emission of the four PEMS test.

Table 5. Information of MAW and real NOx emission of the four PEMS tests.

	Test 1	Test 2	Test 3	Test 4
MAW Number	12,291	12,513	13,742	10,432
Valid MAW Number	7562	9047	6895	6709
Valid MAW Ratio (%)	61.5	72.3	50.2	64.3
Power Threshold (%)	20	17	20	20
90th cumulative percentile of Valid MAWs NOx emission (g/kWh)	0.068	0.551	0.438	0.337
Real NOx emission of the entire trip (g/kWh)	0.648	0.794	0.811	0.893
Error 90th Valid MAW to Overall (%)	−89.48	−30.59	−45.91	−62.29

The percentage of valid windows was less than 50% when power threshold was 20%, 19%, 18%, respectively in Test 2. So, the final power threshold used to evaluate the NOx emission in Test 2 was 17%.

There was a great difference between the real NOx emission of the entire trip and 90th cumulative percentile of valid MAWs NOx emission (g/kWh). The evaluation results of the four PEMS tests were all lower than the real, the error is −89.48% in Test 1, −30.59% in Test 2, −45.91% in Test 3 and −62.29% in Test 4. Not only that, the evaluated NOx emissions were all lower than the required limit (0.690 g/kWh), but for Test 2, Test 3 and Test 4, the real NOx emissions were higher than 0.690 g/kWh.

Figure 10a shows the distribution, in g/kWh, of the NOx emissions of each MAW versus the average power (% of maximum engine power) of each MAW during Test 3. As it can be seen, the data points in the red rectangle are considered in the final analysis according to the boundary conditions. The higher NOx emissions which are excluded from the final analysis because of the imposed boundary conditions which are mainly concentrated in the MAWs whose average power are higher.

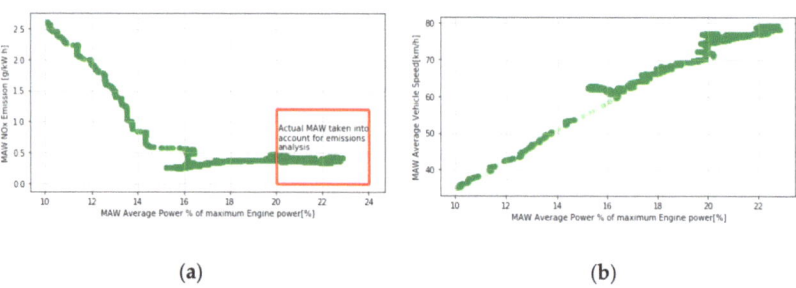

Figure 10. (**a**) MAWs NOx emissions vs. % of maximum engine power of MAWs during Test 3; (**b**) MAWs vehicle speed vs. % of maximum engine power of MAWs during Test 3.

Figure 10b shows the average vehicle speed and average power (% of maximum engine power) of each MAW. The lower the average power of the MAW is, the lower the average vehicle speed of the MAW is. As we know, low vehicle speed occurs mainly in the urban section.

Figure 11 shows the distribution of the urban start position, rural start position, motorway start position, test end position, start position of the first valid MAW and end position of the last valid MAW on the timeline. As it can be seen, the end positions of the last valid MAWs are all located in the rural section of the four PEMS tests, that means the valid MAWs obtained by the rules described in Section 2.5 in this paper (in accordance with China VI, GB17691-2018) just represent the emission characteristics of rural and motorway sections.

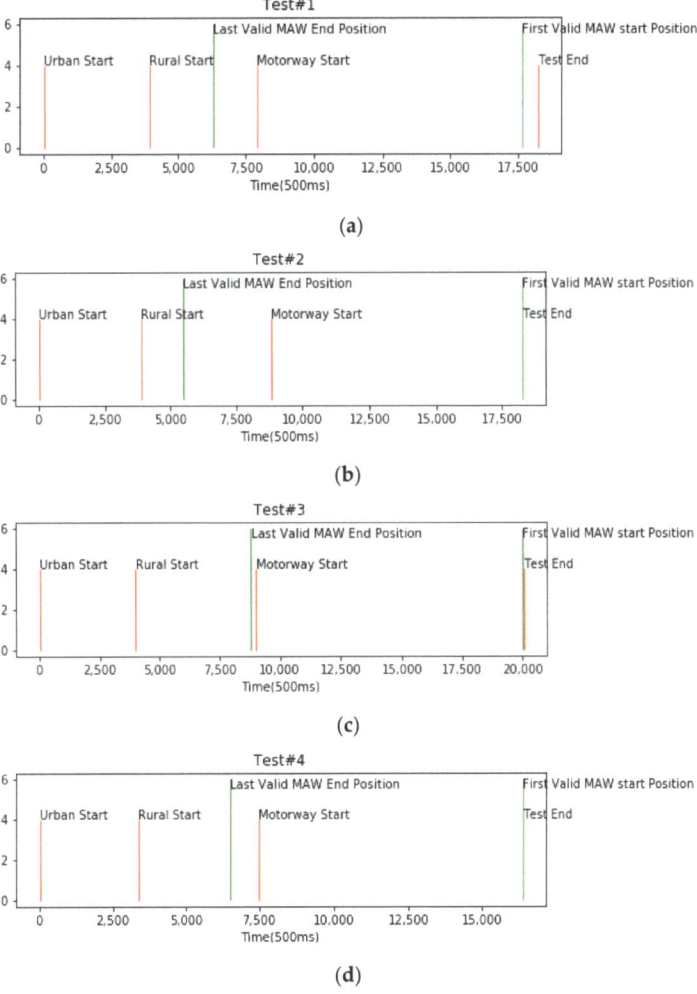

Figure 11. Section and first (last) valid MAW distribution information; (**a**) for Test 1; (**b**) for Table 2. (**c**) for Test 3; (**d**) for Test 4.

So, if we still want to evaluate the NOx emission by MAW method, the data during the urban operation must be taken into account for the evaluation and new rules may be needed. Before that, we have to find main influence factors of the MAW NOx emission.

Figure 12 shows a heatmap that reveals the Pearson correlation coefficient between the parameters' mean value or cumulative value of the MAWs in Test 3.

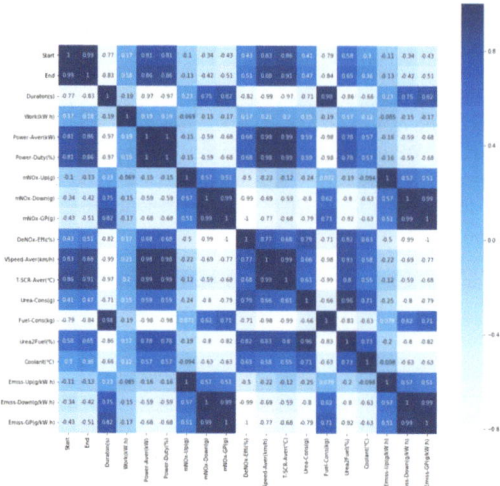

Figure 12. Heatmap of MAW parameters in Test#3 (Pearson correlation coefficient).

Table 6 shows the parameters which have a strong correlation to MAW NOx emission according to the heatmap shown in Figure 12. MAW NOx emission is positively correlated with the window duration, the cumulative mass of NOx emission and the cumulative fuel consumption of the MAW; moreover, it is negatively correlated with the average power, NOx conversion efficiency, average vehicle speed, average SCR temperature, urea consumption and urea-fuel ratio of the MAW.

Table 6. Pearson correlation coefficient of MAW NOx emission.

	Test 1	Test 2	Test 3	Test 4
	MAW NOx emission [g/kWh]			
MAW Duration (s)	0.910	0.739	0.817	0.842
MAW Average Power (kW)	−0.800	−0.590	−0.678	−0.730
MAW Cumulative NOx Mass (g)	1.000	1.000	1.000	1.000
MAW DeNOx Efficiency (%)	−0.999	−0.998	−1.000	−1.000
MAW Average Vehicle Speed (km/h)	−0.876	−0.744	−0.772	−0.775
MAW Average SCR Temperature (°C)	−0.885	−0.641	−0.677	−0.751
MAW Fuel Consumption (kg)	0.995	0.623	0.709	0.807
MAW engine coolant(°C)	-0.871	-0.620	-0.632	-0.672
MAW NOx Emission (g/kWh)	1.000	1.000	1.000	1.000

3.4. The Exploration of Evaluation Methods

The parameters which have a strong correlation to MAW NOx emission including the MAW duration, the MAW average power, the MAW average vehicle speed, etc. (Table 6) may help us clearly distinguish urban section from rural and motorway sections. Therefore, in order to guarantee the objectivity and accuracy of the evaluation, referring to the baseline already described in Section 2.5 in this paper, three different methods (Method 1~3) were applied to extend the regulatory boundary conditions, details were as follows:

Method 1—in accordance with the baseline except power threshold (PT), and setting the PT as 10% for valid MAWs.

Method 2—in accordance with Method 1 except using the 90th cumulative percentile of the valid MAWs emissions as the result of the PEMS test. Instead, according to the average vehicle speed of the MAWs, redefining the MAWs as "Motorway MAWs" (>70 km/h), "Rural MAWs" (≤70 km/h and ≥reference vehicle speed) and "Urban MAWs" (<reference

vehicle speed), where, the reference vehicle speed is equal to the average vehicle speed of the rural section of the PEMS test. More than that, distributing weighting factors to "Motorway MAWs" (0.55), "Rural MAWs" (0.25), and "Urban MAWs" (0.2) according to the required trip share. The final evaluation results shall be calculated by Equation (4):

$$EF = 90\%ileEF_{Urban\ MAW} * 0.2 + 90\%ileEF_{Rural\ MAW} * 0.25 + 90\%ileEF_{Motorway\ MAW} * 0.55 \qquad (4)$$

where: $90\%ile$ is 90th cumulative percentile.

Method 3—in accordance with Method 1 except using the 90th cumulative percentile of the valid MAWs emissions as the result of the PEMS test. Instead, redefining the MAWs according to the average engine coolant temperature of the MAWs as "Cold MAWs" (<reference engine coolant temperature) and "Hot MAWs" (≥reference engine coolant temperature), where, the reference engine coolant temperature is equal to the average engine coolant temperature of the entire trip. At the same time, distributing weighting factor to "Cold MAWs" (0.14), "Hot MAWs" (0.86) referring to the WHTC rules. The final evaluation results shall be calculated by Equation (5):

$$EF = 90\%ileEF_{ColdMAW} * 0.14 + 90\%ileEF_{HotMAW} * 0.86 \qquad (5)$$

where: $90\%ile$ is 90th cumulative percentile.

Due to the uncertainty of the PEMS test procedures, such as trip share, vehicle speed or cumulative work, etc., it is hardly for us to find a constant "Reference Value" to ensure a better universality of Method 2 or Method 3, so, the key point of these two methods were to find a proper "Reference Value". Our data analysis revealed that the average value of a certain section of the test or the entire trip may be a good choice.

Table 7 shows the evaluation results of the Method 1~3, As it can be seen:

Table 7. The results of the Method 1~3.

Method	Parameters	Test 1	Test 2	Test 3	Test 4
——	Real NOx emission of the entire trip (g/kWh)	0.648	0.794	0.811	0.893
Method 1	Power Threshold (%)	10	10	10	10
	NOx emission by Method 1 (g/kWh)	1.43	0.611	0.645	1.203
	Error1 (Method 1 to Real NOx emission)	120.7%	−23.1%	−20.4%	34.6%
	MAW Number	12,291	12,513	13,742	10,432
	Valid MAW Number	12,291	12,513	13,742	10,432
	Valid MAW Ratio	100%	100%	100%	100%
Method 2	Power Threshold (%)	10	10	10	10
	NOx emission by Method 2 (g/kWh)	0.619	0.773	0.833	0.812
	Error2 (Method 2 to Real NOx emission)	−4.5%	−2.7%	2.8%	−9.1%
	Reference Vehicle Speed (km/h)	55.4	60.9	60.9	57.7
Method 3	Power Threshold (%)	10	10	10	10
	NOx emission by Method 3 (g/kWh)	0.661	0.784	0.763	0.955
	Error3 (Method 3 to Real NOx emission)	2.0%	−1.2%	−5.9%	6.9%
	Reference Engine Coolant Temperature (°C)	78.0	80.8	80.4	80.1

For Method 1, when the power threshold (PT) was set as 10%, all of the MAWs of the four PEMS tests were valid, and the data produced during the entire trip was taken into account for the evaluation. Even so, there was also a great difference between the real NOx emission of the entire trip and 90th cumulative percentile of valid MAWs NOx emission (g/kWh), not only that, both positive error and negative error were existed. The error was as high as 120.7% in Test 1, −23.1% in Test 2, −20.4% in Test 3, 34.6% in Test 4. So, we may conclude that just reducing the power threshold (PT) may be not useful enough for an objective evaluation.

For Method 2, the average vehicle speed of rural section of the test was used to distinguish "Urban MAWs" and "Rural MAWs", that because, for N3 category vehicles, there may be no MAW during the urban operation only. As shown in Figure 13, the cumulative work of urban section of each PEMS test was less than the work performed over the WHTC cycle (38.72 kWh). So, for the four PEMS tests, no MAW was merely composed by the urban operation. The less the cumulative work of urban section, the more the data produced in rural section and used to compose the last MAW.

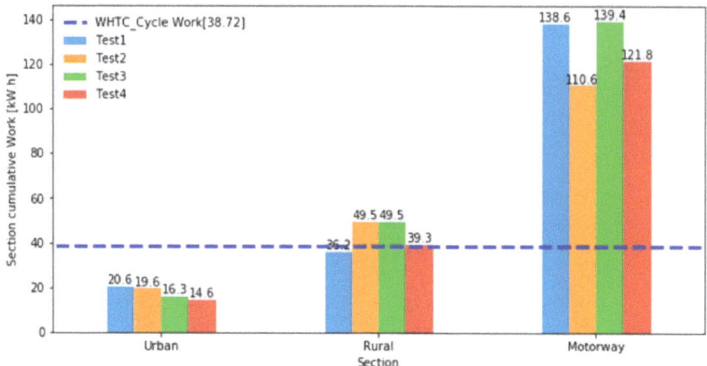

Figure 13. Cumulative work of each section of the four PEMS tests.

Figure 14 shows the distribution of the redefined MAWs of the four PEMS tests by Method 2. As it can be seen, the worse polluting windows were categorized as "Urban MAW", most MAWs were categorized as "Motorway MAW" because of its longest test duration and maximum cumulative work. The error between the real NOx emission of the entire trip and the NOx emission evaluated by Method 2 was −4.5% in Test 1, −2.7% in Test 2, 2.8% in Test 3 and −9.1% in Test 4.

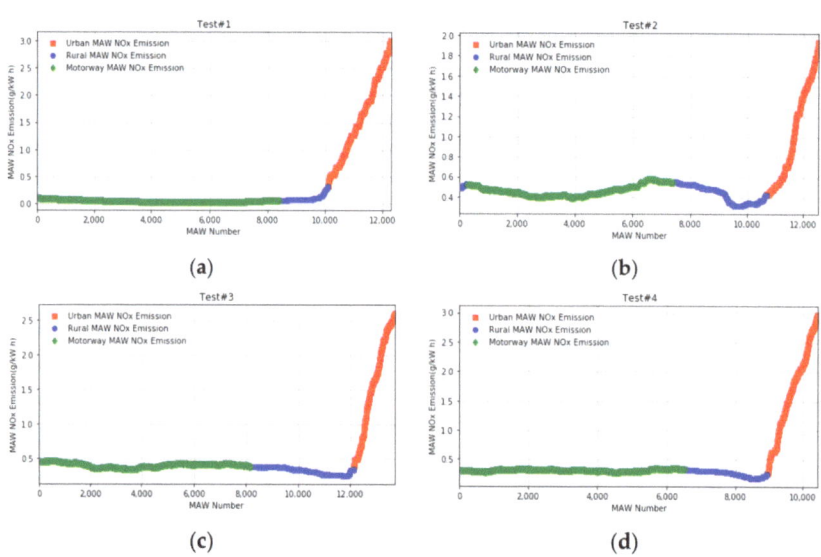

Figure 14. Distribution of the redefined MAWs of the four PEMS tests by Method 2; (**a**) for Table 1. (**b**) for Test 2; (**c**) for Test 3; (**d**) for Test 4.

For Method 3, the average engine coolant temperature of the entire trip was used to distinguish "Hot MAWs" and "Cold MAWs". As shown in Figure 15, the engine coolant temperature generally rose gradually and then stabilized around a certain value after the engine coolant temperature had reached 70 °C for the first time during a PEMS test. Usually, there was no urea solution injection during the "rising period". So, we renamed the "rising period" as "Cold Operation". Via the data analysis, we found that taking the average engine coolant temperature of the entire trip as the "Reference Value" would lead to an exciting result.

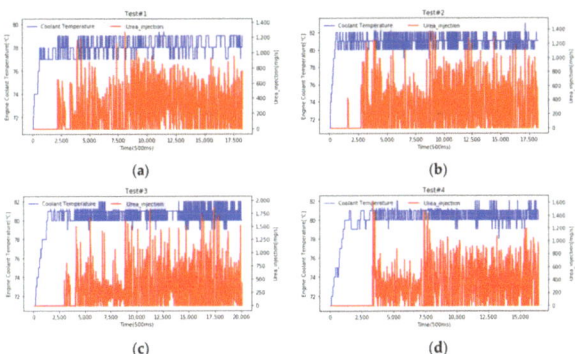

Figure 15. The instantaneous engine coolant temperature and urea solution injection of the entire trip; (**a**) for Test 1; (**b**) for Test 2; (**c**) for Test 3; (**d**) for Test 4.

Figure 16 shows the distribution of the redefined MAWs of the four PEMS tests by Method 3, only a few of MAWs were regarded as "Cold MAWs" which had worse NOx emission. The error between the real NOx emission of the entire trip and the NOx emission evaluated by Method 3 was 2.0% in Test 1, −1.2% in Test 2, −5.9% in Test 3 and 6.9% in Test 4.

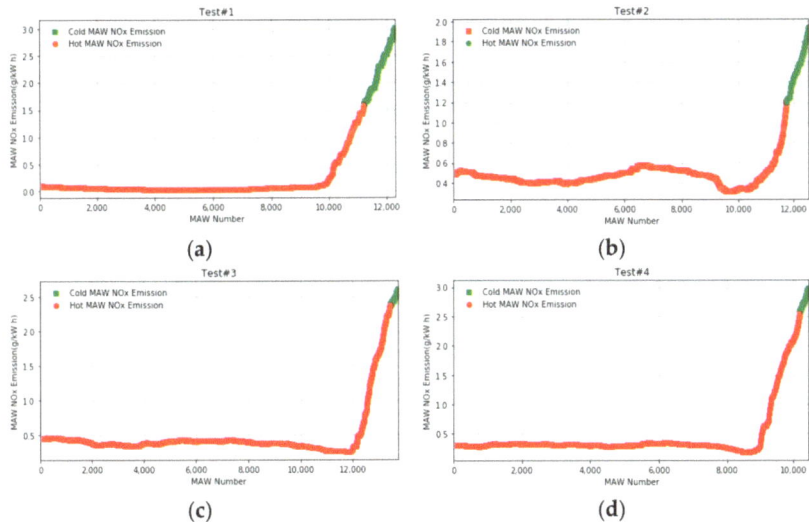

Figure 16. Distribution of the redefined MAWs of the four PEMS tests by Method 3; (**a**) for Test 1; (**b**) for Test 2; (**c**) for Test 3; (**d**) for Test 4.

To sum it up, the performance of Method 2 or Method 3 was much better than Method 1, and the evaluation errors of Method 2 or Method 3 were all within ±10% (Table 7), that meant vehicle speed and engine coolant temperature could be used as the assistant parameters to classify the MAWs for better heavy-duty vehicle real world NOx emission evaluation. At the same time, we may conclude that a method that could represent the characteristics of each section (such as "urban section", "cold section") of a PEMS test shall be adopted for the emissions evaluation.

4. Conclusions and Outlook

This study had presented and analyzed the data produced by four valid PEMS tests of a China VI (step B), N3 category heavy-duty diesel vehicle. The conclusions of the present study can be summarized to the following:

- The highest tailpipe NOx emissions including the cumulative mass and the brake-specific emission were invariably found during urban operation, which was of great concern for urban air quality and human health. This was mainly because the SCR temperature of the catalytic converter was not high enough to ensure the urea solution injection in most of urban operation. Therefore, heating up SCR rapidly may be an effective means to reduce NOx emission in the urban section.
- For N3 category vehicle, the data produced during the urban section of a PEMS test was excluded from emissions analysis by MAW method due to the higher power threshold (20%) required by boundary conditions. Therefore, a lower power threshold should be used or power threshold boundary should be avoided.
- There was a great difference between the real NOx emission of the entire trip and 90th cumulative percentile of valid MAWs NOx emission(g/kWh)whether the power threshold was set as 10% or 20%.
- The 90th cumulative percentile of valid MAWs NOx emission just represents the emission characteristics of certain sections (rural and motorway for N3 category vehicles) of a PEMS test rather than the entire trip. So, average vehicle speed of the MAWs was used to categorize the MAWs into "Urban MAWs", "Rural MAWs" and "Motorway MAWs"; average engine coolant temperature of the MAWs was used to categorize the MAWs into "Hot MAWs" and "Cold MAWs". The evaluation results of the NOx emission of these two kinds of categorized MAWs were close to the real NOx emission, and the errors were all within ±10%.

In this work, we had pointed out the insufficient of the current evaluation method for heavy-duty vehicle real world NOx emission. Future studies should focus on the following aspects:

- The control and evaluation for NOx emission of cold start (engine coolant temperature less than 70 °C) or low load operation conditions.
- More individualized boundary conditions or MAW rules for the real-world NOx emission of each category vehicle, such as the definition of power thresholds (PT), valid MAW, weights factors, etc.
- The real amount (g, g/kW.h or g/km) of heavy-duty diesel vehicles' real world NOx emission, especially the urban section.

Author Contributions: Conceptualization, P.L. and L.L.; methodology, P.L. and L.L.; formal analysis, P.L.; investigation, P.L. and L.L.; resources, L.L.; data curation, P.L.; writing—original draft preparation, P.L.; writing—review and editing, L.L.; visualization, P.L.; supervision, L.L.; project administration, L.L.; funding acquisition, L.L. All authors have read and agreed to the published version of the manuscript.

Funding: This research was funded by National Natural Science Foundation of China (NSFC), grant number 51679176.

Institutional Review Board Statement: Not applicable.

Informed Consent Statement: Not applicable.

Data Availability Statement: Data sharing not applicable.

Acknowledgments: The authors are thankful to all the personnel who either provided technical.

Conflicts of Interest: The authors declare no conflict of interest.

References

1. Anenberg, S.C.; Miller, J.; Minjares, R.; Du, L.; Henze, D.K.; Lacey, F.; Malley, C.S.; Emberson, L.; Franco, V.; Klimont, Z.; et al. Impacts and mitigation of excess diesel-related NOx emissions in 11 major vehicle markets. *Nat. Cell Biol.* **2017**, *545*, 467–471. [CrossRef]
2. Cheng, Y.; He, L.; He, W.; Zhao, P.; Wang, P.; Zhao, J.; Zhang, K.; Zhang, S. Evaluating on-board sensing-based nitrogen oxides (NOX) emissions from a heavy-duty diesel truck in China. *Atmos. Environ.* **2019**, *216*, 116908. [CrossRef]
3. Yang, D.; Zhang, S.; Niu, T.; Wang, Y.; Xu, H.; Zhang, K.M.; Wu, Y. High-resolution mapping of vehicle emissions of atmos-pheric pollutants based on large-scale, real-world traffic datasets. *Atmos. Chem. Phys.* **2019**, *19*, 8831–8843. [CrossRef]
4. Grange, S.K.; Lewis, A.C.; Moller, S.J.; Carslaw, D.C. Lower vehicular primary emissions of NO_2 in Europe than assumed in policy projections. *Nat. Geosci.* **2017**, *10*, 914–918. [CrossRef]
5. Zhang, S.; Zhang, S.; Shaojun, Z.; Liu, H.; Wu, X.; Hu, J.; Walsh, M.P.; Wallington, T.J.; Zhang, K.M.; Stevanovic, S. On-road vehicle emissions and their control in China: A review and outlook. *Sci. Total Environ.* **2017**, *574*, 332–349. [CrossRef]
6. Grigoratos, T.; Fontaras, G.; Giechaskiel, B.; Zacharof, N. Real world emissions performance of heavy-duty Euro VI diesel vehicles. *Atmos. Environ.* **2019**, *201*, 348–359. [CrossRef]
7. Sowman, J.; Box, S.; Wong, A.; Grote, M.; Laila, D.S.; Gillam, G.; Cruden, A.J.; Preston, J.; Fussey, P. In-use emissions testing of diesel-driven buses in Southampton: Is selective catalytic reduction as effective as fleet operators think? *IET Intell. Transp. Syst.* **2018**, *12*, 521–526. [CrossRef]
8. Lisi, L.; Landi, G.; Di Sarli, V. The Issue of Soot-Catalyst Contact in Regeneration of Catalytic Diesel Particulate Filters: A Critical Review. *Catalysts* **2020**, *10*, 1307. [CrossRef]
9. Jiaqiang, E.; Xie, L.; Zuo, Q.; Zhang, G. Effect analysis on regeneration speed of continuous regeneration-diesel particulate filter based on NO 2 -assisted regeneration. *Atmos. Pollut. Res.* **2016**, *7*, 9–17. [CrossRef]
10. Jiaqiang, E.; Zuo, W.; Gao, J.; Peng, Q.; Zhang, Z.; Hieu, P.M. Effect analysis on pressure drop of the continuous regenera-tion-diesel particulate filter based on NO2 assisted regeneration. *Appl. Therm. Eng.* **2016**, *100*, 356–366.
11. Velders, G.J.; Geilenkirchen, G.P.; De Lange, R. Higher than expected NOx emission from trucks may affect attainability of NO_2 limit values in the Netherlands. *Atmos. Environ.* **2011**, *45*, 3025–3033. [CrossRef]
12. Wu, Y.; Zhang, S.J.; Li, M.L.; Ge, Y.S.; Shu, J.W.; Zhou, Y.; Xu, Y.Y.; Hu, J.N.; Liu, H.; Fu, L.X.; et al. The challenge to NOx emission control for heavy-duty diesel vehicles in China. *Atmos. Chem. Phys. Discuss.* **2012**, *12*, 18565–18604.
13. Varella, R.A.; Gonçalves, G.; Duarte, G.; Farias, T. Cold-Running NOx Emissions Comparison between Conventional and Hybrid Powertrain Configurations Using Real World Driving Data. *SAE Tech. Pap. Ser.* **2016**, *1*. [CrossRef]
14. Varella, R.A.; Faria, M.V.; Mendoza-Villafuerte, P.; Baptista, P.C.; Sousa, L.; Duarte, G.O. Assessing the influence of boundary conditions, driving behavior and data analysis methods on real driving CO_2 and NOx emissions. *Sci. Total Environ.* **2018**, *658*, 879–894. [CrossRef] [PubMed]
15. Degraeuwe, B.; Weiss, M. Does the New European Driving Cycle (NEDC) really fail to capture the NOX emissions of diesel cars in Europe? *Environ. Pollut.* **2017**, *222*, 234–241. [CrossRef] [PubMed]
16. Liu, D.; Lou, D.; Liu, J.; Fang, L.; Huang, W. Evaluating Nitrogen Oxides and Ultrafine Particulate Matter Emission Features of Urban Bus Based on Real-World Driving Conditions in the Yangtze River Delta Area, China. *Sustainability* **2018**, *10*, 2051. [CrossRef]
17. Giechaskiel, B.; Gioria, R.; Carriero, M.; Lähde, T.; Forloni, F.; Perujo, A.; Martini, G.; Bissi, L.M.; Terenghi, R. Emission Factors of a Euro VI Heavy-duty Diesel Refuse Collection Vehicle. *Sustainability* **2019**, *11*, 1067. [CrossRef]
18. European Commission. Commission Regulation (EU) 2016/427 of 10 March 2016 Amending Regulation (EC) No 692/2008 as Regards Emissions from Light Passenger and Commercial Vehicles (Euro 6). *Off. J. Eur. Union* **2016**, *82*, 1–98.
19. Vlachos, T.G.; Bonnel, P.; Perujo, A.; Weiss, M.; Villafuerte, P.M.; Riccobono, F. In-Use Emissions Testing with Portable Emissions Measurement Systems (PEMS) in the Current and Future European Vehicle Emissions Legislation: Overview, Underlying Principles and Expected Benefits. *SAE Int. J. Commer. Veh.* **2014**, *7*, 199–215. [CrossRef]
20. Giechaskiel, B.; Clairotte, M.; Valverde-Morales, V.; Bonnel, P.; Kregar, Z.; Franco, V.; Dilara, P. Framework for the assessment of PEMS (Portable Emissions Measurement Systems) uncertainty. *Environ. Res.* **2018**, *166*, 251–260. [CrossRef]
21. Mendoza-Villafuerte, P.; Suarez-Bertoa, R.; Giechaskiel, B.; Riccobono, F.; Bulgheroni, C.; Astorga, C.; Perujo, A. NOx, NH_3, N_2O and PN real driving emissions from a Euro VI heavy-duty vehicle. Impact of regulatory on-road test conditions on emissions. *Sci. Total Environ.* **2017**, *609*, 546–555. [CrossRef] [PubMed]
22. Tan, Y.; Henderick, P.; Yoon, S.; Herner, J.; Montes, T.; Boriboonsomsin, K.; Johnson, K.C.; Scora, G.; Sandez, D.; Durbin, T.D. On-Board Sensor-Based NOx Emissions from Heavy-Duty Diesel Vehicles. *Environ. Sci. Technol.* **2019**, *53*, 5504–5511. [CrossRef] [PubMed]
23. Varella, R.A.; Duarte, G.; Baptista, P.; Sousa, L.; Villafuerte, P.M. Comparison of Data Analysis Methods for European Real Driving Emissions Regulation. *SAE Tech. Pap. Ser.* **2017**, *1*, 0997. [CrossRef]

Article

Finding Optimal Stations Using Euclidean Distance and Adjustable Surrounding Sphere

Athita Onuean [1,2], Hanmin Jung [1,2,*] and Krisana Chinnasarn [3,*]

1. Data & High Performance Computing Science, University of Science and Technology, Daejeon 34113, Korea; athita@kisti.re.kr
2. Korea Institute of Science and Technology Information, Daejeon 34141, Korea
3. Faculty of Informatics, Burapha University, Chon Buri 20131, Thailand
* Correspondence: jhm@kisti.re.kr (H.J.); krisana@it.buu.ac.th (K.C.); Tel.: +82-42-869-1772 (H.J.); +66-81-590-5009 (K.C.)

Citation: Onuean, A.; Jung, H.; Chinnasarn, K. Finding Optimal Stations Using Euclidean Distance and Adjustable Surrounding Sphere. *Appl. Sci.* **2021**, *11*, 848. https://doi.org/10.3390/app11020848

Received: 31 December 2020
Accepted: 15 January 2021
Published: 18 January 2021

Publisher's Note: MDPI stays neutral with regard to jurisdictional claims in published maps and institutional affiliations.

Copyright: © 2021 by the authors. Licensee MDPI, Basel, Switzerland. This article is an open access article distributed under the terms and conditions of the Creative Commons Attribution (CC BY) license (https://creativecommons.org/licenses/by/4.0/).

Abstract: Air quality monitoring network (AQMN) plays an important role in air pollution management. However, setting up an initial network in a city often lacks necessary information such as historical pollution and geographical data, which makes it challenging to establish an effective network. Meanwhile, cities with an existing one do not adequately represent spatial coverage of air pollution issues or face rapid urbanization where additional stations are needed. To resolve the two cases, we propose four methods for finding stations and constructing a network using Euclidean distance and the k-nearest neighbor algorithm, consisting of Euclidean Distance (ED), Fixed Surrounding Sphere (FSS), Euclidean Distance + Fixed Surrounding Sphere (ED + FSS), and Euclidean Distance + Adjustable Surrounding Sphere (ED + ASS). We introduce and apply a coverage percentage and weighted coverage degree for evaluating the results from our proposed methods. Our experiment result shows that ED + ASS is better than other methods for finding stations to enhance spatial coverage. In the case of setting up the initial networks, coverage percentages are improved up to 22%, 37%, and 56% compared with the existing network, and adding a station in the existing one improved up by 34%, 130%, and 39%, in Sejong, Bonn, and Bangkok cities, respectively. Our method depicts acceptable results and will be implemented as a guide for establishing a new network and can be a tool for improving spatial coverage of the existing network for future expansions in air monitoring.

Keywords: Euclidean Distance; spatial coverage; air quality monitoring network; sustainability monitoring

1. Introduction

Air quality monitoring networks (AQMN) are established as tools that determine policies and strategies for achieving air quality standards. A plan for designing an AQMN depends on objectives such as urban planning, environmental policies, and budget. Generally, designing an AQMN is done by environmental authorities or governmental organizations based on empirical judgments. An expert group assesses various criteria to make their decisions, such as budget and population thresholds. These play a significant role in determining the required number and location of monitoring stations. For example, Thailand and South Korea set up air monitoring stations in government office areas to reduce the cost of installation, maintenance, and the safety of the devices [1,2]. Guidelines of the USA and Australia suggest installing new air quality monitoring stations based on population size [3,4]. In such cases, monitoring stations are often considered in an ad hoc fashion.

The critical issues in AQMN designing are separated into a setting up network for allocating optimum stations and optimizing the existing network to better reflect air quality in the area. There are different methods to design a new network, all of which must comply with the Environmental Protection Agency (EPA) guideline or local government regulations [5]. Furthermore, various studies mainly consider pollution data, population density, land-use regression, distance to major roads, and high-risk observation regions to identify new air quality monitoring stations [6–10]. Some studies investigated pollution indicators in the area surrounding stations [11–13]. Such a zone is called a sphere of influence (SOI) and the original algorithm was developed by Liu et al. in 1986. Their proposed method uses air pollution data to determine spatial coverage and the number of air monitoring stations [14].

Optimization, evaluation, and revision of existing AQMN to meet changing local area pollution levels has been an important research topic over the past few decades [15–18]. The main cause regarding the distribution of air pollution and emission source changed caused by urbanization. The rapid expansion of the urban areas in developing countries of Asia leads to air quality monitoring issues such as excessive numbers of stations in urban regions and a lack of adequate numbers of stations in rural areas [19]. Consequently, several approaches have suggested adding or removing stations by considering various constraints such as population density, historical pollution data, terrain conditions, budget, and health impact. They use statistical analytics, weight criteria, holistic approaches, data simulation, and pure measurement for AQMN optimization [20–22]. Moreover, some studies apply terrain maps, heat maps, gridded synthetic assessment, and graphical information systems (GIS) to show high pollution concentration areas. They analyze pollution criteria and combine them with spatial statistics to determine suitable areas to recommend for station location sites [23–27].

However, studies of optimal designs or revision of AQMN in less developed countries (LDCs) are relatively scarce. Such studies in the literature are mostly related to minimizing air pollution's health impact [28,29]. The LDCs have different issues than in developed countries, such as limited budgets to establish the stations, lack of historical pollution data, and requirements that demand air pollution monitoring cover vast areas. All of these constraints are critical factors for AQMN design [30]. Therefore, recommending the station location to cover a prospective land use expansion has an important key role for sustainable development of air pollution control and helps assess air pollution's spatial variability for better monitoring of air quality [31].

Previous studies attempted to design air monitoring networks by considering specific air pollution concentration, cost, population data, and other parameters, while spatial coverage was ignored. Such a design methodology causes station locations to have high density in some particular regions, especially in urban areas. It does not distribute coverage to rural areas with a sufficient number of stations. The incremental building of air quality monitoring in urban areas can make better monitoring of the specific areas. However, it leads to an increase in installation and maintenance costs, which is one of the main constraints for designing a dense air quality monitoring network. Furthermore, none of the methods in previous studies mentioned designing an AQMN without historical pollution data or city characteristics data. Furthermore, it is important to consider that the air monitoring networks designed are not only for use today but also for urban expansion in the future.

We organized the rest of the paper as follows. Problem formulation is introduced in Section 2. We describe the study areas in Section 3. In Section 4 proposes four methods for finding the best next station, which includes algorithms and equations. In Section 5, the evaluation criteria will be explained. The results of the proposed methods are compared and discussed in Section 6. Finally, Section 7 draws our conclusion and suggests possible future work.

2. Problem Formulation

The studied problem is based on finding the next stations to achieve maximum spatial coverage. We apply the concept of Euclidean Distance and the k-nearest neighbor algorithm (k-NN) to calculate the distance between station neighbors. The objectives are set up a network and improve an existing network. In the preparation process, we divide map of study areas into a square of a grid and identify latitude longitude pairs at the centroid of all grids. Such geographic coordinates are used to calculate the distance. Given only a study area map and a specified number of stations, how can one calculate and recommend the stations to achieve maximum spatial coverage? The objective is to find the stations which meet the following constraints:

(i) Achieve maximum spatial coverage while maintaining proper overlapped area between the nearest-neighbors' stations
(ii) Propose methods without using historical pollution and city characteristics data.

The proposed method was tested and demonstrated in three real cities and compared with existing network coverage. This study can provide recommendations for environmental authorities or city planners to select the stations for increasing spatial coverage of air quality monitoring.

3. Study Areas

This study used three cities—Sejong: South Korea, Bangkok: Thailand, and Bonn: Germany—as the primary research areas. The three cities are developed smart city prototypes facing urban expansion shortly. Another reason for our choices is that the three regions have different sizes and shapes. The sizes of the cities in ascending order are Bonn, Sejong, and Bangkok, respectively. Because the size of Sejong is in the middle, we chose Sejong city as the primary implementation, and the other two cities are used as test cases. Our study areas are described in this section.

Sejong city is located almost the middle of Korea on a long, stretching mainly north to south. The population was about 350,000 in October 2020, and the size of Sejong city is 465.23 km^2. Sejong was specifically designed to be a smart city, so it serves as an example of the standard for the other cities experimenting with developing smart city infrastructure. Sejong city has four air quality monitoring stations, as shown with triangle symbols in Figure 1a. For this study, the whole area of the city is divided into 2024 grid cells of 500 × 500 m, as shown in Figure 1d.

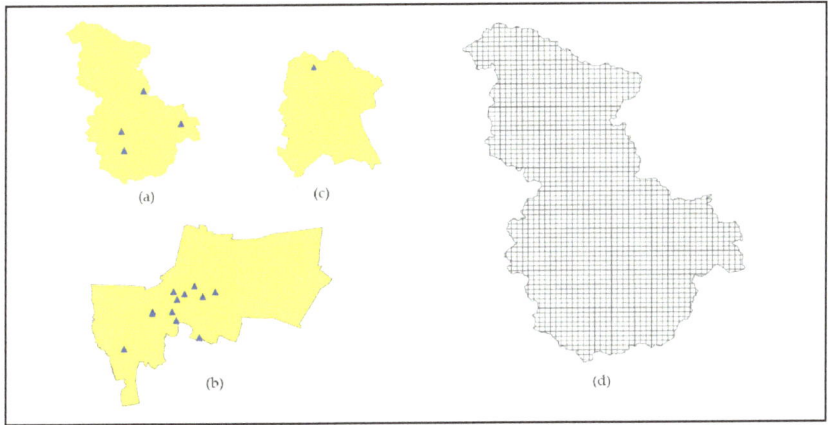

Figure 1. Study areas (**a**–**c**) show administrative boundaries, and triangle symbols show current stations. The map regions (**d**) divided into 2024 grid cells in Sejong city.

Bangkok is the capital city of Thailand. The city is located in almost the middle of Thailand and occupies 1568.7 km^2. Because this city has the highest population density in Thailand, Bangkok city has an air pollution problem from traffic and energy consumption. There are 12 air quality monitoring stations from the pollution control department, which are depicted with the triangle symbols, as shown in Figure 1b. Our experiment divided the area of Bangkok into 7010 grid cells of the same size as we used for Sejong city.

Bonn city is located in western Germany and occupies 141.06 km^2. This city is one of eight major smart cities studied and one of two cities that will now access their brand new 5G network. Bonn has only one air quality monitoring station, which is located in the north of the city. Figure 1c shows the map and location of the air quality monitoring station in Bonn. Furthermore, for our experiment, the whole area of the city is divided into 653 grid cells.

4. Proposed Methods

In this study, finding optimal stations' main target goal is achieving the maximum spatial coverage while still preserving appropriate overlapping areas. The maximum spatial coverage designed can be realized through the optimal placement of the stations in the city. Moreover, maintaining a relevant overlapped area can enhance effectiveness for more reliable and accurate data collection. Accordingly, in this section, we proposed finding the stations based on the Euclidean Distance and the k-nearest neighbor's algorithm (k-NN). The proposed methods framework consists of four main methods and two evaluation criteria, as shown in Figure 2.

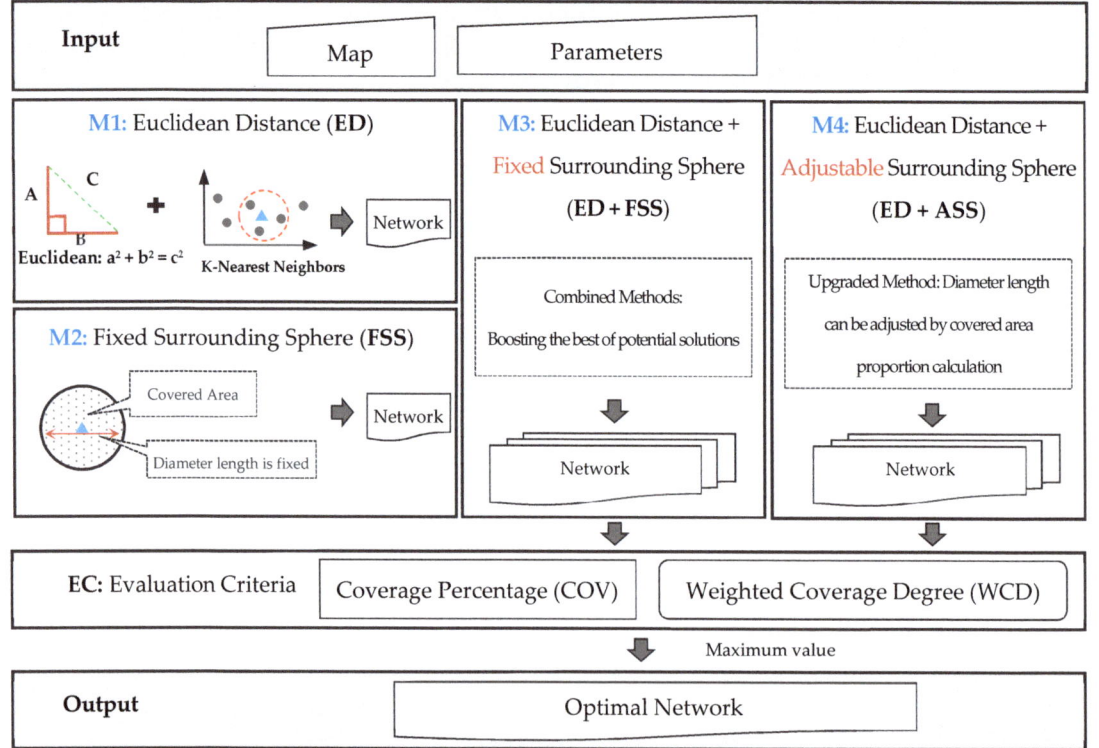

Figure 2. The framework of proposed methods.

In the proposed framework, our input consists of a map and parameters. The map of the study area is divided into a square of continue grid. A centroid of each square grid is a point that consists of latitude-longitude coordinates. For simple understanding throughout this paper, we use the term "location" (L_i), $i = 1 \ldots N$ to represent the centroid of a grid. The parameters consisted of (i) a grid index at the center of the map, (ii) a specified number of stations, and (iii) the diameter length of the surrounding sphere of each station as described in Section 6.1. Next, we give stations in area A, which consist of η stations. The value $S = \{S_j, \ldots, S_\eta\}$ is a list of stations. Each station $S_j, j = 1 \ldots \eta$ is at a location in our study area. All of the input will use to calculate by our proposed methods.

M1: Euclidean Distance (ED) and M2: Fixed Surrounding Sphere are the initial method for finding the next stations. The M3 method is a combination of previous methods. The M4 is an upgraded version of M3. Our evaluation criteria, coverage percentage (COV), and weighted coverage degree (WCD) are used to evaluate the results from the methods then return the optimal network. The comprehensive methods are described in the following sections.

4.1. Euclidean Distance (ED)

In this study, we apply the Euclidean Distance function to calculate distance from location (L_i) to three nearest neighbor stations $\{NS_1, NS_2, NS_3\}$, as illustrated in Figure 3. We use k-NN ($k = 3$) because the nearby stations can exchange data reliability with existing stations. Let NS_1, NS_2, and NS_3 be members of the set of nearest neighboring stations of L_i, where L_i is a position to calculate ED_i. We calculate the Euclidean Distance (ED_i) at any L_i as:

$$ED_i = \sqrt{\left(D_{L_i,\ NS_1}\right)^2 + \left(D_{L_i,NS_2}\right)^2 + \left(D_{L_i,NS_3}\right)^2} \quad (1)$$

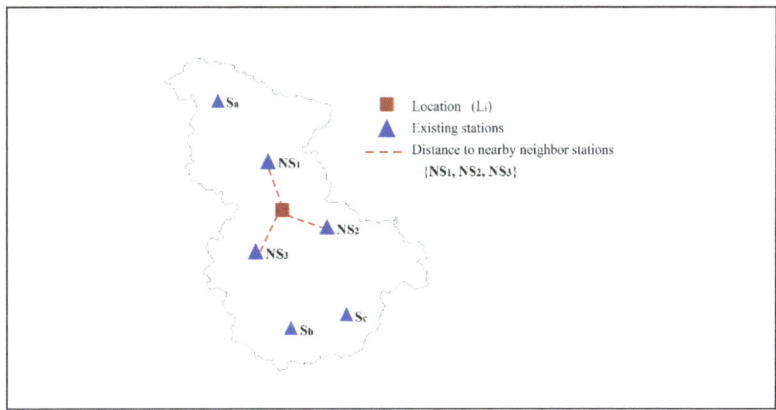

Figure 3. The location of L_i and its nearest neighbor stations NS_1, NS_2, and NS_3.

ED_i denotes Euclidean Distance at L_i, where i is an index of the location, and each parenthesized value is the distance from L_i to NS_1, NS_2, and NS_3, respectively. If $k < 3$, then we adapt the equation by using only existing stations. For example, if $k = 2$, then D_{L_i,NS_3} is equal to zero. Thus, ED_i is calculated using only D_{L_i,NS_1} and D_{L_i,NS_2}.

In order to calculate and find the next station, the following definitions are used. Let S denote a list of stations in a network and let L_t be a candidate station to be evaluated for the possibility of it being the next station. Thus, the current station network is $S + L_t$. Here, we calculate ED_i according to Equation (1). Subsequently, we sum ED_i values as:

$$TED_{L_t} = \sum_{i=0}^{N} ED_i \quad (2)$$

TED_{L_t} denotes the total of ED_i, where L_t is a candidate for the next station, and N is the number of locations. The candidate with the lowest TED_{L_t} will be defined as the additional station.

4.2. Fixed Surrounding Sphere (FSS)

The fixed surrounding sphere (FSS) method was inspired by the original sphere of influence (SOI) [12]. We applied such an idea for determining the area surrounding each station without using pollution concentration data. The constant value is identified to a diameter length of FSS with reference to city air quality monitoring network designing in Seoul city [2]. In Seoul's existing air monitoring network, the 25 stations are located approximately 5 km away from each other. They are not located close to the road, high emission concentration sources, or high-density population areas. On the other hand, these stations are distributed throughout the city. Consequently, we predefined a fixed diameter (dm_s) of FSS to divide the covered and non-covered areas under stations. As shown in Figure 4, the triangle symbols represent stations. The surrounding covered area of the stations is illustrated as a circle shape in Figure 4a and a pie shape in Figure 4b by determining a diameter length. The areas outside represent the non-covered areas. The shapes depend on the position of stations on the map. However, such areas are defined as covered areas, although the shapes of the areas are different.

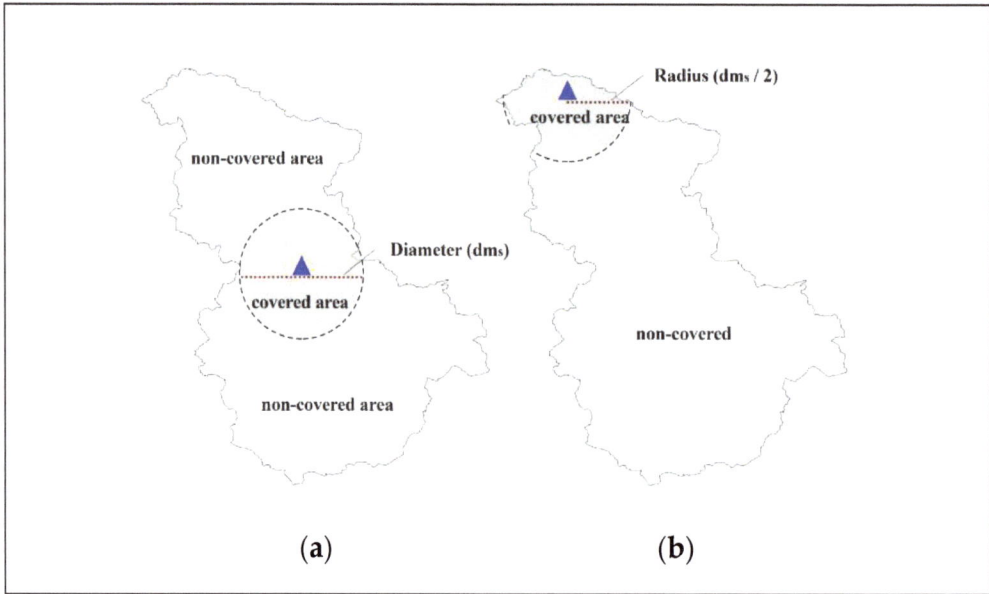

Figure 4. The (**a**) circle and (**b**) pie shapes indicate the covered areas by stations at the different positions on the map by determining a diameter length.

The procedure to classify the covered and non-covered areas under a station are described in this subsection. Let station S_j be a location of station and the diameter of S_j be equal to dm_s. The location (L_i) is covered (monitored) by a station S_j when the distance from L_i to S_j is less than $\frac{dm_s}{2}$ and also such L_i will be members of all covered areas for A_{sj}. Equation (3) represents the probability p(L_i,S_j) that a location is covered by station S_j.

$$p(L_i, S_j) = \begin{cases} 1: \text{"covered area"}, \ if \ distance(L_i, S_j) \leq \frac{dm_s}{2} \\ 0: \text{"non} - \text{covered area"}, \ otherwise \end{cases} \in A_{Sj} \quad (3)$$

i is the index of location from 0 to $N-1$, where N represents the number of locations in area A, and j provides an index of stations in S.

The Algorithm 1 Fixed Surrounding Sphere (FSS) can be described as follows:

Algorithm 1. Fixed Surrounding Sphere (FSS)

Input: Map; a specified number of stations; index of center of map;
1:   Identify the first station at center of map and insert into station list $S = \{S1\}$.
2:     **while** a specified number of stations is not satisfied **do**
3:       **while** all L_t in the non-covered area have not been tried **do**
4:         Select candidate station (L_t) and we temporarily append L_t to the station list S.
5:         Classify covered and non-covered areas of S with Equation (3).
6:         Calculate the total of the non-covered area and pass value to the #poor variable.
7:         Remove the temporarily added L_t from S.
8:       **end while**
9:       #poor values are compared.
10:      The L_t with the smallest value of #poor will be chosen as the next location of station.
11:      L_t is permanently appended to the station list S.
12:    **end while**
13: **return** List of stations S. # *The output is a list of stations with the first station at S1. For examples:*
14:             # $S = \{S1, L_a, L_b, ..., L_\eta\}$

4.3. Euclidean Distance + Fixed Surrounding Sphere (ED + FSS)

This method combines the concepts Euclidean Distance (ED) and Fixed Surrounding Sphere (FSS). However, the difference between this method and the previous one is the location of the first station. For ED and FSS, the first stations are set at the center of the map, while for ED + FSS, all map locations are tried as a first station. The output of this method is a multi-list of stations in which the first stations are different. Consequently, further evaluation criteria have been introduced to evaluate the best network with a maximum of spatial coverage.

We divide ED + FSS into two processes: (i) finding the next stations and (ii) evaluating a maximum coverage percentage while still preserving an appropriate overlapped area for the network. The first process, finding the next stations, is described with Algorithm 2 Euclidean Distance + Fixed Surrounding Sphere (ED + FSS) as follows:

Algorithm 2. Euclidean Distance + Fixed Surrounding Sphere (ED + FSS)

Input: Map; a specified number of stations;
1:   **For each** L_i in map **do**
2:      select L_i and insert into station list, $S = \{L_i\}$.
3:      **while** a specified number of stations is not satisfied **do**
4:         **while** all L_t in the non-covered area have not been tried **do**
5:            Select candidate station (L_t) and we temporarily append L_t to the station list S.
6:            Classify covered and non-covered areas of S with Equation (3).
7:            Calculates TED_{Lt} with Equation (2) and stores the current TED_{Lt} value in a list.
8:            Remove the temporarily added L_t from S.
9:         **end while**
10:        Compare TED_{Lt} value in a list
11:        The L_t with the smallest value of TED_{Lt} will be chosen as the next location of station.
12:        L_t is permanently appended to the station list S.
13:     **end while**
14:     Store station list S into multi-list
15: **end for**
16: **return** Multi-list of stations.
17:        # *The output is multi-list of stations. For examples:*
18:        # $\{S_i = \{L_i, L_a, L_b, ..., L_\eta\}, S_{i+1} = \{L_{i+1}, L_c, L_d, ..., L_\eta\}, ..., S_N = \{L_N, L_e, L_f, ..., L_\eta\}\}$

After we obtain a multi-list of stations, the next process is to evaluate the maximum coverage percentages of all S_x. The ED + FSS will use two criteria, Coverage Percentage (COV) and Weighted Coverage Degree (WCD), for selecting the best network. The descriptions of COV and WCD criteria are outlined in Section 5.

4.4. Euclidean Distance + Adjustable Surrounding Sphere (ED + ASS)

This method is an upgrade from ED + FSS. We change from using a fixed diameter to an adjustable diameter, which depends on a station's covered area proportion. Our proposed method considers economic benefits based on deployment costs and station location with the highest spatial coverage for a specified number of economically feasible stations. The covered area proportion can be adjusted as needed. For example, if the environmental authority has a budget limit for establishing monitoring stations, an arbitrary ratio can be predefined as 50% or 70%. On the other hand, if they require high spatial resolution monitoring or an unlimited budget, the ratio can be predefined as 10% or 30%.

As a warmup to ED + ASS, we select a station at the center of the map (S_{center}) for calculating the length of the diameter. Next, we calculate ED_i from all locations (L_i) on map to the S_{center} with Equation (1). We pass all the ED_i values into a list. Subsequently, we sort the obtained list in ascending order, which means that locations closer to the station with lower ED_i values will be at the beginning of the list. The cut off position corresponds to an arbitrary ratio, as mentioned above. We calculate the cut off position value with Equation (4).

$$Cut\ off\ position = round(ratio \times length(list)) \qquad (4)$$

ratio is predefined covered area proportion of the station, length (list) is equal N where N represents the total number of locations in area.

In the sorted list, we access the index of the list at the cut off position and get the value of that ED_i value. The value of ED_i is multiplied by two and defined as a diameter length of the ED + ASS method. Once, we have obtained the diameter length, and then we can continue the procedure of establishing the station network and evaluating the global maximum coverage of the station network in the same way as we did with ED + FSS.

5. Evaluation Criteria

In ED + FSS and ED + ASS, there is a multi-list of station networks with different first locations that must be processed to evaluate the maximum coverage percentage. Consequently, in order to find the best station network, the evaluation criteria are designed and described next.

5.1. Coverage Percentage (COV)

The covered area of stations (A_S) can be determined using Equation (3). Given the list of stations S located in a study area, we can assess the k-coverage when a location is covered by at least k different stations. The parameter k is called Coverage Degree. It means at least k stations cover each location in the study area. There are previous studies of wireless network coverage that have discussed the required k value of a network. Such studies explain a proper value of k that depends on the application. For example, an application requires $k = 1$ in a monitoring environment in which fault tolerance is not important. Meanwhile, $k > 1$ should be used when stronger monitoring is required, such as in an industrial or dangerous chemical region. Furthermore, in cases requiring fault tolerance, $k \geq 3$ is required. Therefore, it is clear that the station networks with higher k-coverage are more reliable [32,33].

Suppose that there are four stations. The circle shapes represent covered areas of a station and × symbols represent L_i for coverage degree assessment. If × symbols are within the covered areas of one station, then we define such L_i as a C1. If × symbol lies within the covered areas of two, three, and four stations, then it is denoted by C2, C3, and C4 as shown in Figure 5a–d, respectively. The area outside the circle is defined as a C0, which means it is a non-covered area.

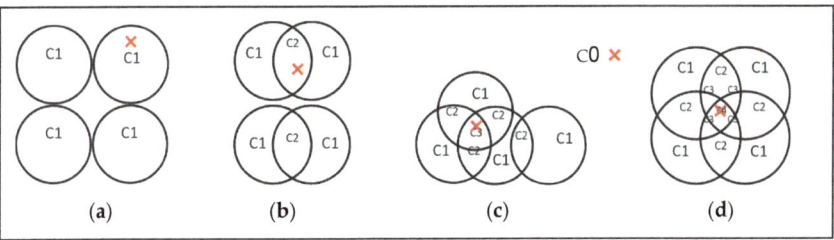

Figure 5. The example of coverage degree, (a–d) indicate the covered areas by one, two, three, and four stations.

We calculate the coverage percentage using the count L_i in each of the coverage degrees. The coverage percentage (COV) is used to evaluate results in our proposed methods and can be calculated with Equation (5).

$$COV_{percentage} = 100 \times \frac{\sum_{i=1}^{k} C_i}{N} \quad (5)$$

C_i denotes the summation of L_i in coverage degree, i is k-coverage number, k indicates a number of stations in the network, and N is total number of locations in study area A.

5.2. Weighted Coverage Degree (WCD)

Weighted coverage degree (WCD) is an additional criterion. It is used whenever COV cannot give a unique answer to the best network. The WCD corresponds to number of stations and k-coverage. For example, if the study area has four stations, here the k-coverage $k = 4$ (coverage degree: C0, C1, C2, C3, C4) and the weight has five values of coverage degree. We calculate the weight value by employing the Divide and Conquer concept. The $\sum W_i$ is a value equal to one. The W_k is calculated from $\sum W_i$ divided by two. The next weight value at W_{k-i} will decrease from the previous by half, which means $W_{k-i} = W_k$ divided by two. The weight calculation is done continuously until the last weight at W_0 is set equal to W_1. Table 1 is an example of weight value generation when the coverage degree is equal to 4. The WCD can be calculated by coverage degree weighting with Equation (6).

$$WCD = 100 \times \frac{\sum_{i=0}^{k} W_i C_i}{N} \tag{6}$$

Table 1. Example of weight generation.

Coverage Degree	C4	C3	C2	C1	C0	C4
Weight, $\sum W_i = 1$	W4 = 0.5	W3 = 0.25	W2 = 0.125	W1 = 0.0625	W0 = 0.0625	W4 = 0.5

W_i is a weight value and $\sum W_i = 1$, $i \in \{0, 1, 2, \ldots, k\}$, k indicates a number of stations in the network, and the weights are associated to Coverage Degree (C_i).

6. Results and Discussion

This section explains the parameters used in the experiment and compares each method's pros and cons. In scenario 1, setting up a network by considering four cases as following, (i) spatial coverage, (ii) performance of coverage percentage versus the number of the stations added incrementally, (iii) coverage percentage versus a specified number of stations, and (iv) flexibility to apply our methods to different cities. In scenario 2, finding an additional station to improve the current network is evaluated.

6.1. Experimental Parameter Settings

The parameter settings are shown in Table 2. The area size (A) is the number of locations in study areas. The centers of the maps are located at indices 1056, 3516, and 218 for Sejong, Bangkok, and Bonn. The specified number of stations is equal to the existing stations in the cities. The diameter length of the M3: ED + FSS is a fixed value of ten kilometers and the M4: ED + ASS is predetermined. We consider the proportion of covered area from reasonable based on the number of existing stations, as shown in Figure 6. Finally, we determined that 30% is a proper value for our experiment.

Table 2. Experiment Parameter Setting.

Parameters	Sejong	Bangkok	Bonn
Area size (A)	2024	7010	653
Index of the center of the map	1056	3516	218
Number of stations	4	12	1
Diameter length of FSS (dm_s)	10 km.	10 km.	10 km.
Diameter [1] length of ASS (dm_s)	14.2 km.	25.4 km.	8.6 km.

[1] Diameter calculation is based on proportion with a covered area of 30% for the cities.

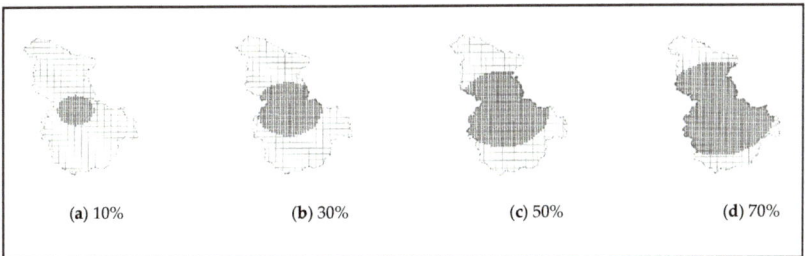

(a) 10% (b) 30% (c) 50% (d) 70%

Figure 6. Depicts the proportion of covered area in 10%, 30%, 50%, and 70% (**a**–**d**), respectively.

6.2. Assessment of Scenario 1: Setting Up a Network

6.2.1. Spatial Coverage

Table 3 shows M1: ED, M2: FSS and compares them with the existing stations in Sejong city. In Table 3, triangle symbols depict existing stations, and circles represent the spatial coverage. Let us consider a result in M1: ED, where we defined the first station at the center and used four stations as input parameters. The output from M1: ED is shown in Table 3 (b). Three stations are located near each other, but one station is located far away from its neighbors. We found significant inefficient spatial coverage in that case because almost all covered areas overlap. The COV of M1: ED is 57%, which is a 16% decrease in the current value, and WCD shows a value of 12.58, which increased 48% when compared with the current value. Next, consider the result in M2: FSS; we used the input parameters as same as M1: ED except adding a fixed diameter of ten kilometers to be used in the calculation. The result shows that most stations are located apart from the first station at the center, as shown in Table 3 (c). As a result, M2: FSS achieves the best coverage percentage up to 91% and an increase of 34% compared to the current value. On the other hand, the WCD value shows 7.44, which decreased by 13%.

Table 3. The existing stations and the results of stations, coverage percentage, and weighted coverage degree of two methods: M1: ED and M2: FSS.

	Current		M1: ED		M2: FSS	
Map of Sejong city and Stations	(a)		(b)		(c)	
1. Coverage Degree & Weight	C0: 655 C1: 774 C2: 527 C3: 67 C4: 1	W0: 0.0625 W1: 0.0625 W2: 0.125 W3: 0.25 W4: 0.5	C0: 877 C1: 526 C2: 44 C3: 508 C4: 69	W0: 0.0625 W1: 0.0625 W2: 0.125 W3: 0.25 W4: 0.5	C0: 183 C1: 1513 C2: 299 C3: 29 C4: 0	W0: 0.0625 W1: 0.0625 W2: 0.125 W3: 0.25 W4: 0.5
2. Coverage Percentage (COV)	68%		57% (−16%)		91% (+34%)	
3. Weighted Coverage Degree (WCD) value	8.52		12.58 (+48%)		7.44 (−13%)	

The M1: ED shows large overlapping areas which make a strengthened area with neighboring stations. The area of overlap can make data more reliable in case of data verification between neighboring stations and also produce a network which is fault tolerant. We denominate such overlap areas as confidence areas because they can enhance data reliability. However, this method shows spatial coverage weaknesses. In contrast, M2: FSS shows achieving good spatial coverage which can cover an area up to 91% in the case of four stations in Sejong city. For the spatial coverage assessment, it is possible to conclude that M2: FSS is better than M1: ED.

6.2.2. Performance of Coverage Percentage Versus the Number of Stations Added Incrementally

Our previous result when establishing four stations using M2: FSS shows a high coverage percentage of about 91%, which is better than M1: ED. In this case, we compared the performance of M2: FSS and M3: ED + FSS in terms of coverage percentage versus the number of stations added incrementally. The specified number of stations is seven. In Figure 7, the plot graph shows a comparison coverage percentage (COV) as the number of stations increases. The dashed line with square symbols indicates M2: FSS that shows coverage percentage increasing sharply and degrading from stations numbered 4 to 7. In contrast, the dotted line with triangle symbols indicates M3: ED + FSS and offers a relatively stable increasing trend from stations numbered 1 to 7. We conclude that M3: ED + FSS shows a coverage increasing trend better than that of M2: FSS.

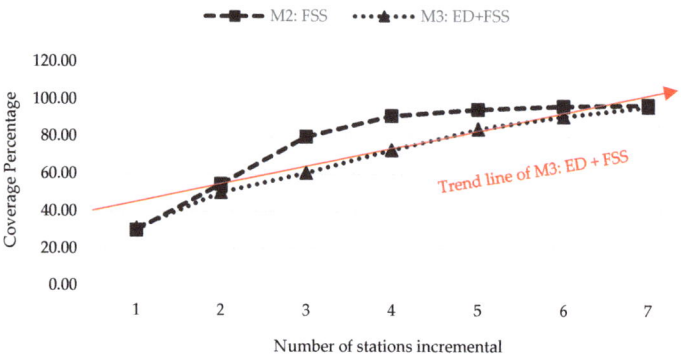

Figure 7. Plot of historical coverage percentage versus number of stations in Sejong.

Next, we investigated the additional stations of M2: FSS. As shown in Table 4 (a), light color of triangle symbols indicated the three additional stations which are located close to the border.

Although the M2: FSS achieves excellent spatial coverage (98%) nonetheless, the additional stations cause loss of area coverage, which shows the transparent area inside circles. On the other hand, all seven stations from the M3: ED + FSS method are located inside the city. These stations provide both sufficient spatial coverage (94%) and right overlapping area, as clearly shown in Table 4 (b). According to the result, we can assert a performance of M3: ED + FSS in achieving spatial coverage without loss of area coverage while still preserving the overlapped areas. The nearby stations can enhance strength to neighboring stations, which makes the network more robust and its data more reliable.

Table 4. The results of stations, coverage percentages, and weighted coverage degrees of two methods: M2: FSS and M3: ED + FSS.

		M2: FSS		M3: ED + FSS	
Map of Sejong city and Stations		(a)		(b)	
1. Coverage Degree & Weighted	C0: 49 C1: 767 C2: 856 C3: 290 C4: 62 C5: 0 C6: 0 C7: 0	W1: 0.0078125 W1: 0.0078125 W2: 0.015625 W3: 0.03125 W4: 0.0625 W5: 0.125 W6: 0.25 W7: 0.5		C0: 121 C1: 520 C2: 757 C3: 584 C4: 42 C5: 0 C6: 0 C7: 0	W1: 0.0078125 W1: 0.0078125 W2: 0.015625 W3: 0.03125 W4: 0.0625 W5: 0.125 W6: 0.25 W7: 0.5
2. Coverage Percentage (COV)		98%		94%	
3. Weighted Coverage Degree (WCD) value		1.61		1.86	

6.2.3. Coverage Percentage Versus a Specified Number of Stations

In this experimental case, we compared the performance of M3: ED + FSS and M4: ED + ASS concerning coverage percentage versus a specified number of stations. The location of station results from two methods is shown in Table 5. This part of our experiment aims to consider the flexibility of finding stations when we predefine the number of stations.

Table 5. Location of stations within the cities and coverage percentage.

	4 Stations	5 Stations	6 Stations	7 Stations
M3: ED + FSS	COV: 76%	COV: 87%	COV: 91%	COV: 94%
M4: ED + ASS	COV: 83%	COV: 93%	COV: 97%	COV: 99%

We will evaluate the coverage percentage and distribution of the stations in the study area. The results based on M3: ED + FSS show that the stations in similar aligned positions seem like a straight line in all four cases. On the other hand, results for M4: ED + ASS show a better balance of stations. The stations are readjusted whenever the number of stations changes. Furthermore, in four cases of a specified number of stations, the M4: ED + ASS has a higher coverage percentage and is better than M3: ED + FSS. Clearly, M4: ED + ASS is better than M3: ED + FSS to better cover the percentage and balance of stations.

6.2.4. Flexibility to Apply Our Methods for Different Cities

The M4: ED + ASS shows good spatial coverage and balance of stations based on a specified number of stations. For this section, we compared our method's capability to apply to different sizes of cities and evaluated balancing and distribution of the stations in cities. We used M3: ED + FSS and M4: ED + ASS in Bangkok, Sejong, and Bonn. The cities in that list are in descending order by size.

Table 6 (bkk-1), (sj-1), and (bo-1) shows the existing stations and current coverage percentages. There are twelve stations in Bangkok, four stations in Sejong, and one station in Bonn. The result of M3: ED + FSS in Table 6 (bkk-2) shows an imbalance of stations. On the other hand, the stations from M4: ED + ASS in Table 6 (bkk-3) are evenly distributed over the city. One benefit of balancing locations and overlapping areas is that the network can better support the future urban expansion and provide better air pollution monitoring. Although in Table 6, (sj-3) and (bo-3) cannot clearly show the balancing of stations when compared with Table 6 (sj-2) and (bo-2), the COV values of M4: ED + ASS are higher than M3: ED + FSS and show significantly increased rates of 56%, 22%, and 37%, in Bangkok, Sejong, and Bonn, respectively. These results allow us to conclude that our M4: ED + ASS is more flexible than M3: ED + FSS for designing new station networks in any city size.

Table 6. Locations of stations within the cities and coverage percentages: (bkk-1), (sj-1), (bo-1) depict current stations, (bkk-2), (sj-2), (bo-2) depict station networks using M3: ED + FSS, and (bkk-3), (sj-3), (bo-3) depict station networks improved using M4: ED + ASS, in Bangkok, Sejong, and Bonn, respectively.

Country/Methods	Bangkok, Thailand (1568.7 km^2)	Sejong, Korea (465.23 km^2)	Bonn, Germany (141.06 km^2)
Current	(bkk-1) COV: 64%	(sj-1) COV: 68%	(bo-1) COV: 27%
M3: ED + FSS	(bkk-2) COV: 99% (+55%)	(sj-2) COV: 76% (+12%)	(bo-2) COV: 33% (+22%)
M4: ED + ASS	(bkk-3) COV: 100% (+56%)	(sj-3) COV: 83% (+22%)	(bo-3) COV: 37% (+37%)

6.3. Assessment of Scenario 2: Finding Additional Stations to Improve the Current Network

For this section, we applied M4: ED + ASS to add one station into the network. Then, we evaluated the results with our proposed criteria: COV and WCD. In Table 7, triangle symbols indicate the existing stations, and star symbol indicates the additional station. The additional stations show the performance of our proposed method and evaluation criteria. They can improve coverage percentages by 39%, 34%, and 130% in Bangkok, Sejong, and Bonn.

Table 7. Location of an additional station and resulting coverage percentages in Bangkok, Sejong, and Bonn.

Bangkok, Thailand	Sejong, Korea	Bonn, Germany
COV: 89% (+39%)	COV: 91% (+34%)	COV: 62% (+130%)

One of the key limitations of our study is the appropriate ratio of spatial coverage satisfaction versus confidence area. As shown in Figure 8, the dashed line plot graph compares coverage percentage and the number of stations. The 6th stations show the most coverage percentage of 90%. Increasing the number of stations from 7 up to 12 does not significantly increase coverage. Thus in future work, a change of strategy is needed after achieving the spatial coverage to increase the overlapped regions.

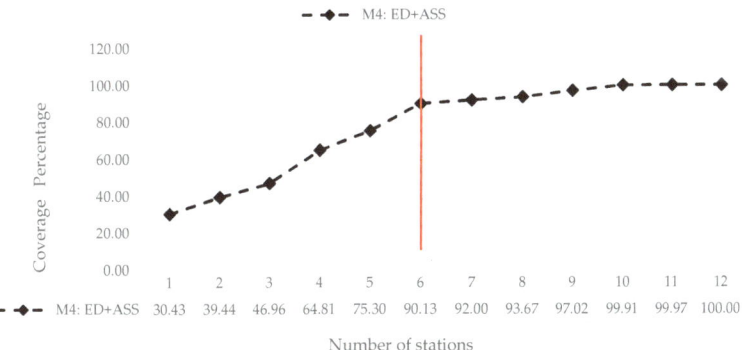

Figure 8. Plot of coverage percentage versus number of stations in Bangkok.

7. Conclusions

Increasing spatial coverage of AQMN gives effective environmental management. Especially in setting up a network in cities that lack historical pollution data, adding stations in cities with an existing network that face urbanization can make it challenging to design adequate spatial coverage and effective AQMN. Our experimental results proved the ability of the proposed method to provide maximum spatial coverage based on a given number of stations. The results showed that ED + ASS achieved effectiveness both in enhance spatial coverage and proper confidence area. In setting up a network in the cities, such a method suggested a balance of station locations and achieved maximum spatial coverage. The stations are neither located close to the border, nor are they close together. However, they are evenly distributed over the city. These results indicated that if the city planner considers economic benefits and the investment costs for constructing a network, the proposed method can give an excellent. In adding stations in cities, the results showed an optimal for the next station and enhanced the spatial coverage. Our proposed method showed effectiveness for expanding the existing networks as well as setting up AQMN.

In future work, we will consider the ratio of spatial coverage satisfaction versus confidence area as we mentioned in our experiment. We will modify our methods by changing strategies after achieving the spatial coverage to increase the overlapped regions of confidence areas. Moreover, we will integrate the historical pollution information, the city characteristics data, and land-use to find stations for better air pollution monitoring.

Author Contributions: Formal analysis, A.O.; methodology, A.O., H.J.; software, A.O.; validation, H.J., K.C.; investigation, A.O., H.J., writing–original draft preparation, A.O.; writing–review & editing, A.O., H.J., K.C.; supervision, H.J., K.C. All authors have read and agreed to the published version of the manuscript.

Funding: This research received no external funding.

Conflicts of Interest: The authors declare no conflict of interest.

References

1. Thailand State of Pollution Report 2019, Pollution Control Department Ministry of Natural Resources and Environment. Available online: https://www.pcd.go.th/publication/8013/ (accessed on 15 August 2020).
2. Seoul Solution, Air Pollution Monitoring Network. Available online: https://www.seoulsolution.kr/sites/default/files/policy/%ED%99%98%EA%B2%BD_2_p23_Air%20Pollution%20Monitoring%20Network.pdf (accessed on 3 August 2020).
3. EPA: United States Environmental Protection Agency. Ambient Monitoring Guidelines for Prevention of Significant Deterioration (PSD). USA; 1987. Available online: https://www.epa.gov/nsr/ambient-monitoring-guidelines-prevention-significant-deterioration (accessed on 1 August 2020).
4. National Environment Protection, Review of Air Quality Monitoring Network Design. Australian; 2019. Available online: https://www.environment.nsw.gov.au/reszearch-and-publications/publications-search/review-of-air-quality-monitoring-network-design (accessed on 1 August 2020).
5. Gulia, S.; Nagendra, S.S.; Khare, M.; Khanna, I. Urban air quality management—A review. *Atmos. Pollut. Res.* **2015**, *6*, 286–304. [CrossRef]
6. Wu, H.; Reis, S.; Lin, C.; Heal, M.R. Effect of monitoring network design on land use regression models for estimating residential NO_2 concentration. *Atmos. Environ.* **2017**, *149*, 24–33. [CrossRef]
7. Piersanti, A.; Ciancarella, L.; Cremona, G.; Righini, G.; Vitali, L. Application of a land cover indicator to characterize spatial representativeness of air quality monitoring stations over Italy. *Air Pollut. Model. Appl. XXIV* **2016**, 625–628. [CrossRef]
8. Min, K.D.; Kwon, H.J.; Kim, K.; Kim, S.Y. Air pollution monitoring design for epidemiological application in a densely populated city. *Int. J. Environ. Res. Public Health* **2017**, *14*, 686. [CrossRef] [PubMed]
9. Pigliautile, I.; Marseglia, G.; Pisello, A.L. Investigation of CO_2 variation and mapping through wearable sensing techniques for measuring pedestrians' exposure in urban areas. *Sustainability* **2020**, *12*, 3936. [CrossRef]
10. Marseglia, G.; Medaglia, C.M.; Ortega, F.A.; Mesa, J.A. Optimal alignments for designing urban transport systems: Application to Seville. *Sustainability* **2019**, *11*, 5058. [CrossRef]
11. Baldauf, R.W.; Wiener, R.W.; Heist, D.K. Methodology for siting ambient air monitors at the neighborhood scale. *J. Air Waste Manag. Assoc.* **2002**, *52*, 1433–1442. [CrossRef]
12. Mofarrah, A.; Husain, T. A holistic approach for optimal design of air quality monitoring network expansion in an urban area. *Atmos. Environ.* **2010**, *44*, 432–440. [CrossRef]
13. Mofarrah, A.; Husain, T.; Alharbi, B.H. Design of urban air quality monitoring network: Fuzzy based multi-criteria decision making approach. *Air Qual. Monit. Assess. Manag.* **2011**, *11*, 25–39.
14. Liu, M.K.; Avrin, J.; Pollack, R.I.; Behar, J.V.; McElroy, J.L. Methodology for designing air quality monitoring networks: I. Theoretical aspects. *Environ. Monit. Assess.* **1986**, *6*, 1–11. [CrossRef]
15. Lin, B.; Zhu, J. Changes in urban air quality during urbanization in China. *J. Clean. Prod.* **2018**, *188*, 312–321. [CrossRef]
16. Vadrevu, K.; Ohara, T.; Justice, C. Land cover, land use changes and air pollution in Asia: A synthesis. *Environ. Res. Lett.* **2017**, *12*, 120201. [CrossRef]
17. Choung, Y.J.; Kim, J.M. Study of the Relationship between Urban Expansion and PM10 Concentration Using Multi-Temporal Spatial Datasets and the Machine Learning Technique: Case Study for Daegu, South Korea. *Appl. Sci.* **2019**, *9*, 1098. [CrossRef]

18. Marseglia, G.; Vasquez-Pena, B.F.; Medaglia, C.M.; Chacartegui, R. Alternative fuels for combined cycle power plants: An analysis of options for a location in India. *Sustainability* **2020**, *12*, 3330. [CrossRef]
19. Health Effects Institute, Outdoor Air Pollution and Health in the Developing Countries of Asia: A Comprehensive Review. Available online: https://www.healtheffects.org/publication/outdoor-air-pollution-and-health-developing-countries-asia-comprehensive-review (accessed on 10 October 2020).
20. Zheng, J.; Feng, X.; Liu, P.; Zhong, L.; Lai, S. Site location optimization of regional air quality monitoring network in China: Methodology and case study. *J. Environ. Monit.* **2011**, *13*, 3185–3195. [CrossRef]
21. Henriquez, A.; Osses, A.; Gallardo, L.; Resquin, M.D. Analysis and optimal design of air quality monitoring networks using a variational approach. *Tellus B Chem. Phys. Meteorol.* **2015**, *67*, 25385. [CrossRef]
22. Benis, K.Z.; Fatehifar, E. Optimal design of air quality monitoring network around an oil refinery plant: A holistic approach. *Int. J. Environ. Sci. Technol.* **2015**, *12*, 1331–1342. [CrossRef]
23. Kazemi-Beydokhti, M.; Abbaspour, R.A.; Kheradmandi, M.; Bozorgi-Amiri, A. Determination of the physical domain for air quality monitoring stations using the ANP-OWA method in GIS. *Environ. Monit. Assess.* **2019**, *191*, 299. [CrossRef]
24. Li, T.; Zhou, X.C.; Ikhumhen, H.O.; Difei, A. Research on the optimization of air quality monitoring station layout based on spatial grid statistical analysis method. *Environ. Technol.* **2018**, *39*, 1271–1283. [CrossRef]
25. Alsahli, M.M.; Al-Harbi, M. Allocating optimum sites for air quality monitoring stations using GIS suitability analysis. *Urban Clim.* **2018**, *24*, 875–886. [CrossRef]
26. Yoo, E.C.; Park, O.H. Optimization of air quality monitoring networks in Busan using a GIS-based decision support system. *J. Korean Soc. Atmos. Environ.* **2007**, *23*, 526–538. [CrossRef]
27. Shareef, M.M.; Husain, T.; Alharbi, B. Optimization of air quality monitoring network using GIS based interpolation techniques. *J. Environ. Prot.* **2016**, *7*, 895–911. [CrossRef]
28. Liu, S.; Wei, Q.; Failler, P.; Lan, H. Fine particulate air pollution, public service, and under-five mortality: A cross-country empirical study. *Healthcare* **2020**, *8*, 271. [CrossRef] [PubMed]
29. IQAir, 2019 World Air Quality Report Region & City PM2.5 Ranking. Available online: https://www.iqair.com/world-most-polluted-cities/world-air-quality-report-2019-en.pdf (accessed on 30 October 2020).
30. Awe, Y.; Hagler, G.; Kleiman, G.; Klopp, J.; Pinder, R.; Terry, S. *Filling the Gaps: Improving Measurement of Ambient air Quality in Low and Middle Income Countries*; World Bank: Washington, DC, USA, 2017; Available online: http://pubdocs.worldbank.org/en/425951511369561703/Filling-the-Gaps-White-Paper-Discussion-Draft-November-2017.pdf (accessed on 30 October 2020).
31. Desa, U.N. Transforming Our World: The 2030 Agenda for Sustainable Development. Available online: https://sdgs.un.org/2030agenda (accessed on 12 January 2021).
32. Patra, R.R.; Patra, P.K. Analysis of k-coverage in wireless sensor networks. *Int. J. Adv. Comput. Sci. Appl.* **2011**, *2*, 91–96.
33. Huang, C.F.; Tseng, Y.C. The coverage problem in a wireless sensor network. *Mob. Netw. Appl.* **2005**, *10*, 519–528. [CrossRef]

Article

Pollution Characteristics of Particulate Matter ($PM_{2.5}$ and PM_{10}) and Constituent Carbonaceous Aerosols in a South Asian Future Megacity

Afifa Aslam [1], Muhammad Ibrahim [1], Imran Shahid [2], Abid Mahmood [1], Muhammad Kashif Irshad [1], Muhammad Yamin [3], Ghazala [1], Muhammad Tariq [1] and Redmond R. Shamshiri [4,*]

1. Department of Environmental Sciences & Engineering, Government College University Faisalabad, Faisalabad 38000, Pakistan; afifaaslam22@gcuf.edu.pk (A.A.); mibrahim@gcuf.edu.pk (M.I.); drabid@gcuf.edu.pk (A.M.); kashifirshad@gcuf.edu.pk (M.K.I.); ghazala@gcuf.edu.pk (G.); tariqnazir159@gmail.com (M.T.)
2. Environmental Science Centre, Qatar University, Doha P.O. Box 2713, Qatar; ishahid@qu.edu.qa
3. Department of Farm Machinery & Power, Faculty of Agricultural Engineering & Technology, University of Agriculture, Faisalabad 38040, Pakistan; yamin529@uaf.edu.pk
4. Leibniz Institute for Agricultural Engineering and Bioeconomy, Max-Eyth-Allee 100, 14469 Potsdam-Bornim, Germany
* Correspondence: rshamshiri@atb-potsdam.de; Tel.: +49-176-2290-3563

Received: 29 October 2020; Accepted: 7 December 2020; Published: 11 December 2020

Abstract: The future megacity of Faisalabad is of prime interest when considering environmental health because of its bulky population and abundant industrial and anthropogenic sources of coarse particles (PM_{10}) and fine airborne particulate matter ($PM_{2.5}$). The current study was aimed to investigate the concentration level of $PM_{2.5}$ and PM_{10}, also the characterization of carbonaceous aerosols including organic carbon (OC), elemental carbon (EC) and total carbon (TC) in $PM_{2.5}$ and PM_{10} samples collected from five different sectors (residential, health, commercial, industrial, and vehicular zone). The data presented here are the first of their kind in this sprawling city having industries and agricultural activities side by side. Results of the study revealed that the mass concentration of $PM_{2.5}$ and PM_{10} is at an elevated level throughout Faisalabad, with ambient $PM_{2.5}$ and PM_{10} points that constantly exceeded the 24-h standards of US-EPA, and National Environment Quality Standards (NEQS) which poses harmful effects on the quality of air and health. The total carbon concentration varied between 21.33 and 206.84 $\mu g/m^3$, and 26.08 and 211.15 $\mu g/m^3$ with an average of 119.16 ± 64.91 $\mu g/m^3$ and 124.71 ± 64.38 $\mu g/m^3$ for $PM_{2.5}$ in summer and winter seasons, respectively. For PM_{10}, the concentration of TC varied from 34.52 to 289.21 $\mu g/m^3$ with an average of 181.50 ± 87.38 $\mu g/m^3$ (for summer season) and it ranged between 44.04 and 300.02 $\mu g/m^3$ with an average of 191.04 ± 87.98 $\mu g/m^3$ (winter season), respectively. No significant difference between particulate concentration and weather parameters was observed. Similarly, results of air quality index (AQI) and pollution index (PI) stated that the air quality of Faisalabad ranges from poor to severely pollute. In terms of AQI, moderate pollution was recorded on sampling sites in the following order; Ittehad Welfare Dispensary > Saleemi Chowk > Kashmir Road > Pepsi Factory, while at Nazria Pakistan Square and Allied Hospital, higher AQI values were recorded. The analysis and results presented in this study can be used by policy-makers to apply rigorous strategies that decrease air pollution and the associated health effects in Faisalabad.

Keywords: particulate matter; aerosols; vehicular exhaust; industrial activity; WHO; NEQS; US-EPA

1. Introduction

In many developing countries, increasing industrialization and overpopulation becomes the reason for escalating air pollution [1]. According to various researches conducted in high-income countries situated in the Asian region, the level of many air pollutants is normally beyond the ambient air quality standards and WHO guidelines. In many developing countries, the use of non-renewable fuel like biomass and diesel is associated with the increasing level of air pollution at the regional level. Airborne particulate matter is abundant in the atmosphere and is the foremost indicator of the quality of air in a specified area. Chemical composition, concentration and size of particulate matter varied widely and are delimited universally under acceptable standards built on size elements ranging from $PM_{2.5}$ to PM_{10} to Total Suspended Particles (TSP), while PM_4 was also identified as the respirable size fraction [2]. Particulate matter instigates from a diversity of anthropogenic (e.g., rapid industrialization, agricultural activities, refineries, waste incineration, biomass burning, motor vehicles, utilities, brick kiln, industrial emissions power plants, factories, large population and heavy traffic) are responsible for bad air quality in the cities due to elevated levels of gaseous and particulate pollutants [3] and natural (e.g., dust storm and sea spray) sources, besides secondary formation processes. However, also mineral dust transport from deserted areas is considered a significant source for regional pollution in Asia [4]. Henceforth, for the air quality management and epidemiological studies, the assessment of the concentration of atmospheric particulate matter (PM) and its associated toxic constituents is a prerequisite [5]. It was consistently confirmed by epidemiological studies that there is a strong association between ambient particulate matter comprising toxic components and cardiovascular- and respiratory-related upsurges in mortality and morbidity, particularly in urban areas [6]. This connection has been revealed to be stronger for $PM_{2.5}$ rather than for PM_{10} or total suspended particles since $PM_{2.5}$ can infiltrate deep into the alveolar areas of the human lungs [6]. The transport and distribution of particulate matter in the atmosphere are distinctly allied with meteorological parameters such as air temperature, relative humidity, atmospheric pressure, wind direction, speed, and rainfall [7]. In various parts of the world, different monitoring programs on atmospheric PM have been directed which exposed varied instabilities and disproportions among the trace element constituents and particulate matter [8].

Components of carbonaceous aerosol, elemental carbon (EC) and organic carbon (OC), account for a large element of atmospheric particulate matter and, on average, subsidize 20–35% of coarse particulate and 20–45% of fine particulate [9]. Carbonaceous aerosols have a chief role in the interactions of light-particles within the atmosphere and are one of the significant components of fine and coarse particulate matter; they are therefore associated with the negative climatic and environmental impacts and the worsening in public health and air quality [10]. Elemental carbon is often used as a substitute for black carbon (BC) and is discharged into the atmosphere mostly through the processes of combustion [11]. Elemental carbon is primarily accountable for the absorption of light in the atmosphere, which sturdily influence the radiative balance of the earth [12]. The six main sources of elemental carbon have been recognized using organic tracers as coal combustion, biomass burning, vehicle exhaust, cigarette smoke, cooking and vegetative detritus [13]. Carbonaceous aerosols were found dominant in $PM_{2.5}$ (which is attained from agricultural waste and wood-fuel burning) and have a strong effect on the decline in visibility and air-quality and also stimulates radiative forcing on a regional scale [14].

In Pakistan, control of air pollution has not yet become a democratic issue because of a lack of suitable information for policy and decision-makers, though some infrequent reports that identify airborne particulate matter as a great health and environmental concern in urban regions of Pakistan are present [15]. Generally, the concentration of particulate matter is many folds higher than the acceptable limits documented by the World Health Organization (WHO), National Environmental Quality Standards (NEQS) and the United States Environmental Protection Agency (US-EPA). According to the World Bank [16], the annual burden of health because of particulate matter was 1% of the GDP and is accountable for 700 deaths among children and 22,000 premature deaths amongst adults in Pakistan. However, due to the absence of air quality management competencies, the country is suffering from the deterioration of air quality. Evidence from many international bodies and governmental organizations has indicated that air pollution is a momentous risk to the health of residents, environment and quality of life [17]. According to a study directed by the World Health Organization, Bombay, Calcutta and Tehran were found to be the most contaminated cities in Asia [18]. Similarly, Faisalabad (the textile city of Pakistan) is also highlighted to be an extremely polluted city in this study. Due to increased industrialization and construction of commercial zones and rapid urbanization, the atmosphere of the city is getting worst day by day [19]. This state of concern stimulated us to conduct a comprehensive study on the status of air pollution in Faisalabad. As a result of the burning issue of air pollution and associated health impacts, a study was planned to examine the quality of air in Faisalabad city for which 12 different sites were selected and categorized as residential, commercial, industrial, and health centers.

Keeping in view the facts discussed above, the present study was conducted with the following objectives: (a) to measure the quality of air with its allied consequences within varying activity zones of Faisalabad city; (b) to compare the ambient air quality of Faisalabad with air pollution indexes such as NEQS -Pakistan, National Ambient Air Quality Standards (NAAQS)-US-EPA and WHO; and (c) to provide an opportunity to conduct additional studies on source identification, impact assessment, and trend analysis for this zone. It is expected that the current study will be supportive for designing and establishing emission regulations and abatement strategies in the future.

2. Methodology

2.1. Study Area and Sampling Sites

Faisalabad is the third-largest city of Pakistan and a major industrial hub (dominated by textile and chemical industries); consequently, the air quality of the city is a major environmental problem. It covers an area of 1230 km^2 and is occupied by more than four million people. The summer season is very hot with a humid climate while a cold winter (falls to 0 °C some days) is experienced by the Faisalabad city. The climate of the city touches extreme hotness and humidity during summer and cold during winter. The sampling sites were banquets around Faisalabad and its vicinities. Twelve sampling sites were nominated based on current anthropogenic activities accountable for atmospheric pollution, and the dominant direction of wind for pollutant dispersion and distribution in the area. The locations were selected based on the zones in the city. The selected locations are comprised of medical units, residential areas, commercial areas, industrial areas, and automobiles rich areas. The average wind speed of 3–6 km in winter and 6–13 km in summer was observed. The map showing the locations on the Faisalabad (Figure 1) represents the coordinates of the location within Faisalabad geography.

2.2. Data Collection

Data of meteorological parameters were obtained from Agromet. The PM concentrations were determined by the first author herself. We took the samples from all the locations and then measurements were made in the Lab. We took sample readings sector-wise and readings for all the residential sites were taken at the same time. A similar trend was followed for commercial, industrial, health, and automobile sites. At Provincial and Federal EPAs, Data Logging systems retrieve the data about the quality of ambient air from air monitoring stations with the help of data processing software. The seasonal average was intended to find out the difference in the mass concentration of $PM_{2.5}$ in summer and winter seasons. For this study, 12 discrete sampling sites under five diverse sectors (residential, health, commercial, industrial, and automobile vehicles) were selected for the evaluation of $PM_{2.5}$ and PM_{10} with the help of high air volume sampler. The interpretations were taken at three diverse times (morning, noon, and evening) daily from November 1 to December 31 for winter and from May 1 to June 30 for summer. It should be noted that wind speed and direction influence the rate of diffusion of pollution. The temperature inversion is also directly linked to solar radiation making the air softer, hence the air converts into fog because pollutants and dust are no longer raised from the surface. This can become a problem in metropolises where numerous pollutants exist. The data of temperature and radiation used in this study was collected from the Agricultural Metrological Cell Agromet Bulletin.

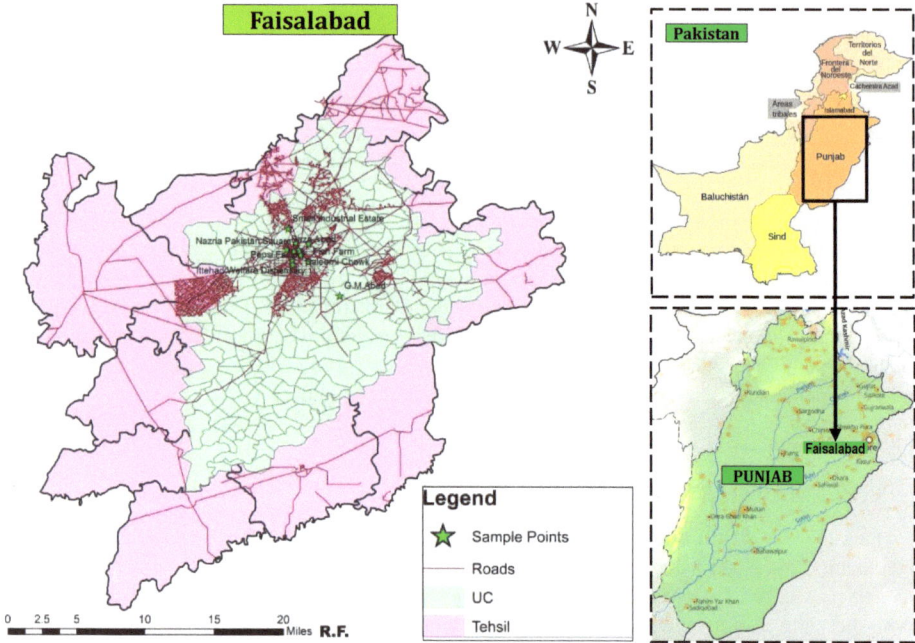

Figure 1. Study sites within the Faisalabad city and location of the city within the political map of Pakistan.

2.3. Sample Analysis (Chemical, Gravimetric, and Carbonaceous Aerosols)

After sample collection, the filter papers were kept in exact environmental conditions at a relative humidity of 30–40% and temperature of 20–23 °C for 24 h as per the US-EPA standard. Before mass analysis, the filter paper with fine and coarse particulates samples was equilibrated for 24 h in silica gel desiccators to abolish the effect of humidity and to attain accurate particulate matter measurements. The $PM_{2.5}$ and PM_{10} masses of each sample were determined gravimetrically by deducting the initial average mass of the blank filter from the final average mass of the sampled filter. Gravimetric analysis is the determination of particulate concentration based on weight difference. Individual filters (Teflon®, 46.2 mm) were weighed on an electronic micro-balance pre and post field sampling. Particulate matter <2.5 µm was collected from ambient air on the filters throughout the sample duration of 24 h. The net variances between pre- and post-sampling filter weights were used to estimate the mass concentration in the ambient air of the city. After post weighing, filters can be stored for a minimum of one year. Using the post-sample and pre-sample filter weights, the total filter mass gain ($PM_{2.5}$) and the concentration of PM_{10} were respectively calculated from Equation (1) and Equation (2):

$$PM_{2.5} = (M_{Post} - M_{Pre})(10^3)/M_{Pre} \qquad (1)$$

$$PM10 = (W_f - W_i)(106)/V \qquad (2)$$

Here $PM_{2.5}$ is the total mass gain in µg, M_{Post} is the post sample filter weight in mg, M_{Pre} is the pre-sample filter weight in µg, W_f is the filter paper weight, W_i is the initial mass of filter paper, and V is the total air sampled in m^3. In the current study, elemental carbon was determined by a two-step combustion method described by [20]. Filters were heated for 2 h at 340 °C in an oxygen atmosphere to remove organic carbon (OC). The calibration procedure was done using tartaric acid dyed in aluminum foil. While, total carbon was determined by a combustion method, where all material on the filter is combusted in pure oxygen at 1000 °C and the resulting CO_2 is measured by non-dispersive IR photometry (NDIR, Maihak) [21].

2.4. Air Quality and Pollution Index

An Air Quality Index is defined as a complete scheme that converts the weighed values of parameters related to individual air pollution (e.g., the concentration of pollutant) into a sole number or set of numbers [22]. Air Quality Index (AQI) is a tool to detect the present scenario of air quality. AQI was calculated based on the arithmetic mean of the ratio of the concentration of pollutants to the standard value of that pollutant such as PM_{10}, $PM_{2.5}$, NO_2, and SO_2. The average is then multiplied by 100 to arrive at the AQI index. The pollutant AQI and the pollution index (PI) of the potentially noxious element were respectively derived from Equations (3) and (4):

$$AQI = (W * C/Cs) \quad (3)$$

$$PI = Cn/Bn \quad (4)$$

where W is the pollutant weighted, C is the observed value ($PM_{2.5}$, PM_{10}, SO_2 and NO_2), Cs is the CPCB standard for the residential area [23], B_n is the background concentration, and C_n is the measured concentration of the element. It should be noted that in Pakistan, the National Air Quality Index is followed. Moreover, it should be highlighted that the pollution index of the potentially contaminated elements is the ratio between the concentration of toxic elements and the reference background concentration of the consistent elements obtained from a previous published study [24].

3. Results

3.1. Mass Concentration of $PM_{2.5}$ and PM_{10} in Winter Season

The results of the present investigations in Faisalabad city for which 12 different sites were selected and categorized as residential, commercial, industrial and health centers are presented (Figure 2). The concentration of $PM_{2.5}$ had the following decreasing order in the air samples collected near medical centers of Mian Trust Hospital (38.50 ± 0.30 µg/m^3), Ittehad Welfare Dispensary (37.35 + 0.45 µg/m^3), Allied Hospital (36.65 ± 0.27 µg/m^3). While, in residential areas, the highest concentration of fine particulate was found in Ghulam Muhammad Abad (39.1 ± 0.50 µg/m^3) followed by Raza Abad (35.2 ± 0.23 µg/m^3), Saleemi Chowk (33.83 ± 0.74 µg/m^3), Kashmir Road (32.25 ± 0.14 µg/m^3). In the selected commercial areas, Nazria Pakistan (43.63 ± 0.59 µg/m^3) was the most polluted site of the city. The average concentration of $PM_{2.5}$ in small industrial estate was recorded as (37.93 ± 0.19 µg/m^3) followed by Pepsi factory area (37.17 ± 0.62 µg/m^3); while in and near transport station (35.08 ± 0.61 µg/m^3) was analyzed. In contrast, the highest concentration level of PM_{10} was found in the samples collected from Nazria Pakistan (800.85 ± 0.93 µg/m^3) followed by Mian Trust Hospital (586.6 ± 3.88 µg/m^3), Allied Hospital (584.62 ± 3.41 µg/m^3), Small Industrial Estate (469.1 ± 0.57 µg/m^3) and Ghulam Muhammad Abad (440.2 ± 0.10 µg/m^3); while the lowest concentration was analyzed in the ambient air of Saleemi Chowk (280.18 ± 0.12 µg/m^3). The average particulate concentration can be compared with guideline values proposed by WHO, US-EPA and NEQS. During the present study, the concentration of $PM_{2.5}$ was within the guideline value of the US-EPA and NEQS (35 µg/m^3) but still higher than the WHO guideline value of 25µg/m^3 [25] for 24 h average, while coarse particles were drastically exceeding the safe limits of all the quality standards. The composition of particulate matter is strongly reliant with its sources, i.e., anthropogenic or natural (Figure 2).

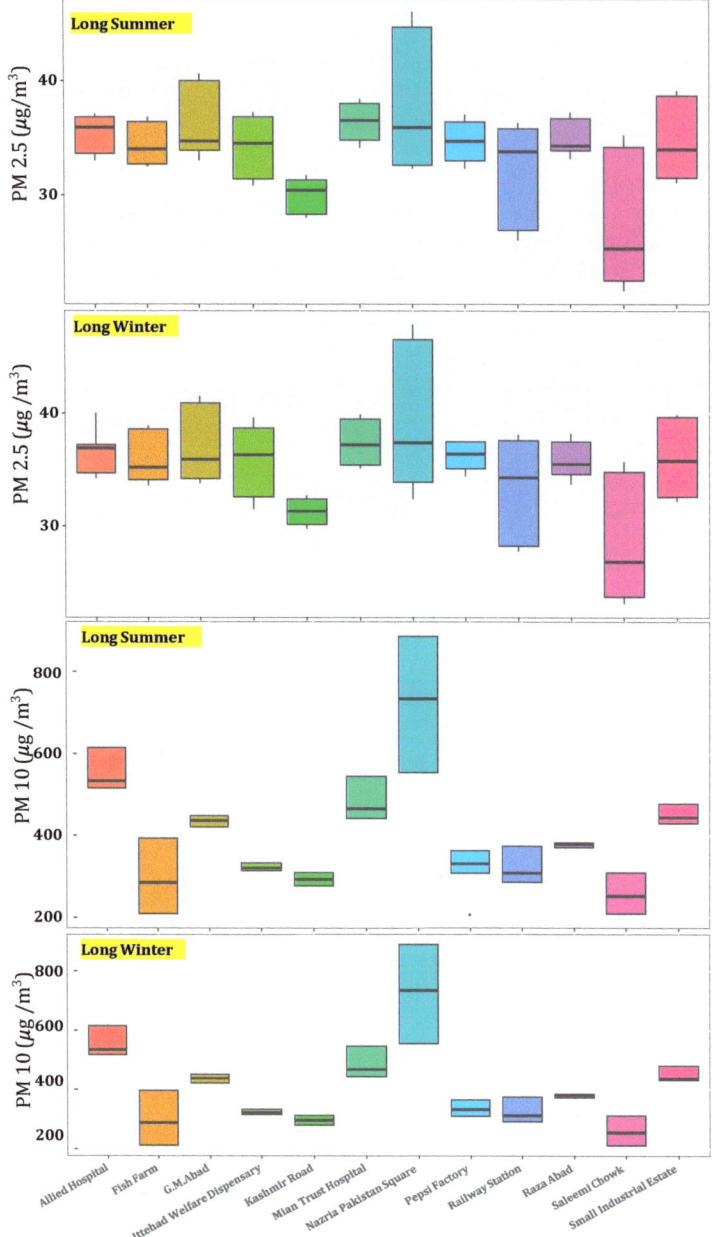

Figure 2. Average mass concentrations of $PM_{2.5}$ and PM_{10} in long summer and long winter in Faisalabad city.

3.2. Mass Concentration of $PM_{2.5}$ and PM_{10} in Summer Season

Figure 2 shows the mass concentration of fine ($PM_{2.5}$) and coarse (PM_{10}) particles calculated from the samples collected during the winter (Dec 2016 to Jan 2017) and summer (May to June 2017) seasons from different selected areas of Faisalabad. It was clear from the results that the concentration levels of both particulates were lower in the summer season as compared to the winter season. According to the results, it was cleared that the contamination level that the concentration level of both particulates was lower in the summer season as compared to the winter season. According to the results, the highest concentration of $PM_{2.5}$ was found in the air samples collected from Nazria Pakistan (42.5 ± 0.57 µg/m^3) followed by a Ghulam Muhammad Abad (38.4 ± 0.34 µg/m^3), Mian Trust Hospital and Small Industrial Estate air samples (36.9 ± 0.34 µg/m^3). While, the mass volume of $PM_{2.5}$ was lower in the air samples of Station Chowk (33.5 ± 0.35 µg/m^3), Raza Abad (34.2 ± 0.28 µg/m^3), Saleemi Chowk (31.8 ± 0.94 µg/m^3) and Kashmir Road (31.2 ± 0.21 µg/m^3) when compared with US-EPA and NEQS guidelines rather than WHO safe limits. The rest of the areas were slightly higher in $PM_{2.5}$ concentrations than US-EPA and NEQS safe limits but still highly polluted if compared with WHO guidelines. Table 1 represents the concentration values of coarse particles obtained after analysis. The decreasing order was followed as Nazria Pakistan (800.6 ± 2.16 µg/m^3), Allied Hospital (477.4 ± 3.7 µg/m^3), Mian Trust Hospital (477.3 ± 0.82 µg/m^3), Small Industrial Estate (465.4 ± 1.16 µg/m^3), Ghulam Muhammad Abad (440.1 ± 0.13 µg/m^3), Raza Abad (380.1 ± 0.08 µg/m^3) with the lowest value obtained at Kashmir Road (297.3 ± 0.50 µg/m^3). According to the results, the mass concentration of PM_{10} is exceeding the safe guidelines of all the selected air quality standards throughout the study area.

Table 1. (National Air Quality Index, CPCB, October 2014).

Category	Range
Good	0–50
Satisfactory	51–100
Moderately Polluted	101–200
Poor	201–300
Very Poor	301–400
Severe	401–500

3.3. Seasonal Impact on $PM_{2.5}$ and PM_{10} Concentration

Figure 2 represents the average concentration of $PM_{2.5}$ and PM_{10} in the long summer and winter seasons during the study period. Figure 2 shows the highest fine particulates concentration in a commercial area on the average 39.18 ± 4.70 µg/m^3 and 40.73 ± 2.9 µg/m^3 with the lowest obtained concentration in the residential sector ranging from 33.91 ± 3.27 µg/m^3 and 35.1 ± 2.93 µg/m^3 for summer and winter season, respectively. The WHO safe limit for $PM_{2.5}$ is 25 µg/m^3 and for PM_{10} is 50 µg/m^3 (WHO, 2005). Similarly, the US-EPA and NEQS safe limit for $PM_{2.5}$ is 35 µg/m^3 and PM_{10} is 150 µg/m^3 [26,27]. Coarse particles (PM_{10}) were similar in trend as shown by $PM_{2.5}$ with decreasing trend as commercial areas, hospital areas, industrial areas, automobile station and residential areas were in the range of 575.19 ± 66.26 µg/m^3, 499.81 ± 148.62 µg/m^3, 409.63 ± 59.44 µg/m^3, 379.63 ± 0.81 µg/m^3 and 350.98 ± 74.29 µg/m^3 in winter and 573.14 ± 321.64 µg/m^3, 427.52 ± 86.30 µg/m^3, 405.63 ± 84.49 µg/m^3, 349.41 ± 74.48 µg/m^3 and 349.46 ± 74.49 µg/m^3 in the summer season, respectively. Correspondingly, Figure 2 showed a strong positive correlation between $PM_{2.5}$ and PM_{10} in both winter and summer seasons on average.

3.4. Analysis of Carbonaceous Aerosols in fine ($PM_{2.5}$) and Coarse Particulate (PM_{10}) Samples

The concentration level of carbonaceous aerosols is presented in Figures 3–5 for summer and winter seasons. It was clear that concentration of EC and OC was higher in winter (Figures 3 and 4) which was quite similar with previous studies. According to the results, EC was found in low concentration on average in the samples of fine particles collected from Kashmir Road (8.56 ± 1.86 µg/m^3) in summer season while highest EC contamination was found in the ambient air of Nazria Pakistan (89.67 ± 1.52 µg/m^3). A similar trend was found for OC with the lowest concentration in the $PM_{2.5}$ samples collected from Kashmir Road (19.93 ± 0.42 µg/m^3) categorized as one of the residential sites, while the highest values were obtained in the ambient air of Nazria Pakistan (178.4 ± 3.51 µg/m^3) nominated as the busiest commercial zone of Faisalabad with a variety of businesses (Figure 4).

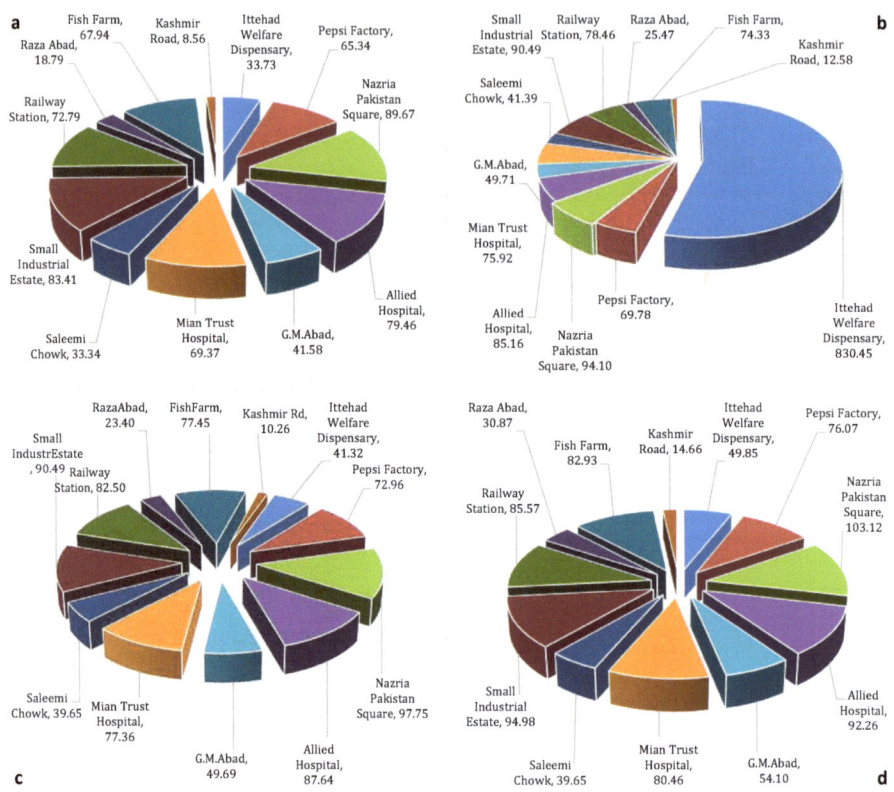

Figure 3. Concentration of carbonaceous species—elemental carbon (EC) in PM$_{2.5}$ in summer (**a**); PM$_{2.5}$ in winter (**b**); PM$_{10}$ in summer (**c**); PM$_{10}$ in winter (**d**). All units are expressed in µg/m^3.

When considering TC for the investigated sites (Figure 5), it was observed that Nazria Pakistan was heavily contaminated (268.08 ± 5.03 µg/m^3) followed by Small Industrial Estate (248.23 ± 5.79 µg/m^3), Fish Farm (235.68 ± 5.02 µg/m^3), Allied Hospital (224.34 ± 3.62 µg/m^3), Vehicular station (223.21 ± 3.82 µg/m^3), Mian Trust Hospital (202.77 ± 8.42 µg/m^3) and Pepsi Factory (184.84 ± 4.27 µg/m^3) with positive OC/EC correlation which indicates the common source of emission of TC in these zones. While, residential areas (G.M Abad, Saleemi Chowk, Raza Abad and Kashmir Road) were less contaminated (114.71 ± 2.76 µg/m^3; 64.97 ± 3.58 µg/m^3; 51.16 ± 3.14 µg/m^3; and 28.49 ± 2.27 µg/m^3, respectively), as compared to the other sites indicating negative OC/EC correlation. While samples of fine particulates collected in the winter season from the same investigating sites were analyzed for carbonaceous aerosol concentration. The levels of EC and OC were higher in winter as compared to the summer season. This may be due to the more wood and fossil fuel burning to warm up the surroundings as well as extra consumption of diesel and petrol by vehicles to warm up the engines in sizzling cold weather. The concentration of TC was much higher in the ambient air of Nazria Pakistan (277.5 ± 4.9 µg/m^3) followed by Small Industrial Estate (262.01 ± 3.68 µg/m^3), Fish Farm (248.07 ± 5.28 µg/m^3), Allied Hospital (237.13 ± 2.89 µg/m^3), Vehicular Station (234.4 ± 4.08 µg/m^3), Mian Trust Hospital (215.97 ± 8.82 µg/m^3), Pepsi Factory (194.27 ± 4.30 µg/m^3) and Ittehad Welfare Dispensary (142.58 ± 7.69 µg/m^3). On the other side, mix community of domestic zones showed less concentration of both EC and OC but still falls in contamination categories that are not safe for human health. Kashmir Road was detected with the least concentration of TC (38 ± 3.43 µg/m^3) while Ghulam Muhammad Abad was higher in TC (130.04 ± 2.6 µg/m^3) concentration in the residential zone. A positive OC/EC correlation was observed in the maximum of the investigating sites in the winter season. Samples of coarse particulate were also analyzed to evaluate the concentration level of carbonaceous aerosols collected from the investigating sites of Faisalabad city for summer and winter seasons, respectively.

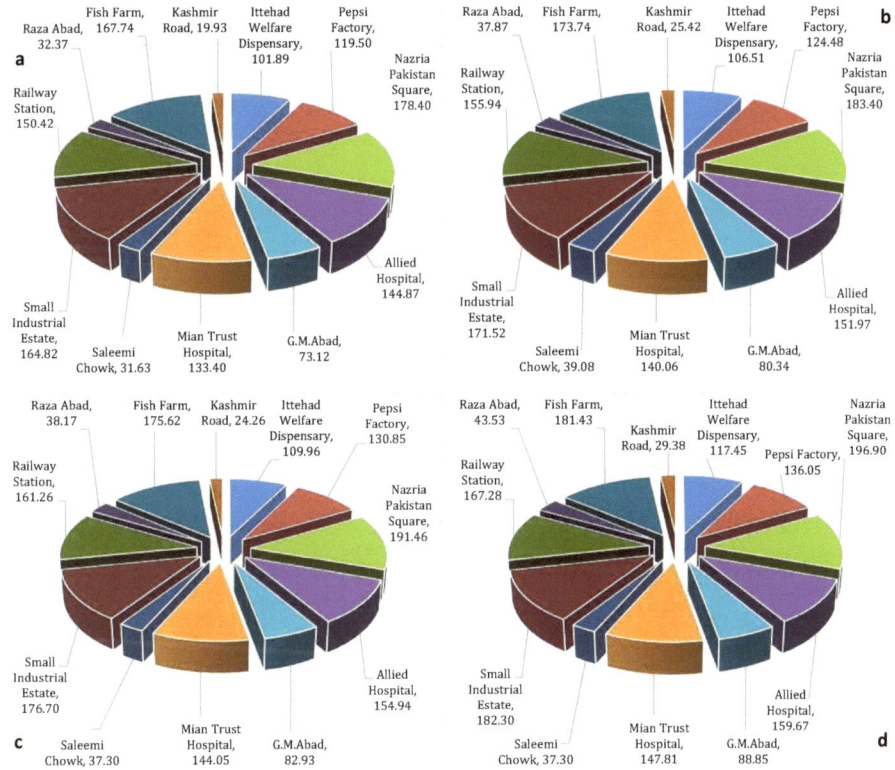

Figure 4. Concentration of carbonaceous species—organic carbon (OC) in PM$_{2.5}$ in summer (**a**); PM$_{2.5}$ in winter (**b**); PM$_{10}$ in summer (**c**); PM$_{10}$ in winter (**d**). All units are expressed in µg/m^3.

We found similar trends of concentration level as experienced with aerosols available in fine particulates but higher in concentration than observed in PM$_{2.5}$ samples. Commercial areas of Faisalabad were enriched with TC (289.21 ± 2.75 µg/m^3 and 300.02 ± 3.25 µg/m^3 for Nazria Pakistan and 253.06 ± 5.59 µg/m^3 and 264.36 ± 4.16 µg/m^3 for Fish Farm) at an elevated level among all the sites. Coarse particulate samples collected from the Small Industrial Estate were also found to be extremely high (267.19 ± 4.28 µg/m^3 and 277.28 ± 3.16 µg/m^3) after Nazria Pakistan followed by Vehicular Station (243.75 ± 3.66 µg/m^3 and 252.85 ± 2.9 µg/m^3), Allied Hospital (242.58 ± 5.24 µg/m^3 and 251.93 ± 4.44 µg/m^3), Mian Trust Hospital (221.4 ± 6.51 µg/m^3 and 228.27 ± 5.53 µg/m^3) and Ittehad Welfare Dispensary (151.28 ± 12.09 µg/m^3 and 167.31 ± 4.57 µg/m^3) for summer and winter seasons, respectively. Concentration level of carbonaceous aerosol was higher in Ghulam Muhammad Abad (132.62 ± 3.4 µg/m^3 and 142.95 ± 3.65 µg/m^3) while considering the residential zone of the city followed by Saleemi Chowk (76.95 ± 4.87 µg/m^3 and 76.95 ± 3.9 µg/m^3), Raza Abad (61.57 ± 3 µg/m^3

and 74.4 ± 2.32 µg/m³) and Kashmir Road (34.52 ± 2.02 µg/m³ and 44.04 ± 1.97 µg/m³) for both the seasons accordingly.

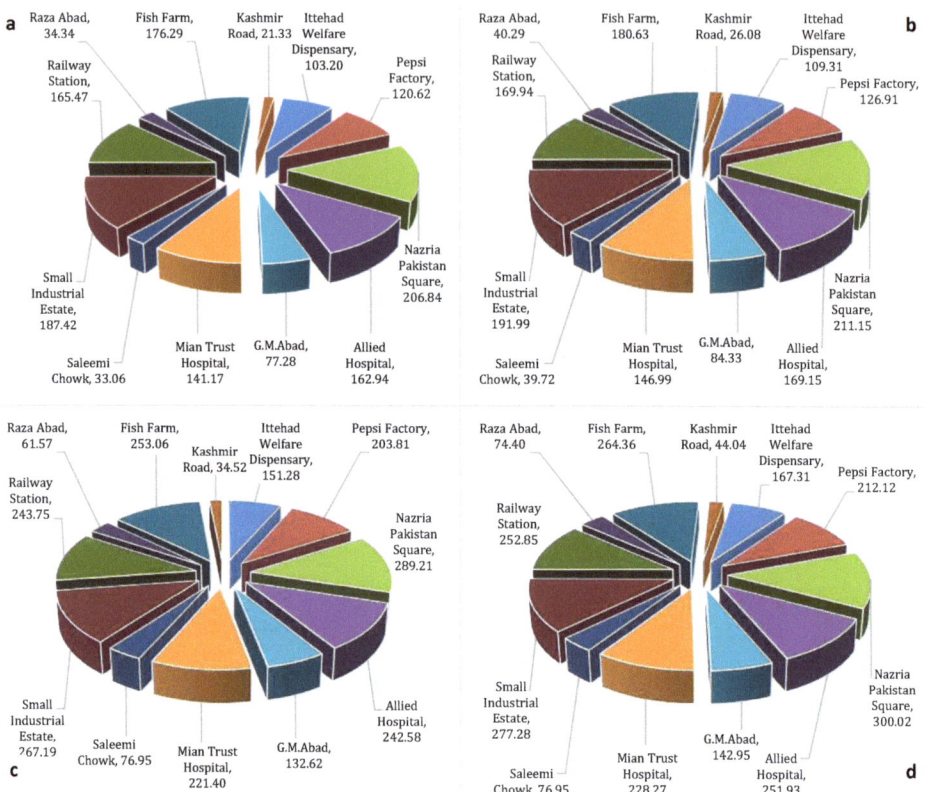

Figure 5. Concentration of carbonaceous species—total carbon (TC) in $PM_{2.5}$ in summer (**a**); $PM_{2.5}$ in winter (**b**); PM_{10} in summer (**c**); PM_{10} in winter (**d**). All units are expressed in µg/m³.

3.5. Air Quality and Pollution Index

The air temperature of the study area fluctuated between 18–25 °C in winter which is considered a typical range while the trend in May–June 2017, as shown in Figure 6, showed an increase to 37.2 °C on average. Figure 6 also demonstrates the trend of relative humidity with a mean value of that varies from 39.45% in summer to 60.4% in winter. Table 1 presented the categories of air quality according to the AQI while Figure 7 illustrates the AQI index of selected sites of Faisalabad city with detrimental outcomes. It was found that the ambient air of Faisalabad city ranges from moderately polluted with the sequence of Ittehad Welfare Dispensary > Saleemi Chowk > Kashmir Road > Pepsi Factory, while severely polluted air was found in the vicinity of Nazria Pakistan Square, followed by Allied Hospital. The air quality of Mian Trust Hospital and the Small Industrial Estate was categorized as very poor besides GM Abad, Raza Abad and Station Chowk where the air quality was poor. The overall air quality of Faisalabad city was not good for health and other activities that require urgent attention from Government institutes and ministries involved in making and implementing policies to safeguard the environment.

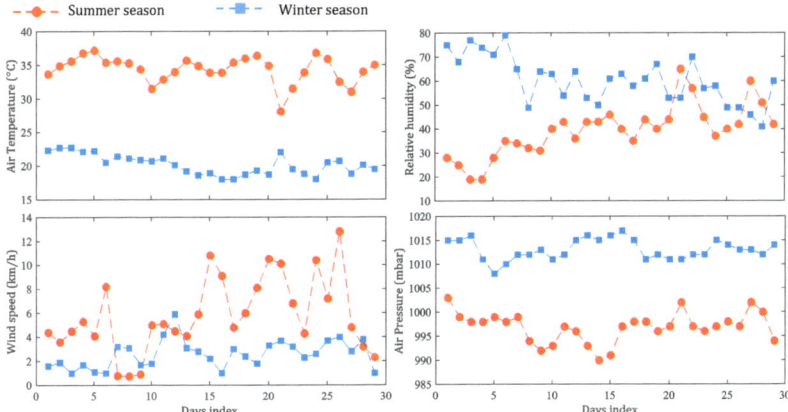

Figure 6. Relation of weather parameters in summer and winter seasons.

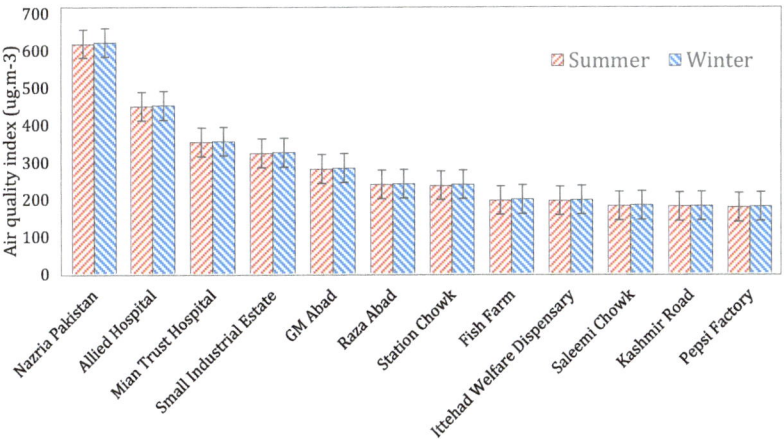

Figure 7. Air Quality Index (AQI) of Faisalabad city after analysis.

The average values of pollution index (PI) for each potential toxic element at selected sites of Faisalabad city for both seasons have been shown in Figure 8. In some residential areas, PI of $PM_{2.5}$ was found in the average level of pollution. While the PI of PM_{10} was estimated for the same areas showed a middle level of pollution $1 < PI \leq 2$, and sample site which is located near Nazria Pakistan Square suggested a high level of environmental pollution $PI > 4$. The $PM_{2.5}$ concentrations for almost all the sampling sites also showed a low level of environmental pollution of $PI \geq 1$. At Saleemi Chowk, Fish Farm, and Kashmir Road, the PI of PM_{10} showed a low level of environmental pollution $PI \leq 1$, while samples collected in the vicinities of Ittehad Welfare Dispensary, Pepsi Factory, Station Chowk, Raza Abad, GM Abad, and Mian Trust Hospital showed the middle level of pollution ($2 < PI \leq 3$) in the environment. It should be noted that the samples of Allied Hospital and Nazria Pakistan showed the highest environmental pollution level of $3 < PI < 4$. This can be due to the toxic elements in urban dust which accumulate and originate mainly from traffic, paint, and many other nonspecific urban sources in the megacity.

Appl. Sci. **2020**, *10*, 8864

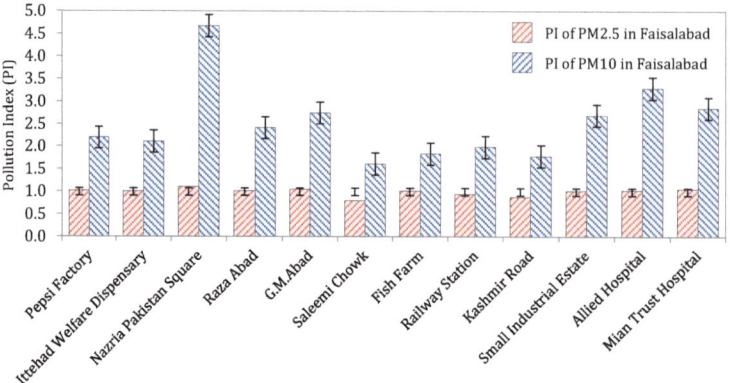

Figure 8. Pollution index flow diagram of $PM_{2.5}$ and PM_{10} in Faisalabad.

In order to show the difference between the mass concentrations of $PM_{2.5}$ and PM_{10} were statistically significant between the 12 locations, four sets of null hypotheses with H_0: $\mu(Location1) = \mu(Location2) = \ldots = \mu(Location12)$ were tested against the alternative hypotheses that the means of mass concentrations in the 12 locations were not equal. Based on the very small p-value that resulted from the one-way analysis of variance tests ($p < 0.0001$), all null hypotheses were rejected at any significant level and we concluded that the difference between locations was statistically significant. This result has been also shown in Figure 9 by means of the four scatter plots that demonstrate a visual comparison between the mean values of $PM_{2.5}$ and PM_{10} data in the 12 locations with respect to the daytime (morning, afternoon, evening) and season (summer or winter). Taking plot labeled (c) of Figure 9 as an example (mean of PM_{10} data in summer), the difference between morning concentrations of $PM_{2.5}$ in the 12 locations is clearly visible. It should be noted that for 61 days of winter the total number of measurements was 549. That is 61 days multiplied by 3 daytime (morning, afternoon, and evening) multiplied by 3 replications for each time.

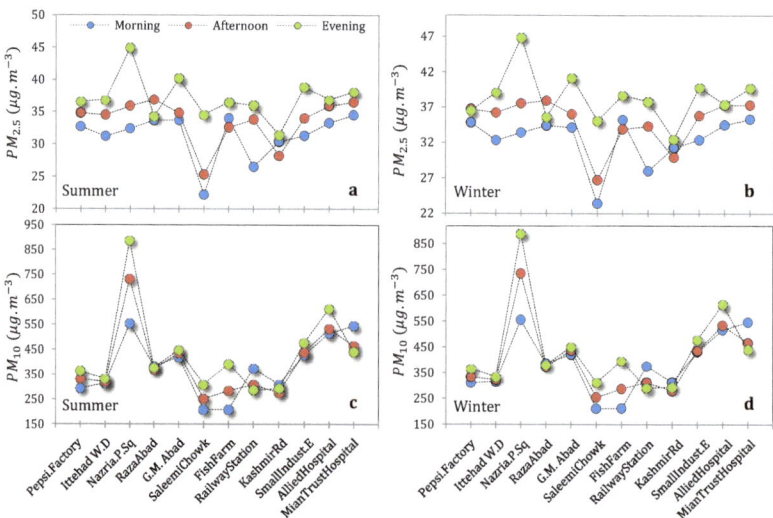

Figure 9. A comparison between mean values of $PM_{2.5}$ and PM_{10} data with respect to the daytime (morning, afternoon, evening) and season (summer or winter) in the 12 locations for (**a**): $PM_{2.5}$ in summer, (**b**): $PM_{2.5}$ in winter, (**c**) PM_{10} in summer, (**d**) PM_{10} in winter.

4. Discussion

Usually, the sources of gaseous pollutants are measured in three categories: natural emission mechanism, combustion sources and industrial manufacturing processes. Industrial sources of particulates, like steel, heavy traffic loads, indiscriminate burning of solid wastes and cement factories are the main sources of PM [2,28] besides gaseous and noise pollutants. The increasing air pollution is, after losses of properties, crops and increased health care costs. Airborne particulate matter is abundant in the atmosphere and varies extensively (temporally and spatially) in size, chemical composition and concentration. Emissions of particulate matter towards air are the focal environmental challenges for the transport and industrial sectors [4]. According to the WHO report, South Asia has developed as one of the most polluted zones in the globe because of its increasing population and rapid industrialization. Acquaintance to PM leads to more appointments to the emergency room or doctor. Health effects include premature deaths with existing lung and heart diseases, lung damage, coughing, aggravated asthma, wheezing and shortness of breath. Specifically, in the dry and cold season in major urban areas of Pakistan, people of almost all ages suffer from throat infections as reported [2,4,17]. In Pakistan, like the other developing countries, the emissions from vehicles have been conquered by emissions from poorly maintained and old vehicles that subsidize to heightened mass concentrations of carbon monoxide and fine particulates [17,28].

The current study focused on particulate volume has reported the highest concentration of $PM_{2.5}$ and PM_{10} in a commercial area on the average 39.18 ± 4.70 µg/m^3, 573.14 ± 321.64 µg/m^3 and 40.73 ± 2.9 µg/m^3, 575.19 ± 225.66 µg/m^3 for summer and winter, respectively (Figure 2). While, the lowest concentration of $PM_{2.5}$ and PM_{10} was obtained in the residential sector ranging $(33.91 \pm 3.27$ µg/m$^3)$, $(35.1 \pm 2.93$ µg/m$^3)$ and $(349.46.75 \pm 74.49$ µg/m$^3)$, $(350.98 \pm 74.29$ µg/m$^3)$ for summer and winter season correspondingly with a strong positive correlation between $PM_{2.5}$ and PM_{10} in both seasons on average. According to the guidelines, the reference value for $PM_{2.5}$ and PM_{10} are WHO (25 µg/m^3 and 50 µg/m^3) [17,25], NEQS and US-EPA (35 µg/m^3, 150 µg/m^3) [26,27] and most samples examined in the present study had values higher than the reference values. Elemental carbon is discharged from a variety of ignition procedures, categorized as a short-lived climate forcer that put up to atmospheric warming and also allied with human mortality and morbidity [28]. Common sources of atmospheric primary and secondary organic carbon antecedents are biomass burning, vehicular exhaust, biogenic emission and industrial emissions [29]. During the winter season, a higher level of pollutants especially the mass concentration of $PM_{2.5}$ persists in the ambient air of Faisalabad, owing to reduced atmospheric dispersion due to high relative humidity. Similarly, it was observed that $PM_{2.5}$ and PM_{10} sources were frequently localized as depicted by high concentrations at low wind speeds, mostly by the emissions from road vehicles [2,29]. This demonstrates the fact that $PM_{2.5}$ and PM_{10} concentrations were lower in summer than in winter (Figure 2) due to an increase in wind speed and temperature.

Prior studies conducted in the carbonaceous aerosols were assessed to account for about 50–60% of the total mass of $PM_{2.5}$ in metropolises in Jordan, Israel and Palestine [30]. Not unexpectedly, since production and processing of oil was prevalent transversely in the Middle East, substantial oil burning was valued to contribute 18% to total mass of PM_{10} and 69% to the total mass of $PM_{2.5}$ in Jeddah, Saudi Arabia [31]. Likewise, in Faisalabad, Pakistan, the quality of air not only reflects the impact of regional and local dust but also momentous local sources which include numerous industries and a heavy traffic weight. In municipal areas, the higher concentrations of $PM_{2.5}$ and PM_{10} are symbolic of the higher density of traffic as presented in the current study (Figures 2 and 4). Additionally, the burden of particulates is higher in the daytime than nighttime one, demonstrating more urban activities throughout day time. In Faisalabad, the textile industry, the topographical configuration and the geographical location make the problem of air pollution so perilous that it is very crucial to study it (Figure 5). The current study aimed at finding out whether or not the situation of air pollution in Faisalabad was previously seriously abundant to warrant the establishment of a regular air quality management system through which intercession measures can be premeditated and executed.

The analysis result of ambient air samples of selected sites of Faisalabad city displays that the level of particulate matter in most of the areas of the city is above the indorsed levels of the WHO, NEQS and US-EPA. Most of the city's commercial and residential areas are within the sensitive zone with the maximum concentrations of PM, which is constant with their proximity to the city's industrial areas.

When compared with the other studies conducted in other cities, it was found that the $PM_{2.5}$ level at Industrial Estate I-10 and IJP Road has reached the critical level (>35 µg/m^3) whereas at Industrial Estate I-9 it was moderate to the high level (31.9 µg/m^3 to 41.1 µg/m^3) [27]. While, the mean concentration of $PM_{2.5}$ and PM_{10} for Peshawar city during the study period has been calculated to be respectively 172 µg/m^3 and 480 µg/m^3 [32]. A similar high mass concentration of particulate matter was observed by [33] at Lahore, Pakistan and documented that the average $PM_{2.5}$ mass was 190 µg/m^3, and ranged from 89 µg/m^3 to 476 µg/m^3, far over US-EPA standards. Much higher PM_{10} mass concentration was experienced in Faisalabad when compared with other megacities [34], In addition, the PM_{10} concentrations were quite higher than the annual mean PM_{10} concentrations in Eastern Mediterranean and Africa [35,36] (WHO. Ambient (outdoor) 2014), Malaysia [34] and Bogota, Egypt, Los Angeles and Mexico [34]. It was also identified that PM_{10} is the dominant pollutant in the index value [37]. While, according to the results obtained after the analysis of particulate matters samples, the highest concentration of elemental carbon was 103.12 ± 1.46 µg/m^3 and the highest concentration of organic carbon was 196.9 ± 1.79 µg/m^3. While, 300.02 ± 3.25 µg/m^3 was the highest TC concentration found in the samples of coarse particulate matter collected in the vicinity of Nazria Pakistan (Figure 3). When compared, it was found that these concentrations are comparatively higher than in other metropolises in the areas like Punjab, India (116 µg/m^3), Hangzhou, China (119 µg/m^3), Kolkatta, India (197 µg/m^3), New Delhi, India (219 µg/m^3) and Lahore, Pakistan (233 µg/m^3) [38–40]. For elemental carbon, a large number of sources are identified, e.g., biomass and coal-fired power plant, two-stroke vehicles, fossil fuel burning, diesel engines and low burning efficiency. Elemental carbon is also utilized as a tracer for vehicular emission [17,40]. It was stated by [41] that diesel and gasoline motor vehicles and traffic exhaust are key sources of elemental carbon, followed by biomass burning. Organic carbon can be released straight from sources identified as primary carbon as a result of biomass and fossil combustion or can be formed as a result of a chemical reaction recognized as secondary organic carbon [42]. Temperature means are also under the normal limit but the increasing trend shows the alarming state of affairs and the same case is with radiations. Relative humidity has a value that is normal and considered healthy but a decreasing trend precedes the deterioration of ambient air quality. AQI and PI indicated that the ambient air quality of Faisalabad city falls from poor to severely polluted categories which are not safe to breathe and perform our daily activities.

5. Conclusions

Studying particle matters with aerodynamic diameters below 10 µm and 2.5 µm have received research attention for atmospheric pollution characteristics due to their severe effects on the human health issue. In this paper, we studied PM_{10} and $PM_{2.5}$ and highlighted that atmospheric pollution has become a significant issue as a result of growing industries in the megacity of Faisalabad, leading to the increased risk factors for chronic respiratory diseases in elderly and accelerated loss of lung function in newborns. To determine the pollution characteristics of particular matter, as well as the source and factors affecting them, we concentrated our study on 12 different sites that were selected and categorized as residential, commercial, industrial and health centers. Results of our study showed that the PM concentrations measured during current study periods (Dec 2016–Jan 2017) at various zones of Faisalabad were surprisingly higher than summer (May–June 2017). The enormous difference between fine ($PM_{2.5}$) and coarse (PM_{10}) particulate specifies that Faisalabad is inclined by a high loading of "coarse" particulate dust. Commercial areas are heavily polluted with fine and coarse particulate pollution. The average levels of pollution for fine and particulate matter were recorded as 39.18 ± 4.70, 573.14 ± 321.64 and 40.73 ± 2.9, 575.19 ± 225.66 during summer and winter, respectively (values in µg/m^3). The average $PM_{2.5}$ and PM_{10} concentrations were higher as compared to other major cities like

Islamabad, Lahore, and Peshawar. The quality of ambient air of Faisalabad has deteriorated beyond the safe limits set by WHO, US-EPA and NEQS. We also concluded that carbonaceous aerosols are in higher concentration in the air of the study sites. The air quality of Faisalabad city ranges from poor to severely polluted category which is highly unsafe for human health. These demands for an effort to introduce appropriate pollution control and management plans such as plantation and green belts for the betterment of civic life. A sustainable solution to improve air quality in Faisalabad would be to reduce emission by replacing high-energy consuming industries with renewable and clean energy sources, besides other strategies that reduce the use of fossil energy. Future studies may involve the use of wavelet analysis to explore the temporal characteristics of $PM_{2.5}$ and PM_{10}, or to investigate the relationship between meteorological factors and PM_{10}.

Author Contributions: A.A., M.I., I.S. and A.M. conceived and designed the experiments; A.A., M.T. and A.M. performed the experiments. M.K.I., G., M.Y., M.T. used software; R.R.S. worked on the final review and editing, figures and analysis; A.A., M.T., R.R.S. performed formal analysis and writing the draft; M.I., A.M. and G. did project funding and finding acquisition; M.I., I.S., A.M. and G. performed supervision of the experiments. All authors have read and agreed to the published version of the manuscript.

Funding: This research was funded through HEC-NRPU, grant number 5635 and the data presented here are a part of a Ph.D. thesis at GCUF. The authors further acknowledge the financial support provided by IERI-GIST project.

Acknowledgments: The authors acknowledge the financial support by the Open Access Publication Fund of the Leibniz Association, Germany, the partial research funding and editorial supports from Adaptive AgroTech Consultancy International and the administrative supports from Benjamin Mahns at the Leibniz Institute for Agricultural Engineering and Bioeconomy in Potsdam, Germany.

Conflicts of Interest: The authors declare that the research was conducted in the absence of any commercial or financial relationships that could be construed as a potential conflict of interest.

References

1. Hamid, A.; Akhtar, S.; Atique, S.A.; Huma, Z.; Mohay Uddin, S.G.; Asghar, S. Ambient air quality & noise level monitoring of different areas of Lahore (Pakistan) and its health impacts. *Pol. J. Environ. Stu.* **2019**, *28*, 623–629.
2. Javed, W.; Wexler, A.S.; Murtaza, G.; Iqbal, M.M.; Zhao, Y.; Naz, T. Chemical characterization and source apportionment of atmospheric particles across multiple sampling locations in Faisalabad, Pakistan. *Clean Soil Air Water* **2016**, *44*, 753–765. [CrossRef]
3. Gurjar, B.R.; Butler, T.M.; Lawrence, M.G.; Lelieveld, J. Evaluation of emissions and air quality in megacities. *Atmos. Environ.* **2008**, *42*, 1593–1606. [CrossRef]
4. Shahid, I.; Kistler, M.; Mukhtar, A.; Ghauri, B.M.; Cruz, C.R.S.; Bauer, H.; Puxbaum, H. Chemical characterization and mass closure of PM10 and PM2.5 at an urban site in Karachi, Pakistan. *Atmos. Environ.* **2016**, *128*, 114–123. [CrossRef]
5. Cheng, F.J.; Lee, K.H.; Lee, C.W.; Hsu, P.C. Association between particulate matter air pollution and hospital emergency room visits for pneumonia with septicemia: A retrospective analysis. *Aerosol Air Qual. Res.* **2019**, *19*, 345–354. [CrossRef]
6. Correia, A.W.; Pope III, C.A.; Dockery, D.W.; Wang, Y.; Ezzati, M.; Dominici, F. The effect of air pollution control on life expectancy in the United States: An analysis of 545 US counties for the period 2000 to 2007. *Epidemiology* **2013**, *24*, 23. [CrossRef]
7. Pakbin, P.; Hudda, N.; Cheung, K.L.; Moore, K.F.; Sioutas, C. Spatial and temporal variability of coarse (PM10–2.5) particulate matter concentrations in the Los Angeles area. *Aerosol. Sci. Technol.* **2010**, *44*, 514–525. [CrossRef]
8. Leghari, S.K.; Zaidi, M.A.; Ahmed, M.; Sarangzai, A.M. Assessment of suspended particulate matters level and role of vegetation in ambient air of North-East Balochistan, Pakistan. *Fuuast J. Biol.* **2013**, *3*, 37–43.
9. Dinoi, A.; Cesari, D.; Marinoni, A.; Bonasoni, P.; Riccio, A.; Chianese, E.; Tirimberio, G.; Naccarato, A.; Sprovieri, F.; Andreoli, V.; et al. Inter-comparison of carbon content in $PM_{2.5}$ and PM_{10} collected at five measurement sites in southern Italy. *Atmosphere* **2017**, *8*, 243. [CrossRef]

10. Choomanee, P.; Bualert, S.; Thongyen, T.; Salao, S.; Szymanski, W.W.; Rungratanaubon, T. Vertical variation of carbonaceous aerosols with in the $PM_{2.5}$ fraction in Bangkok, Thailand. *Aerosol. Air Qual. Res.* **2020**, *20*, 43–52. [CrossRef]
11. Ji, D.; Zhang, J.; He, J.; Wang, X.; Pang, B.; Liu, Z.; Wang, L.; Wang, Y. Characteristics of atmospheric organic and elemental carbon aerosols in urban Beijing, China. *Atmos. Environ.* **2016**, *125*, 293–306. [CrossRef]
12. Li, C.; Chen, P.; Kang, S.; Yan, F.; Hu, Z.; Qu, B.; Sillanpää, M. Concentrations and light absorption characteristics of carbonaceous aerosol in $PM_{2.5}$ and PM_{10} of Lhasa city, the Tibetan Plateau. *Atmos. Environ.* **2016**, *127*, 340–346. [CrossRef]
13. Li, Y.C.; Yu, J.Z.; Ho, S.S.; Schauer, J.J.; Yuan, Z.; Lau, A.K.; Louie, P.K. Chemical characteristics and source apportionment of fine particulate organic carbon in Hong Kong during high particulate matter episodes in winter 2003. *Atmos. Res.* **2013**, *120*, 88–98. [CrossRef]
14. Bisht, D.S.; Tiwari, S.; Dumka, U.C.; Srivastava, A.K.; Safai, P.D.; Ghude, S.D.; Chate, D.M.; Rao, P.S.; Ali, K.; Prabhakaran, T.; et al. Tethered balloon-born and ground-based measurements of black carbon and particulate profiles within the lower troposphere during the foggy period in Delhi, India. *Sci. Total Environ.* **2016**, *573*, 894–905. [CrossRef]
15. Colbeck, I.; Nasir, Z.A.; Ali, Z. The state of ambient air quality in Pakistan-a review. *Environ. Sci. Poll. Res.* **2010**, *17*, 49–63. [CrossRef]
16. World Bank and Institute for Health Metrics and Evaluation. *The Cost of Air Pollution: Strengthening the Economic Case for Action*; World Bank Group: Washington, DC, USA, 2016.
17. Niaz, Y.; Zhou, J.; Iqbal, M.; Nasir, A.; Dong, B. Ambient air quality evaluation: A comparative study in China and Pakistan. *Pol. J. Environ. Stu.* **2015**, *24*, 1723–1732. [CrossRef]
18. Gunawardana, C.; Goonetilleke, A.; Egodawatta, P.; Dawes, L.; Kokot, S. Source characterization of road dust based on chemical and mineralogical composition. *Chemosphere* **2012**, *87*, 163–170. [CrossRef]
19. Shahid, M.A.K.; Hussain, K.; Awan, M.S. Characterization of Solid Aerosols related to Faisalabad Environment and their probable sources. *Coden Jnsmac.* **2012**, *52*, 09–29.
20. Cachier, H.; Bremond, M.P.; Buat Ménard, P. Determination of atmospheric soot carbon with a simple thermal method. *Tellus B* **1989**, *41*, 379–390. [CrossRef]
21. Shahid, I.; Kistler, M.; Shahid, M.Z.; Puxbaum, H. Aerosol Chemical Characterization and Contribution of Biomass Burning to Particulate Matter at a Residential Site in Islamabad, Pakistan. *Aeros. Air Qual. Res.* **2019**, *19*, 148–162. [CrossRef]
22. Ott, W.R. *Environmental Indices: Theory and Practices*; Ann Arbor Science Publishers Inc.: Ann Arbor, MI, USA, 1978.
23. Central Pollution Control Board (CPCB). *Guidelines for National Ambient Air Quality Monitoring*; Series: NAAQM/25/2003- 04; Central Pollution Control Board: Delhi, India, 2009.
24. Chen, T.B.; Zheng, Y.M.; Lei, M.; Huang, Z.C.; Wu, H.T.; Chen, H.; Fan, K.K.; Yu, K.; Wu, X.; Tian, Q.Z. Assessment of heavy metal pollution in surface soils of urban parks in Beijing, China. *Chemosphere* **2005**, *60*, 542–551. [CrossRef]
25. World Health Organization. *Mental Health: Facing the Challenges, Building Solutions: Report from the WHO European Ministerial Conference*; WHO Regional Office Europe: Geneva, Switzerland, 2005.
26. Abbas, M.; Tahira, A.; Jamil, S. Air quality monitoring of particulate matter (PM2.5 & PM10) at Niazi and Daewoo bus station, Lahore. *FUUAST J. Biol.* **2017**, *7*, 13–18.
27. Hassan, M.; Malik, A.H.; Waseem, A.; Abbas, M. Air pollution monitoring in urban areas due to heavy transportation and industries: A case study of Rawalpindi and Islamabad. *J. Chem. Soc. Pak.* **2013**, *35*, 1623.
28. Huang, X.H.; Bian, Q.J.; Louie, P.K.K.; Yu, J.Z. Contributions of vehicular carbonaceous aerosols to $PM_{2.5}$ in a roadside environment in Hong Kong. *Atmos. Chem. Physics.* **2014**, *14*, 9279–9293. [CrossRef]
29. Weinhold, B. Global bang for the buck: Cutting black carbon and methane benefits both health and climate. *Environ. Health Perspect.* **2012**, *120*, A245. [CrossRef]
30. Abdeen, Z.; Qasrawi, R.; Heo, J.; Wu, B.; Shpund, J.; Vanger, A.; Sharf, G.; Moise, T.; Brenner, S.; Nassar, K.; et al. Spatial and temporal variation in fine particulate matter mass and chemical composition: The Middle East consortium for aerosol research study. *Sci. World J.* **2014**, *878704*. [CrossRef]
31. Khodeir, M.; Shamy, M.; Alghamdi, M.; Zhong, M.; Sun, H.; Costa, M.; Chen, L.-C.; Maciejczyk, P. Source apportionment and elemental composition of $PM_{2.5}$ and PM_{10} in Jeddah City, Saudi Arabia. *Atmos. Pollut. Res.* **2012**, *3*, 331–340. [CrossRef]

32. Alam, K.; Rahman, N.; Khan, H.U.; Haq, B.S.; Rahman, S. Particulate matter and its source apportionment in Peshawar, Northern Pakistan. *Aerosol Air Qual. Res.* **2015**, *15*, 634–647. [CrossRef]
33. Husain, L.; Dutkiewicz, V.A.; Khan, A.J.; Ghauri, B.M. Characterization of carbonaceous aerosols in urban air. *Atmos. Environ.* **2007**, *41*, 6872–6883. [CrossRef]
34. Safar, Z.S.; Labib, M.W. Assessment of particulate matter and lead levels in the Greater Cairo area for the period 1998–2007. *J. Advan. Res.* **2010**, *1*, 53–63. [CrossRef]
35. World Health Organization. *Ambient (Outdoor) Air Quality and Health*; Fact sheet No. 313; 2014; Available online: https://www.who.int/news-room/fact-sheets/detail/ambient-(outdoor)-air-quality-and-health (accessed on 25 April 2020).
36. Ul-Saufie, A.; Yahya, A.; Ramli, N.; Hamid, H. Future PM_{10} concentration prediction using quantile regression models. In *International Conference on Environmental and Agriculture Engineering*; IACSIT Press: Singapore, 2012; Volume 37.
37. Nigam, S.; Rao, B.P.S.; Kumar, N.; Mhaisalkar, V.A. Air quality index-A comparative study for assessing the status of air quality. *Res. J. Eng. Technol.* **2015**, *6*, 267–274. [CrossRef]
38. Cao, J.; Shen, Z.; Chow, J.C.; Qi, G.; Watson, J.G. Seasonal variations and sources of mass and chemical composition for PM10 aerosol in Hangzhou, China. *Particuology* **2009**, *7*, 161–168. [CrossRef]
39. Awasthi, A.; Agarwal, R.; Mittal, S.K.; Singh, N.; Singh, K.; Gupta, P.K. Study of size and mass distribution of particulate matter due to crop residue burning with seasonal variation in rural area of Punjab, India. *J. Environ. Monit.* **2011**, *13*, 1073–1081. [CrossRef] [PubMed]
40. Alam, K.; Mukhtar, A.; Shahid, I.; Blaschke, T.; Majid, H.; Rahman, S.; Khan, R.; Rahman, N. Source apportionment and characterization of particulate matter (PM10) in urban environment of Lahore. *Aerosol. Air Qual. Res.* **2014**, *14*, 1851–1861. [CrossRef]
41. Ghauri, B.; Lodhi, A.; Mansha, M. Development of baseline (air quality) data in Pakistan. *Environ. Monit. Assess.* **2007**, *127*, 237–252. [CrossRef]
42. Seinfeld, J.H.; Pandis, S.N. *Atmospheric Chemistry and Physics: From Air Pollution to Climate Change*; John Wiley & Sons: Hoboken, NJ, USA, 2016.

Publisher's Note: MDPI stays neutral with regard to jurisdictional claims in published maps and institutional affiliations.

© 2020 by the authors. Licensee MDPI, Basel, Switzerland. This article is an open access article distributed under the terms and conditions of the Creative Commons Attribution (CC BY) license (http://creativecommons.org/licenses/by/4.0/).

Article

Chemical and Biological Characterization of Particulate Matter (PM 2.5) and Volatile Organic Compounds Collected at Different Sites in the Los Angeles Basin

Arthur K. Cho [1,2,*], Yasuhiro Shinkai [3], Debra A. Schmitz [1], Emma Di Stefano [1], Arantza Eiguren-Fernandez [4], Aline Lefol Nani Guarieiro [5], Erika M. Salinas [6], John R. Froines [1] and William P. Melega [2]

1. Department of Environmental Health Sciences, School of Public Health, University of California Los Angeles, Los Angeles, CA 90095-1772, USA; dschmitz@ucla.edu (D.A.S.); emmadeste@gmail.com (E.D.S.); jfroines@ucla.edu (J.R.F.)
2. Department of Molecular and Medical Pharmacology, University of California Los Angeles, Los Angeles, CA 90095-1772, USA; WMelega@mednet.ucla.edu
3. Environmental Biology Laboratory, Faculty of Medicine, University of Tsukuba, 1-1-1 Tennodai, Tsukuba, Ibaraki 305-8575, Japan; ya_shinkai@md.tsukuba.ac.jp
4. Aerosol Dynamics, Inc., Berkeley, CA 94710, USA; arantza@aerosol.us
5. Universidade Federal da Bahia, Instituto de Química, 40170290 Salvador-BA, Brasil; alinelefol@gmail.com
6. División de Ciencias Básicas e Ingeniería, Universidad Autónoma Metropolitana, Azcapotzalco, Av San Pablo Xalpa 150, Mexico; eri_salinas@hotmail.com
* Correspondence: acho@mednet.ucla.edu

Received: 4 April 2020; Accepted: 3 May 2020; Published: 7 May 2020

Abstract: Background: Most studies on air pollution (AP) exposure have focused on adverse health effects of particulate matter (PM). Less well-studied are the actions of volatile organic compounds (VOCs) not retained in PM collections. These studies quantified chemical and biological properties of both PM2.5 and VOCs. Methods: Samples were collected near the Port of Los Angeles (Long Beach, LB), railroads (Commerce, CM), and a pollution-trapping topography-site (San Bernardino, SB). Quantitative assays were conducted: (1) chemical—prooxidant and electrophile content, (2) biological—tumor necrosis factor-α (TNF-α) and heme oxygenase-1 (HO-1) expression (3), VOC modulation of PM effects and (4), activation of the antioxidant response element (ARE) using murine RAW 264.7 macrophages. Results: SB site samples were the most potent in the chemical and biological assays, followed by a CM railroad site. Only PM2.5 exhibited significant proinflammatory responses. VOCs were more potent than PM2.5 in generating anti-inflammatory responses; further, VOC pretreatment reduced PM-associated TNF-α expression. VOCs significantly increased ARE activation compared to their corresponding PM2.5 which remained at background levels. Conclusion: Ambient VOCs are major contributors to adaptive responses that can modulate PM effects, in vitro, and, as such, need to be included in comprehensive assessments of AP.

Keywords: PM2.5; VOC; volatile organic compounds; prooxidants; electrophiles; tumor necrosis factor alpha; hemeoxygenase-1; ambient air; antioxidant response element; murine RAW 264.7 macrophages

1. Introduction

Epidemiological and clinical studies have firmly established associations between ambient air pollution (AP) levels and adverse health effects (see for examples, [1–4]). To advance our understanding of the relationships between AP exposure and disease processes, chemical and biological studies are needed to identify and characterize underlying molecular and cellular mechanisms. An essential element of those studies is the use of quantitative methodologies that allow for the comparison of results across other AP studies conducted in different local environments and geographical regions.

Our prior studies have been focused on characterizing reactive chemicals in AP that interact with cellular targets to elicit responses such as inflammation. In ambient air, these chemicals are differentially distributed between particle and vapor phases as obtained in AP sample collections. The particulate matter phase (typically PM2.5) contains mostly prooxidants such as redox active metals together with higher molecular weight quinones and humic-like substances (HULIS) [5,6] together with inorganic electrophiles such as arsenic and zinc. The vapor phase contains mostly volatile organic compounds (VOCs) that include quinones [7–9] and electrophilic carbonyls [10]. At the functional level, prooxidants catalyze the reduction of oxygen to reactive oxygen species and electrophiles react with nucleophilic functions, such as thiolate and amino groups, to form covalent bonds (see Graphical Abstract).

Surprisingly, in recent years, research on biological assessments of AP has largely focused only on the consequences of PM exposure without addressing the potential biological contributions of corresponding VOCs. To address this deficiency, the present studies were designed to generate quantitative data on both PM2.5 and VOCs of ambient AP samples for evaluating their distinct chemical reactivities and biological effect profiles.

In AP biological studies, the consequences of chemical exposure upon cellular components involves induction of oxidative stress and/or protein modifications that at low levels activate cytoprotective mechanisms but at higher levels initiate proinflammatory responses [11]. To understand the relationship between levels of reactive chemicals and this continuum of cellular responses, in this study we determined in PM2.5 and VOC fractions of ambient AP samples the prooxidant content based on their ability to transfer electrons from dithiothreitol to oxygen [12] and electrophile content, based on their ability to inactivate glyceraldehyde 3-phosphate dehydrogenase (GAPDH) [13]. The same fractions were then subjected to cellular assays designed to determine relative potencies of their biological responses.

The AP samples for this study were collected at three regions that neighbor railyards (Long Beach (LB), Commerce (CM), and San Bernardino (SB)) in the Los Angeles Basin with the objective of assessing regional differences in the chemical and biological actions of AP [14]. The chemical study results show distinct distributions of reactive chemicals: prooxidants in the PM2.5 and the electrophiles in the VOCs. The biological study results show that the samples with high reactive chemical content were the most potent in eliciting biological responses: PM2.5 promoting an inflammatory response and VOCs promoting an anti-inflammatory response.

Additionally, a second set of AP samples were collected at CM at sites 0.03–1 mile from its railyards to assess their chemical and biological actions as a function of proximity to the emission source. Those CM local samples exhibited profiles similar to the three regions' data, i.e., high chemical reactivity was associated with potent biological effects; furthermore, distance-dependent chemical and biological reactivities were also observed.

The pro- and anti-inflammatory biological responses to the PM2.5 and VOC samples raised the question of whether additive or antagonistic interactions occur following ambient AP exposures (i.e., PM2.5 + VOCs). We could not address this issue directly insofar as the concentrations of PM2.5 and VOC extracts precluded simultaneous exposure studies, however, we did conduct a two phase exposure in which RAW 264.7 macrophages were pre-exposed to VOCs, followed 24 h later by exposure to PM2.5. The results show that pre-exposure to VOCs reduces the magnitude of an inflammatory response to subsequent PM2.5. Based on prior studies showing that compounds contained in VOCs can activate the Nrf2-ARE pathway [15], the current samples were similarly tested. VOCs, but not PM2.5, activated this pathway, suggesting a cellular mechanism for their anti-inflammatory actions. Collectively, our in vitro studies show that VOCs contain reactive chemicals capable of inducing significant anti-inflammatory effects and, as such, need to be included with PM analyses to enable comprehensive studies of ambient AP.

2. Materials and Methods

2.1. Materials

Rabbit muscle glyceraldehyde-3-phosphate dehydrogenase (GAPDH), nicotinamide adenine dinucleotide (NAD$^+$), ethylenediaminetetraacetic acid (EDTA), glyceraldehyde-3-phosphate (GAP), dithiothreitol (DTT), 5,5'-dithiobisbis-(2-dinitrobenzoic acid (DTNB), and diethylenetriaminepentaacetic acid (DTPA) were purchased from Sigma-Aldrich (St. Louis, MO, USA). Other reagents were of the highest grade available and purchased from Fisher Scientific (Pittsburgh, PA, USA). Murine macrophage cell line RAW 264.7 cells were purchased from American Type Cell Culture (Manassas, VA, USA). Dulbecco's modified Eagle medium (DMEM) and penicillin/streptomycin were purchased from Life Technologies (Carlsbad, CA, USA). Fetal bovine serum was purchased from Gemini Bio-Products (Sacramento, CA, USA). The ELISA kits for TNF-α were purchased from BD Biosciences Pharmingen (San Diego, CA, USA) and the kits for HO-1 were purchased from Enzo Life Sciences (Farmington, NY, USA). Lipofectamine LTX transfection reagent was obtained from Life Technologies (Carlsbad, CA, USA).

2.2. Sample Collection and Extraction

2.2.1. PM2.5

Medium-volume samplers (Tish, Cleves, OH) were located in the selected locations; PM2.5 and vapor samples were collected continuously for 48 h as one time collections. The LB and SB samples were collected in early summer (June). The CM samples were collected continuously over a 5 d period in late spring (May). Teflon-coated glass fiber filters (Pall Corp, East Hills, NY) were used for PM2.5 collection and XAD-4 resin beds (Acros, Thermo Fisher Scientific) for the vapor-phase. Sampling details and matrix cleaning procedures have been previously published (16). Estimates of the volume equivalent for each vapor sample analyzed were based on the air volume collected divided by the fraction of the total extract used in each analysis. Aqueous suspensions of PM2.5 samples were prepared by sonicating filter punches in cell culture water for 20 min and the corresponding volume of air (m^3/cm^2) calculated to normalize the results. As the mass of the PM2.5 on the filter was not measured, the volume equivalent of air was used to describe the final concentration of the particles in the aqueous suspension, which was 2.2–6.0 m^3/mL.

2.2.2. VOCs

XAD-4 resin beds containing the trapped vapor phase organic components corresponding to each particle sample were extracted by sonication (30 min) with dichloromethane. The suspension was filtered through a 0.45 µm nylon filter (Millipore, Billerica, Massachusetts), the volume reduced, and solvent evaporated into a known volume of dimethyl sulfoxide (DMSO), so the concentration for analysis could be expressed as m^3 per mL of DMSO. The final concentration of the organic extract was approximately 300 m^3/mL of DMSO. Blank XAD-4 resin extracts were prepared as described previously and used as controls [16]. Highly polar compounds such as those with multiple hydroxyl moieties that are associated with the vapor phase would not be extracted with dichloromethane.

2.3. Chemical Assays

2.3.1. DTT Assay

This assay measured the prooxidant content of the sample from its ability to transfer electrons from DTT to oxygen [12–17]. In the procedure, aliquots of vapor-phase and PM2.5 water suspensions were incubated with DTT (Sigma Chemical Co., MO, USA) for times varying from 10 to 30 min. The reaction was quenched at specific times and after addition of DTNB to the complex with the remaining DTT, the absorbance at 412 nm was measured. Rates were calculated averaging duplicate runs, and were blank-corrected. The units were DTT nmol consumed per min per m^3, with 95% confidence intervals derived from regression analysis of rates.

2.3.2. GAPDH Assay

Electrophilic reactivity was measured from the sample's ability to inhibit or inactivate the thiolate enzyme GAPDH, through covalent bonding [13]. In brief, a mixture of 1 unit of rabbit GAPDH was incubated with aliquots of the organic extracts of vapors or water suspensions of particles under argon gas at 25 °C for 120 min. The reaction was then quenched by adding an equal volume of cold DTT solution; GAPDH activity, based on the rate of nicotinamide adenine dinucleotide (NADH) formation, was measured by its absorption at 340 nm. The ability to inactivate the enzyme was expressed in equivalents of N-ethylmaleimide (NEM), the standard electrophile, which was included in each assay as a control. Samples were run in triplicate and values reported as averages ± SEM. The assay provided a measure for electrophile content that was based on structures capable of interacting with the catalytic center of the enzyme and provided a quantitative estimate of the electrophilic content which could be used in comparison studies. The units used were the equivalents of NEM per m^3.

2.4. Cell Culture and Treatment

2.4.1. Cell Viability

Cell viability was determined by the 3-(4,5-dimethylthiazol-2yl)-2,5-triphenyl tetrazolium bromide (MTT) tetrazolium salt colorimetric assay [18]. A 5 mg/mL MTT solution was prepared in phosphate buffered saline and sterilized by filtration through a Steriflip. RAW264.7 cells were exposed to particle and vapor samples in 96-well plates for 16 h then treated with 10 µL of 5 mg/mL MTT for 2 h at 37 °C. The medium was removed and 100 µL DMSO was added to dissolve the formazan. Absorbance was measured at 540 nm on a Biotek Synergy 2 Multi-mode Microplate Reader. To determine suitable sample concentration ranges for these experiments, air sample toxicity was assessed by RAW 264.7 macrophages exposure to concentrations of 1 m^3/mL for 16 h. The loss of cells was normalized to that caused by a blank filter extract for the PM2.5 and a blank XAD resin extract for the vapors. The PM2.5 loss was 113.6% ± 3.9% (SEM for N = 8) and that for XAD was 94.7% ± 3.0% of their respective controls. Based on these results, all subsequent cell experiments used concentrations of 1 m^3/mL or less.

2.4.2. Cell Exposure

Murine RAW 264.7 cells were cultured in DMEM, supplemented by 1% penicillin-streptomycin and 10% FBS as described by Li et al. [11] with slight modifications. Cells were exposed to three concentrations (0.1 to 2.0 m^3 air equivalent/mL) of PM2.5 or VOC extracts in duplicate for 16 h, after which the medium and cells were separated and the cells subjected to lysis to obtain a cell extract. The cell extracts were used in the ELISA assays for HO-1 and the medium used to assay TNF-α. The results were analyzed by linear regression procedures (Graph Pad Prism, (San Diego CA, USA) to determine (a) concentration dependency of the response, and (b), if the slope was significant, its value, which is a measure of the potency of the sample.

2.5. Two Phase Study VOC Pretreatment/PM Exposure

In phase 1, cells were exposed to the VOC at a single concentration in triplicate (1 m^3/mL) together with the relevant controls for 24 h. In phase 2, the medium was removed and replaced with fresh medium containing the challenge agent, or PM2.5, also at 1 m^3/mL for 16 h. Cells and media were then processed for HO-1 and TNF-α analyses

2.5.1. ELISA Assays

The ELISA assays were performed following instructions provided by the manufacturers (HO-1; Enzo Life Sciences; TNF-α, BD Pharmingen). The results reported are the differences between the control and the experimental conditions. Values for HO-1 (ng/mg cell protein) and TNF-α (pg/mg cell protein) were normalized to cell protein.

2.5.2. ARE/EpRE Activation

DNA transfections were performed with Lipofectamine LTX transfection reagent (Life Technologies, Carlsbad, CA) following the manufacturer's instructions performing cell culture in 12-well plates. ARE-luciferase cDNA (1 µg/well) and pRL-TK cDNA (0.1 µg/well) or transfection reagent (2 µL/well) were mixed with serum-free media. Before addition to the cells, the DNA solution and transfection reagent solution were mixed together and incubated for 20 min at room temperature to allow the formation of complexes. The complexes were mixed with the culture media and incubated for 24 h to allow transfection. After transfection, the cells were exposed to the samples as described above and luciferase activity measured in cellular extracts according to the manufacturer's instructions (Dual-Luciferase reporter assay system; Promega, WI, USA) with a multi-mode microplate reader (Synergy 2, Bio-Tek, Winooski, VT, USA).

2.6. Collection Sites

The ambient air samples for this study were collected in three communities with railyards in the Los Angeles Basin: Commerce (CM), Long Beach (LB) and San Bernardino (SB) (Figure 1). CM, with two railyards, is located in the midtown area of Los Angeles. LB is southwest of CM and neighbors the Pacific Ocean and the Port of Los Angeles, which is the largest container port in the United States as measured by container volume and cargo value. Its emission sources would include both local railyards and the Port. The SB site is located in the eastern end of the Basin, approximately 80 miles from the coast and represents a receptor site that receives air parcels that have been subjected to modifications by photochemical and chemical reactions as they move from east to west across the Basin [19], together with emissions from a SB railyard. We previously observed that as the air mass moves east, a trend toward higher levels of 9,10-phenanthroquinone was found in the air samples, consistent with its formation from phenanthrene by atmospheric processes [7].

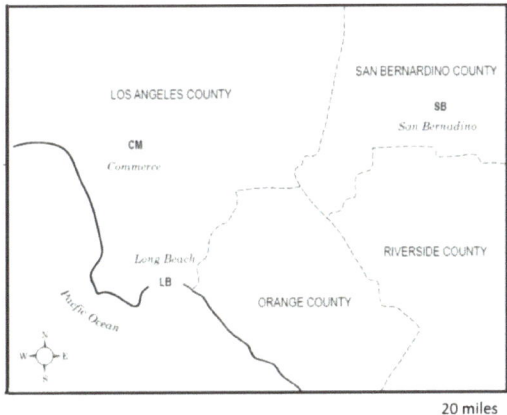

Figure 1. Map of the Los Angeles Basin with locations of the air collection sites: Long Beach (LB), Commerce (CM) and San Bernardino (SB). The Los Angeles Basin contains a coastal area bordered by the San Gabriel Mountains located in northern Los Angeles County and western San Bernardino County.

The second set of collections focused on CM at different distances nearby its railyards to assess ambient air properties as a function of distance from emission sources (Figure 2).

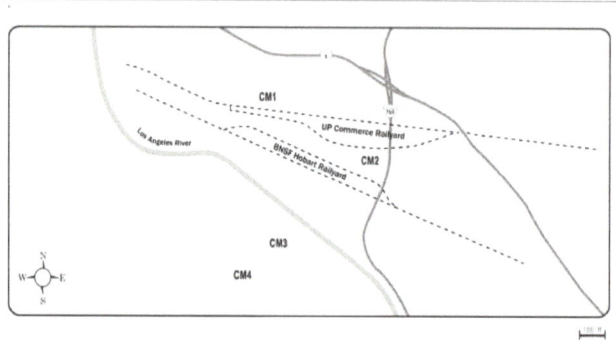

Figure 2. Map of the Commerce railyards with locations of air collection sites.

The sites, CM1 and CM2, were closest to the CM-Union Pacific railyard and downwind from the Burlington Northern and Santa Fe (BNSF) Hobart yard. Sites CM3 and CM4 were both upwind of the BNSF yard, but CM3 was closer to the yards and to a diesel truck processing center. CM4 was considered as a background site, minimally directly impacted by the railyard activities. It should be pointed out that as the samples were collected continuously 24-h/d for 5-d, changes in air movement during that period could have minimized differences due to changes from prevailing onshore to infrequent offshore winds. The distances from the railyards for the CM collections are summarized in Table 1.

Table 1. Location of collection sites at Commerce.

Site	Distance from Railyard (ft)	Emission Environment
CM1	155	Railyard
CM2	400	Railyard (between 2 tracks)
CM3	4440	Diesel truck loading site
CM4	7000	Background

2.7. Biological Reactivities

Two markers were used to assess cellular responses by the RAW 264.7 macrophage cell line: TNF-α as a proinflammatory response [20,21] and HO-1 as an anti-inflammatory response [22]. Cellular responses were determined at three concentrations (0.1, 0.5 and 1.0 m^3/mL) to ensure that the responses compared were linear over the concentrations used. The results were analyzed by linear regression to obtain slopes of the concentration vs. response curves. The results provide an assessment of the concentration-dependency of the response with the slope providing a quantitative assessment of potency in stimulating expression of the target markers, TNF-α and HO-1 (see Supplementary Materials).

2.8. Computational Procedures

All computations were performed with GraphPad Prism 8.12 (San Diego, CA, USA). Linear regression procedures were used to generate the best-fit values, the standard error of the slope and its 95% confidence intervals together with an assessment of the significance of the slope provided by its p value. These procedures generated Pearson correlation coefficients and their respective p values.

3. Results

3.1. Chemical and Biological Reactivities for Basin Sites

The PM and VOC samples from SB were the most reactive compared to those collected at CM and LB. For all three regions, the prooxidant content was mostly associated with the PM2.5 fraction (Figure 3).

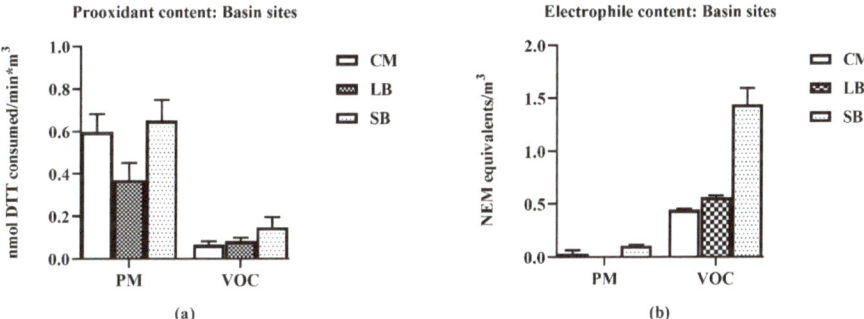

Figure 3. Chemical properties of particulate matter (PM) and volatile organic compound (VOC) air samples collected at CM, LB and SB. For each region's PM and VOC samples, nmol dithiothreitol (DTT) consumed (prooxidant content (**a**)) and N-ethylmaleimide (NEM) equivalents derived from a glyceraldehyde 3-phosphate dehydrogenase (GAPDH)-inhibition assay (electrophile content (**b**)) were measured. The VOCs represent the dichloromethane extracts of XAD resin traps of the vapor phases. Error bars indicate 95% CI (n = 3) for DTT results; SEM (n = 3) for the NEM results (values of the CIs are detailed in supplemental data—Table S1).

In contrast, electrophile content was mostly associated with the VOC fractions; electrophile content of the PM2.5 was close to background levels. SB VOC electrophile content was approximately three-fold higher than those for CM, LB. All PM2.5 samples increased TNF-α expression, with SB being 40-fold more potent than CM and LB (Figure 4). Increases in TNF-α expression were not detected in the VOC samples, i.e., they did not exhibit a concentration-dependent response (data not shown).

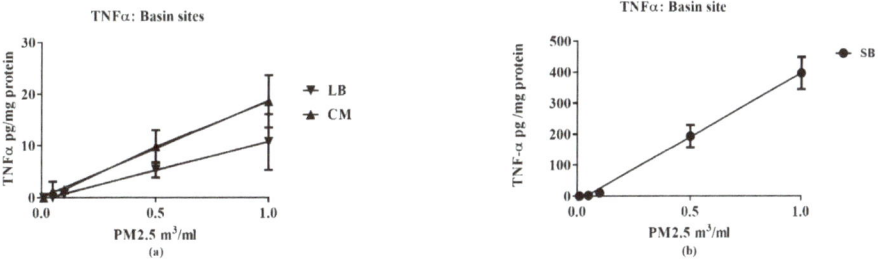

Figure 4. Relative potencies of air samples from CM, LB (**a**) and SB (**b**) to increase TNF-α in RAW 264.7 macrophages. TNF-α expression in the media was measured following exposure to the indicated PM 2.5 concentrations for 16 h. The solid lines represent the best line fit following linear regression analysis of the results (n = 3 for each data point) with potencies indicated by the slopes (values of the regression analysis are detailed in supplemental data—Table S4).

All VOC samples increased HO-1 expression, with SB being three-fold more potent than CM, LB, (Figure 5). The SB PM also increased HO-1 expression but the response was markedly lower than its VOC; concentration-dependent increases in HO-1 expression for CM PM2.5 and LB PM2.5 were not detected.

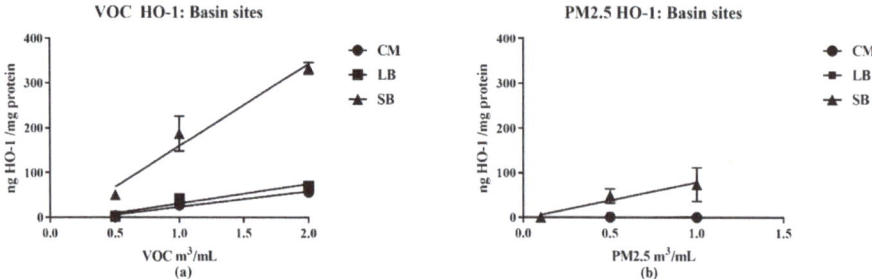

Figure 5. Relative potencies of VOC and PM 2.5 air samples from CM, LB (**a**) and SB (**b**) to increase HO-1 expression in RAW 264.7 macrophages. Cell extracts were measured following exposure to the indicated concentrations of either VOC or its corresponding PM for 16 h. The lines represent the best line fit following linear regression analysis of the results (n = 3 for each data point) with potencies indicated by the slopes (values of the regression analysis are detailed in supplemental data—Table S5).

3.2. Chemical and Biological Reactivities: CM Railroad Sites

The collection sites were located at different distances (0.03 to 1.3 mi from the railyards (Table 1) identified as CM1 to CM4 in increasing distance. The PM2.5 prooxidant and VOC electrophile content decreased as a function of distance from the railroad emission source (Figure 2) with the exception of CM2 which was adjacent to two railyards.

Prooxidant content was mostly associated with PM2.5 samples while electrophile content was mostly associated with the VOC samples. VOC prooxidant and PM2.5 electrophile contents remained close to background levels for all samples (Figure 6).

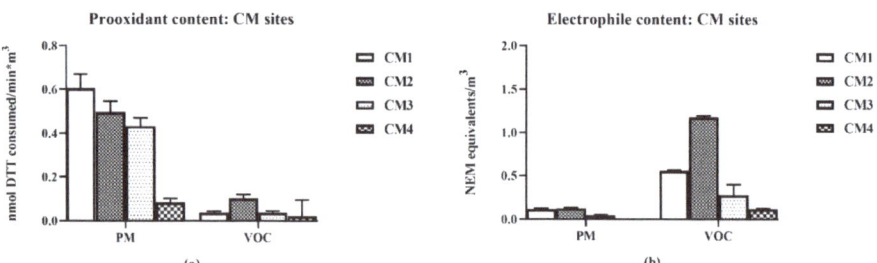

Figure 6. Chemical properties of PM and VOC air samples collected near to the CM railroad area (CM1—closest, CM4 most distant). For each PM and VOC local sample, the nmol DTT consumed (prooxidant content (**a**)) and NEM equivalents (electrophile content (**b**)) were measured. Error bars indicate 95% CI for DTT (n = 3) results and SEM (n = 3) for the GAPDH results; analysis methods as described in Figure 4, (values of the CIs are detailed in supplemental data—Tables S2 and S3).

The biological reactivities paralleled the profiles of the regional samples. Further, PM2.5 TNF-α and VOC HO-1 expression showed a 'distance' pattern, with CM2 being the most potent and CM4 (most distant) being least potent; PM2.5 HO-1 expression remained close to background levels for all samples (Figure 7).

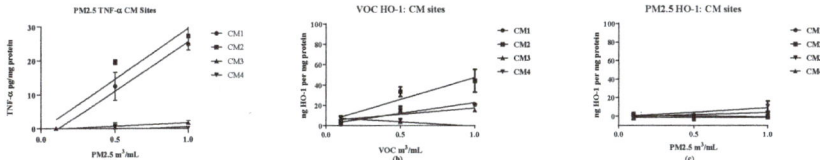

Figure 7. TNF-α expression (**a**) and HO-1 expression (**b**,**c**) in RAW 264.7 macrophages using air samples collected nearby CM railroads (CM1—closest, CM4 most distant). Macrophages were incubated with three concentrations of VOC or PM for 16 h. The lines represent the best line fit following linear regression analysis of the results (n = 3 for each data point) with potencies indicated by the slopes (values of the regression are detailed in supplemental data—Tables S4 and S5).

3.3. SB VOC Pretreatment Effects on SB PM2.5 2.5 Expression of TNF-α and HO-1 in RAW 264.7 Macrophages

The anti-inflammatory properties of the VOC suggested that VOC pretreatment may affect the attenuation of the pro-inflammatory effects of a subsequent PM2.5 challenge. We selected the most potent samples—SB VOC for the pretreatment and SB PM2.5 for the challenge. A sequential exposure experiment was performed in which cells were first exposed to the SB VOC for 24 h. The media was then replaced with fresh media containing control filter extracts or suspensions of the SB PM2.5 and the exposure continued for 16 h. Macrophages and media were analyzed for HO-1 and TNF-α content, respectively (Figure 8).

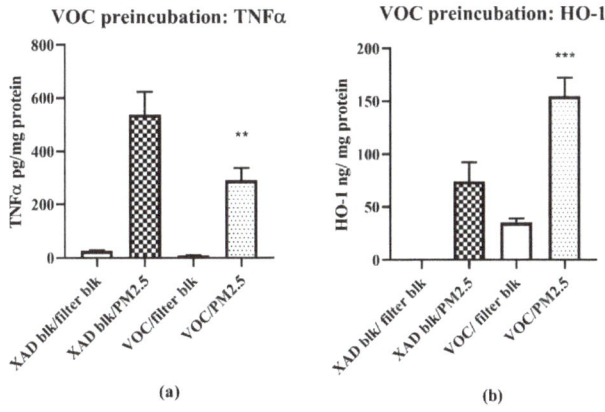

Figure 8. SB VOC pretreatment effects on SB PM 2.5 expression of TNF-α (**a**) and HO-1 (**b**) in RAW 264.7 macrophages. Cells were preincubated with VOC (1 m^3/mL) for 24 h followed by incubation with PM 2.5 (1 m^3/mL) for 16 h; TNF-α—XAD Blk/PM 2.5 vs. VOC/PM 2.5, n = 4; $p^{**} < 0.005$; HO-1—XAD Blk/PM 2.5 vs. VOC/PM 2.5, n = 4; *** $p < 0.001$.

The reason for the two-stage exposure was the limitation in the volume of sample solutions that could be added to the plate cells while maintaining the incubation conditions (see Methods for details). The results show a 54% decrease in the TNF-α response to the PM2.5 challenge following pretreatment to SB VOCs at 1 m^3/mL, n = 4; $p^{****} < 0.005$, together with a significant 200% increase in HO-1 expression (Blank XAD/SB vs. SB VOC/SB 2.5; (n = 4; p < 0.001).

3.4. Activation of the antioxidant Response Element (ARE) in RAW 264.7 Macrophages by Air Samples from CM, LB and SB

In prior studies, we measured an HO-1 response in VOC samples collected at Riverside CA (a sample site nearby SB). Those samples also showed an electrophilic effect on the ARE [23].

Accordingly, we evaluated in this study the three regional samples for their potential to activate the ARE. The results show that all VOC samples significantly increased ARE activation 1.5- to three-fold compared to their corresponding PM2.5 samples which remained at background levels. The SB VOC sample was approximately two-fold more potent than the LB and CM samples (Figure 9).

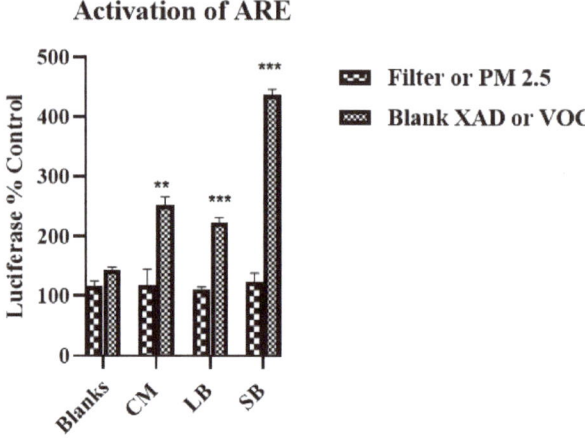

Figure 9. Activation of the antioxidant response element (ARE) in RAW 264.7 macrophages by air samples from CM, LB and SB. Cells transfected with ARE-luciferase cDNA were exposed to either PM2.5 and VOC from CM, LB, SB at 1 m^3/mL for 16 h and ARE activation assessed; (see Methods for ARE procedures). Each value is the mean ± SD, n = 4/sample site; site values normalized to corresponding blanks; CM VOC vs. CM PM2.5, ** p < 0.005; LB VOC vs. LB PM 2.5, SB VOC vs. SB PM2.5, *** p < 0.00001.

4. Discussion

In our prior AP studies on PM2.5 and VOCs, we characterized chemical species that can participate in two reactions associated with acute toxicity, (1) prooxidants that catalyze the formation of hydrogen peroxide and hydroxyl radical, and (2) electrophiles that form covalent bonds with nucleophilic groups such as amino and thiolate found in proteins. Furthermore, we showed that VOC electrophiles associated with AP can activate cellular signaling pathways relevant to adverse health effects [23]. However, VOC biological assessment has received minimal attention in conjunction with PM research despite its apparent contributions to AP toxicology, with the exception of adductomics studies by Rappaport and his associates (e.g., [24]).

For this study, analytical methods we previously developed and validated [14] were used to characterize chemical and biological properties of SB, CM and LB AP samples. The samples were obtained from (1) an aqueous suspension of the particulates, PM2.5, and (2) an organic extract of the VOCs trapped by polystyrene resins placed downstream of the PM2.5 filters which have been shown to contain the majority of the total electrophiles present in ambient air [14]. Based on our previous analyses of seasonal effects on AP properties [14,16], the samples for this study collected in late spring/early summer were likely have relatively higher prooxidant and VOC electrophile content compared to winter collections. Nonetheless, despite such quantitative differences, predominantly PM—TNF-α—and VOC electrophile responses were consistently observed for both seasonal collections. The quantitative nature of the regional data allowed for multiple comparisons between different parameters of AP chemical reactivities in terms of pro- and anti-inflammatory properties, and their distribution between PM 2.5 and VOC phases. Prooxidant reactivity and proinflammatory responses (TNF-α increases) were predominantly associated with PM samples while electrophile reactivity and anti-inflammatory responses (HO-1 increases and ARE activation) were predominantly associated with VOCs. Future studies, both in vitro and in vivo, are needed to determine whether the interactions of these responses are synergistic or antagonistic following concurrent exposure of PM and VOCs. Presently, evidence for a potential interaction is suggested from the results of the two-phase studies we conducted with RAW 264.7 macrophages in which the anti-inflammatory effects of the VOC suppressed the pro-inflammatory effects of the PM2.5.

The following analysis illustrates how the data we generated can be used to explain differences in the response profiles of the two most reactive and potent samples, SB and CM2.

Of particular note, the SB and CM2 samples have comparable chemical reactivities in the DTT assay (0 65, 0.50; Table 2), but the SB pro- and anti-inflammatory potencies as derived from the biological assays were significantly greater (slope values from linear regression analysis; Table 2B). Namely, the SB TNF-α and HO-1 responses were 14 and four times more potent, respectively, than those generated by CM2. We interpret these differences in terms of the chemical content in the samples being collected at different regions of the LA Basin. The CM2 sample consists mostly of railyard traffic emissions, whereas the SB sample has been subjected to atmospheric chemical reactions on the air mass as it moves approximately 80 miles across the Basin. These reactions convert, among other compounds, polynuclear aromatic hydrocarbons to their corresponding quinones as exemplified by 9,10-phenanthroquinone generation (PQ) [7] a potent biological prooxidant associated with PM [25]. Relatively higher levels of such organic prooxidants in the SB PM2.5 sample are suggested by the DTT assay data (Table 2B) that show a larger fraction of the SB PM activity not inhibited by metal chelation (i.e., DTPA-insensitive), in support of this phenomenon of atmospheric formation of organic prooxidants. Analogously, the generation of organic electrophiles such as 1,4-benzoquinone (BQ) likely contribute to the more potent SB anti-inflammatory HO-1 response in the VOC fraction (Table 3). Based on these considerations, we suggest that the higher pro- and anti-inflammatory actions of the SB sample compared to that of CM2 are due to its higher content of organic reactants generated by photochemical processes.

Table 2. Prooxidant content of SB and CM2 samples.

	DTT		DTT + DTPA	
	Mean	95% CI	Mean	95% CI
SB	0.65	0.55–0.75	0.16	0.11–0.20
CM2	0.50	0.47–0.55	0.08	0.03–0.13
SB/CM2 ratio	1.31		1.89	

Prooxidant content, defined by DTT activity (nmols DTT consumed/min*m^3), in the presence and absence of 20 µM, diethylenetriamine pentaacetic acid (DTPA), a metal chelator [16].

Table 3. Biological properties of SB and CM2 samples.

	PM2.5 TNF-α		VOC HO-1	
	Equation	Slope: 95% CI	Equation	Slope: 95% CI
SB	Y = 413.9*X − 16.7	319.9–507.9	Y = 185.2.4*X − 37.8	158.5–212.9
CM2	Y = 29.88*X − 0.34	22.58–37.18	Y = 42.74*X + 4.49	22.76–58.22
SB/CM2 ratio	13.8		4.3	

Slopes of cellular responses to three concentrations of PM2.5 from the indicated PM2.5 samples as a function of cellular protein concentration. Units: Y = TNF-α pg/mg protein; Y = HO-1 ng/mg protein/m^3; X = m^3/mL.

A second consideration in determining relative biological potencies of PM and VOCs is their cellular access upon exposure. PM/metal ions' entry into cells is regulated by endocytosis and transport processes [6,26], whereas quinones such as PQ and BQ are neutral and non-polar compounds that can readily penetrate membranes to promote oxidative stress-related reactions [27]. These biological differences underscore the complex chemical composition of PM, which limits the interpretive value of a general chemical assay such as the DTT assay and necessitate inclusion of parallel biological assays for more definitive evaluation of PM reactivities.

Direct extrapolation of these results to potential toxicological effects in humans, i.e., PM-induced oxidative stress and VOC-induced covalent adducts, remains to be established. In one of the very few direct comparisons of 'whole' diesel exhaust (DE) and VOC components, Campen et al. [28] compared the cardiovascular effects of these two samples in mice and observed major effects consistent with myocardial ischemia, attributable to VOC components. These VOC effects were also seen in vitro using isolated blood vessels. Although some of the chemical components were determined, they did not, however, assay for redox-active species or electrophiles. A second relevant study examined the role of electrophiles that included 1,2-naphthoquinone, a component of VOCs [25] in the actions of DE on mouse lungs [29]. In an intact lung model and in multiple cultured human lung cell models, they showed activation of transient receptor potential ankyrin-1, a pro-inflammatory mediator, on airway C-fibers by a variety of qualitatively and quantitatively different DE particles. These studies demonstrated the actions of VOCs underlying the toxicity of emissions and emphasized the need for further studies that model human exposure conditions.

5. Conclusions

We have shown that quantitative assessments of chemical and biological properties of ambient PM2.5 and VOCs can be used effectively to characterize, compare and contrast AP across different geographical regions, to account for effects of atmospheric modifications on air mass, and to evaluate exposure proximity to an emission source. Of major significance is the consistent characterization of VOCs in terms of their electrophilic chemical reactivity and biological modulation of HO-1 expression and ARE activation. Collectively, these results have human health implications. VOCs, being non-polar volatile species, will readily enter the lungs and penetrate membranes to access all body compartments and thus can potentially contribute to multiple organ effects. Furthermore, electrophilic properties of VOCs can affect the long-term inactivation of protein function resulting from their covalent binding with tissue nucleophiles. Consequently, the effects of low level VOC exposure in AP can accumulate over time and contribute to chronic adverse health effects.

Supplementary Materials: The following are available online at http://www.mdpi.com/2076-3417/10/9/3245/s1, Table S1 Basin prooxidant and electrophile content of PM2.5 and VOC; Table S2 CM1 -CM2 prooxidant and electrophile content of PM2.5 and VOC; Table S3 CM3–CM4 prooxidant and electrophile content of PM2.5 and VOC; Table S4. RAW cell TNF-α response to PM2.5; Table S5 RAW cell HO-1 responses to PM2.5 and VOC.

Author Contributions: Conceptualization, A.K.C., J.R.F., W.P.M.; methodology, Y.S., D.A.S., E.D.S., A.E.-F., A.L.N.G., E.M.S.; software, A.K.C.; validation, A.K.C., W.P.M., D.A.S.; writing—original draft preparation, A.K.C., Y.S., W.P.M.; writing—review and editing A.K.C., W.P.M., Y.S.; supervision, A.K.C.; project administration, A.K.C., J.R.F.; funding acquisition, J.R.F., A.K.C. All authors have read and agreed to the published version of the manuscript.

Funding: This research was funded by the SOUTH COAST AIR QUALITY MANAGEMENT (AQMD) under awards 12865 and 20090799. YS was supported by a grant, "Institutional Program for Young Researcher Overseas Visits" from the Japan Society for the Promotion of Science. EMS was supported by NIH Research Grant # D43TW000623 funded by the Fogarty International Center.

Conflicts of Interest: The authors declare no conflict of interest. The opinions, findings, conclusions, and recommendations are those of the authors and do not necessarily represent the views of AQMD. AQMD, its officers, employees, contractors, and subcontractors make no warranty, expressed or implied, and assume no legal liability for the information in this report. AQMD has not approved or disapproved this report, nor has AQMD passed upon the accuracy or adequacy of the information contained herein.

References

1. Berhane, K.; Chang, C.C.; McConnell, R.; Gauderman, W.J.; Avol, E. Association of Changes in Air Quality With Bronchitic Symptoms in Children in California, 1993–2012. *JAMA* **2016**, *315*, 1491–1501. [CrossRef] [PubMed]
2. Cassee, F.R.; Heroux, M.E.; Gerlofs-Nijland, M.E.; Kelly, F.J. Particulate matter beyond mass: Recent health evidence on the role of fractions, chemical constituents and sources of emission. *Inhal. Toxicol.* **2013**, *25*, 802–812. [CrossRef] [PubMed]

3. Delfino, R.J.; Wu, J.; Tjoa, T.; Gullesserian, S.K.; Nickerson, B.; Gillen, D.L. Asthma morbidity and ambient air pollution: Effect modification by residential traffic-related air pollution. *Epidemiology* **2014**, *25*, 48–57. [CrossRef] [PubMed]
4. McConnell, R.; Wu, W.; Berhane, K.; Liu, F.; Verma, G. Inflammatory Cytokine Response to Ambient Particles Varies due to Field Collection. *Am. J. Respir. Cell Mol. Biol.* **2013**, *48*, 497–502. [CrossRef] [PubMed]
5. Ghio, A.J.; Soukup, J.M.; Madden, M.C. The toxicology of air pollution predicts its epidemiology. *Inhal. Toxicol.* **2018**, *30*, 327–334. [CrossRef]
6. Ghio, A.J.; Madden, M.C. Human lung injury following exposure to humic substances and humic-like substances. *Environ. Geochem. Health* **2018**, *40*, 571–581. [CrossRef]
7. Eiguren-Fernandez, A.; Miguel, A.; Lu, R.; Purvis, K.; Grant, B. Atmospheric formation of 9,10-phenanthroquinone in the Los Angeles air basin. *Atmos. Environ.* **2008**, *42*, 2312–2319. [CrossRef]
8. Cho, A.; Di Stefano, E.; Ying, Y.; Rodriguez, C.E.; Schmitz, D. Determination of Four Quinones in Diesel Exhaust Particles, SRM 1649a and Atmospheric PM2.5. *Aerosol Sci. Technol.* **2004**, *38*, 68–81. [CrossRef]
9. Jakober, C.A.; Riddle, S.G.; Robert, M.A.; Destaillats, H.; Charles, M.J. Quinone emissions from gasoline and diesel motor vehicles. *Environ. Sci. Technol.* **2007**, *41*, 4548–4554. [CrossRef]
10. Jakober, C.A.; Robert, M.A.; Riddle, S.G.; Destaillats, H.; Charles, M.J. Carbonyl emissions from gasoline and diesel motor vehicles. *Environ. Sci. Technol.* **2008**, *42*, 4697–4703. [CrossRef]
11. Li, N.; Kim, S.; Wang, M.; Froines, J.; Sioutas, C.; Nel, A. Use of a stratified oxidative stress model to study the biological effects of ambient concentrated and diesel exhaust particulate matter. *Inhal. Toxicol.* **2002**, *14*, 459–486. [CrossRef] [PubMed]
12. Cho, A.K.; Sioutas, C.; Miguel, A.H.; Kumagai, Y.; Schmitz, D.A. Redox activity of airborne particulate matter at different sites in the Los Angeles Basin. *Environ. Res.* **2005**, *99*, 40–47. [CrossRef] [PubMed]
13. Shinyashiki, M.; Rodriguez, C.E.; Di Stefano, E.W.; Sioutas, C.; Delfino, R.J. On the interaction between glyceraldehyde-3-phosphate dehydrogenase and airborne particles: Evidence for electrophilic species. *Atmos. Environ.* **2008**, *42*, 517–529. [CrossRef]
14. Eiguren-Fernandez, A.; Di Stefano, E.; Schmitz, D.A.; Guarieiro, A.L.N.; Salinas, E.M. Chemical reactivities of ambient air samples in three Southern California communities. *J. Air. Waste Manag. Assoc.* **2015**, *65*, 270–277. [CrossRef] [PubMed]
15. Shinkai, Y.; Nakajima, S.; Eiguren-Fernandez, A.; Stefano, E.D.; Schmitz, D.A. Ambient vapor samples activate the Nrf2-ARE pathway in human bronchial epithelial BEAS-2B cells. *Environ. Toxicol.* **2013**, *25*, 1222–1230. [CrossRef] [PubMed]
16. Eiguren-Fernandez, A.; Miguel, A.H.; Froines, J.R.; Thurairatnam, S.; Avol, E.L. Seasonal and Spatial Variation of Polycyclic Aromatic Hydrocarbons in Vapor-Phase and PM2.5 in Southern California Urban and Rural Communities. *Aerosol Sci. Technol.* **2004**, *38*, 447–455. [CrossRef]
17. Kumagai, Y.; Koide, S.; Taguchi, K.; Endo, A.; Nakai, Y. Oxidation of proximal protein sulfhydryls by phenanthraquinone, a component of diesel exhaust particles. *Chem. Res. Toxicol.* **2002**, *15*, 483–489. [CrossRef]
18. Denizot, F.; Lang, R. Rapid colorimetric assay for cell growth and survival. Modifications to the tetrazolium dye procedure giving improved sensitivity and reliability. *J. Immunol. Methods* **1986**, *89*, 271–277. [CrossRef]
19. Ito, T.; Bekki, K.; Fujitani, Y.; Hirano, S. The toxicological analysis of secondary organic aerosol in human lung epithelial cells and macrophages. *Environ. Sci. Pollut. Res. Int.* **2016**, *26*, 22747–22755. [CrossRef]
20. Sama, P.; Long, T.C.; Hester, S.; Tajuba, J.; Parker, J. The cellular and genomic response of an immortalized microglia cell line (BV2) to concentrated ambient particulate matter. *Inhal. Toxicol.* **2007**, *19*, 1079–1087. [CrossRef]
21. Park, H.S.; Kim, S.R.; Lee, Y.C. Impact of oxidative stress on lung diseases. *Respirology* **2009**, *14*, 27–38. [CrossRef] [PubMed]
22. Wu, L.; Wang, R. Carbon monoxide: Endogenous production, physiological functions, and pharmacological applications. *Pharmacol. Rev.* **2005**, *57*, 585–630. [CrossRef] [PubMed]
23. Iwamoto, N.; Nishiyama, A.; Eiguren-Fernandez, A.; Hinds, W.; Kumagai, Y. Biochemical and cellular effects of electrophiles present in ambient air samples. *Atmos. Environ.* **2010**, *44*, 1483–1489. [CrossRef]
24. Grigoryan, H.; Edmands, W.; Lu, S.S.; Yano, Y.; Regazzoni, L. Adductomics Pipeline for Untargeted Analysis of Modifications to Cys34 of Human Serum Albumin. *Anal. Chem.* **2016**, *88*, 10504–10512. [CrossRef] [PubMed]

25. Eiguren-Fernandez, A.; Miguel, A.; Di Stefano, E.; Schmitz, D.; Cho, A. Atmospheric Distribution of Gas- and Particle-Phase Quinones in Southern California. *Aerosol Sci. Technol.* **2008**, *42*, 854–861. [CrossRef]
26. Su, R.; Jin, X.; Li, H.; Huang, L.; Li, Z. The mechanisms of $PM_{2.5}$ and its main components penetrate into HUVEC cells and effects on cell organelles. *Chemosphere* **2020**, *241*, 125127. [CrossRef]
27. Taguchi, K.; Fujii, S.; Yamano, S.; Cho, A.K.; Kamisuki, S. An approach to evaluate two-electron reduction of 9,10-phenanthraquinone and redox activity of the hydroquinone associated with oxidative stress. *Free Radic. Biol. Med.* **2007**, *43*, 789–799. [CrossRef]
28. Campen, M.J.; Babu, N.S.; Helms, G.A.; Pett, S.; Wernly, J. Nonparticulate components of diesel exhaust promote constriction in coronary arteries from ApoE-/- mice. *Toxicol. Sci.* **2005**, *88*, 95–102. [CrossRef]
29. Deering-Rice, C.E.; Memon, T.; Lu, Z.; Romero, E.G.; Cox, J. Differential Activation of TRPA1 by Diesel Exhaust Particles: Relationships between Chemical Composition, Potency, and Lung Toxicity. *Chem. Res. Toxicol.* **2019**, *32*, 1040–1050. [CrossRef]

© 2020 by the authors. Licensee MDPI, Basel, Switzerland. This article is an open access article distributed under the terms and conditions of the Creative Commons Attribution (CC BY) license (http://creativecommons.org/licenses/by/4.0/).

Article

Impact of the '13th Five-Year Plan' Policy on Air Quality in Pearl River Delta, China: A Case Study of Haizhu District in Guangzhou City Using WRF-Chem

Juanming Zhan [1,2,3], **Minyi Wang** [1,2,3], **Yonghong Liu** [1,2,3], **Chunming Feng** [4], **Ting Gan** [1,2,3], **Li Li** [1,2,3], **Ruiwen Ou** [4] **and Hui Ding** [1,2,3,*]

1 School of Intelligent Systems Engineering, Sun Yat-sen University, Guangzhou 510275, China; zjmxian@126.com (J.Z.); wangmy68@mail2.sysu.edu.cn (M.W.); liu_its@163.com (Y.L.); gt09032206happy@163.com (T.G.); lylee1990@126.com (L.L.)
2 Guangdong Provincial Key Laboratory of Intelligent Transport System, Guangzhou 510275, China
3 Guangdong Provincial Engineering Research Center for Traffic Environmental Monitoring and Control, Guangzhou 510275, China
4 Guangzhou Haizhu District Environmental Monitoring Station, Guangzhou 510220, China; fengcming@126.com (C.F.); sherman_au@126.com (R.O.)
* Correspondence: hui_gnid@126.com

Received: 6 July 2020; Accepted: 27 July 2020; Published: 30 July 2020

Abstract: Due to increasingly stringent control policy, air quality has generally improved in major cities in China during the past decade. However, the standards of national regulation and the World Health Organization are yet to be fulfilled in certain areas (in some urban districts among the cities) and/or certain periods (during pollution episode event). A further control policy, hence, has been issued in the 13th Five-Year Plan (2016–2020, hereafter 13th FYP). It will be of interest to evaluate the air quality before the 13th FYP (2015) and to estimate the potential air quality by the end of the 13th FYP (2020) with a focus on the area of an urban district and the periods of severe pollution episodes. Based on observation data of major air pollutants, including SO_2 (sulphur dioxide), NO_2 (nitrogen dioxide), CO (carbon monoxide), PM_{10} (particulate matter with aerodynamic diameter equal to or less than 10 μm), $PM_{2.5}$ (particulate matter with aerodynamic diameter equal to or less than 2.5 μm) and O_3 (Ozone), the air quality of Haizhu district [an urban district in the Pearl River Delta (PRD), China] in 2015 suggested that typical heavy pollution occurred in winter and the hot season, with NO_2 or $PM_{2.5}$ as the key pollutants in winter and O_3 as the key pollutant in the hot season. We also adopted a state-of-the-art chemical transport model, the Weather Research and Forecasting model coupled with Chemistry (WRF-Chem), to predict the air quality in Haizhu District 2020 under different scenarios. The simulation results suggested that among the emission control scenarios, comprehensive measures taken in the whole of Guangzhou city would improve air quality more significantly than measures taken just in Haizhu, under all conditions. In the urban district, vehicle emission control would account more than half of the influence of all source emission control on air quality. Based on our simulation, by the end of the 13th FYP, it is noticeable that O_3 pollution would increase, which indicates that the control ratio of volatile organic compounds (VOCs) and nitrogen oxides (NO_x) may be unsuitable and therefore should be adjusted. Our study highlights the significance of evaluating the efficacy of current policy in reducing the air pollutants and recommends possible directions for further air pollution control for urban areas during the 13th FYP.

Keywords: 13th Five-Year Plan; pollution episode; WRF-Chem; emission control scenario; PRD

1. Introduction

The quality of ambient air is vital to human health. Air quality management is important for many authorities around the world [1]. The Chinese government has put great effort into mitigating the elevated level of air pollutants in the past decade, especially since the Air Pollution Control Action Plan (APCAP) was issued by the State Council in 2013 [2]. The annual levels of $PM_{2.5}$, PM_{10}, SO_2, and NO_2 decreased by 12%, 11%, 20%, and 5%, respectively, in major cities during 2013–2015 [3,4].

However, the effect was highly heterogeneous both spatially and temporally since the national standards of air quality were violated in some districts/towns in cities and during severe pollution episode events. Observation stations in the urban area of Beijing [5], Shanghai [6], and Guangzhou [7,8] recorded heavy pollution episodes in winter (PM_{10}, $PM_{2.5}$, and NO_2 as key pollutants) and/or hot seasons (summer and autumn, O_3 as key pollutant) [9,10].

To further tackle the air pollution issue, authorities of national and local level have formulated a series of regulations for the 13th Five-Year Plan (13th FYP) [11,12]. Targets have been set such that, by the end of 13th FYP, the emission of SO_2 and NO_2 should decrease 15% compared to that in 2015, and the ratio of heavy pollution days in 2020 should reduce 25% compared to that in 2015 [11]. The three national typical air pollution city clusters (Beijing-Tianjin-Hebei, the Yangtze River Delta and the Pearl River Delta (PRD)) have their own targets. For example, for the cities of the PRD region, concentrations of air pollutants ($PM_{2.5}$, PM_{10}, SO_2, NO_2, CO, and O_3) should meet national secondary air quality standards and the number of heavy pollution day should be zero [11]. Although the emission of the air pollutant from large industrial point sources has been decreasing in the past decade, the emission from traffic has been significantly increasing, which leads to critical challenges in air pollution control. It will be of significance to assess the air quality level in the year before the 13th FYP (2015) and to predict the impact of 13th FYP on air quality by 2020.

Some modelling studies have been conducted to evaluate the potential influence of the 13th FYP policy or the impact of policies on air quality during some major public events. Wang et al. (2016) [13] evaluated the impact of emission control measures on the air quality in the PRD region with WRF-CMAQ (The Weather Research and Forecasting Model-Community Multiscale Air Quality Model) model simulation on emission scenarios (a base case in 2010, two cases in 2020). Liu et al. (2017) [14] assessed the cobenefits (air quality and climate change) of vehicle emission control measures for 2015–2020 in the PRD region. Maji et al. (2018) [15] reported the $PM_{2.5}$-related mortality under air pollution control policies for China 2020. Yang et al. (2016) [16] analysed the effect of the coal control strategy in China on carbon mitigation and pollutants control for 2020 and 2030. Wang et al. (2015) [17] assessed the air quality situation under the pollution control policy of thermal power plants in China for 2020 with MM5-CMAQ (Fifth-Generation Mesoscale Regional Weather Model-Community Multiscale Air Quality Model). Qui et al. (2017) [18] studied the effect of emission control strategies on air quality of Baotou, China. Cai et al. (2017) [19] researched the impact of the 'Air Pollution Prevention and Control Action Plan' on $PM_{2.5}$ in the Jing-Jin-Ji region from 2012–2020 with WRF-CMAQ. Li et al. (2017) [20] estimated the effect of policies in the '13th Five-Year Plan' period on air pollutants emission of China's electric power sector. Wei et al. (2017) [21] analysed the impact of policies in Shanxi province, China. Guo et al. (2016) [22] researched the impact of emission control measures on air quality during APEC (Asia–Pacific Economic Cooperation summit) China 2014 with the Weather Research and Forecasting coupled with Chemistry (WRF-Chem). Xu et al. (2013) [23] evaluated the effect of air pollution control policies on air quality during the 16th Asian Games with CMAQ (Community Multiscale Air Quality Model). Shen et al. (2016) [24] analysed the influence of emission control policies on air quality during China's V-Day parade in 2015. Air quality numerical models (WRF-CMAQ, MM5-CMAQ, WRF-Chem, etc.) and scenario analyses were used widely to evaluate the effectiveness of policies on air quality. These studies only focused on the impact of policies on annual air quality in 2020 or on air quality during some major public events and did not investigate air quality on heavy pollution events.

For heavy pollution, previous works focused on the characteristics and formation of pollution episodes. Tan et al. (2009) [25] investigated the chemical characteristics of haze in Guangzhou

(2002–2003, summer and winter) with PM_{10} samples and gas chromatography-mass spectrometry (GC-MS). Wang et al. (2015) [26] researched the formation process of a severe haze episode in the Yangtze River Delta (2013 winter) based on visibility and meteorological parameters, and backward trajectories of the air mass. Zhang et al. (2015) [27] applied WRF-Chem to simulate a severe haze in Beijing (2013 winter) and discussed the meteorological impacts on haze. Zhan et al. (2017) [7] analysed the spatial and temporal association of $PM_{2.5}$ pollution events between typical cities of PRD (2014 winter). Ding et al. (2004) [28] discussed the effects of sea–land breezes on the transport of air pollution during an ozone episode in PRD (2001 autumn) with the MM5 (Fifth-Generation Mesoscale Regional Weather) model. Shen et al. (2015) [29] researched the source of an ozone episode in PRD (2008 autumn) with the CAMx (Comprehensive Air-quality Model with extensions) model. Zhao et al. (2015) [30] investigated the chemical characteristics of ozone episodes in Shanghai (2010–2013, O_3 peaked in summer) with the differential optical absorption spectroscopy (DOAS) and the hybrid single particle Lagrangian integrated trajectory (HYSPLIT) model. Xu et al. (2008) [31] simulated typical summertime ozone episodes in Beijing (2000) with the WRF-CAMx (Weather Research and Forecasting Model Comprehensive Air-quality Model with extensions) model to analyse the process, and Qu et al. (2014) [32] used the CMAQ-MADRID (Community Multiscale Air Quality Model-Model of Aerosol Dynamics, Reaction, Ionization, and Dissolution) model to evaluate the effects of NO_x and VOCs emissions on ozone pollution in Beijing (2007 summer). According to previous studies, meteorological conditions and emission significantly affected pollution, and numerical models were widely used to analyse the characteristics of pollution. However, the model evaluation of the impact of the 13th FYP on the air quality in 2020 during the pollution episode has not attracted much attention.

The air pollution control strategy in China is now at the new stage that the manufacturing industry in megacities in China (e.g., Guangzhou) is no longer the dominant emission source; control on vehicle emissions is becoming the primary subject [33]; the control strategy is transforming from urban/regional control to district/town grid control [34]. The PRD region is one of the three national typical air pollution city clusters [2]. PRD has experienced the major problem of transferring from haze to complex (haze and photochemical) pollution. Moreover, it is the first city cluster to achieve the goal in the APCAP, i.e., compared with the levels in 2012, the concentration of fine particles ($PM_{2.5}$) in Beijing-Tianjin-Hebei, the Yangtze River Delta, and the Pearl River Delta (PRD) reduced by 25%, 20%, and 15%, respectively, by 2017 [4]. As the centre of PRD and the capital city in Guangdong province, Guangzhou is one of the cities with the poorest air quality in PRD, more comprehensive and more stringent air pollution control measures were implemented in Guangzhou than in other cities of PRD, so that Guangzhou is a role model in air quality management [35]. Haizhu district is an island in the urban area of Guangzhou, with a typical urban landscape (residential and commercial areas) comprising many high rises, shops, residential apartments, highways, and major roads that pass through the urban area, and its air quality was relatively poor [36–38]. It is an ideal testbed to determine how the emission control measures affect air quality in urban areas. Besides, manufacturing industries have been moved out from Haizhu due to the 'limitation of high-pollution production' policy [39,40], which has been a typical trend in cities of China recently [3]. Therefore, our study focused on the Guangzhou central district (Haizhu district) (Figure 1).

In this study, we first compiled an evaluation on the observational data in Haizhu district in 2015 with a focus on the periods of pollution episodes. Then we designed four scenarios based on the 13th FYP regulations, and utilized the WRF-Chem model to evaluate the effect of the emission control measures and the influence of Haizhu policy and Guangzhou (except Haizhu) policy on air quality in Haizhu district 2020 under those scenarios. The research framework is shown in Figure S1 in the Supplementary Materials.

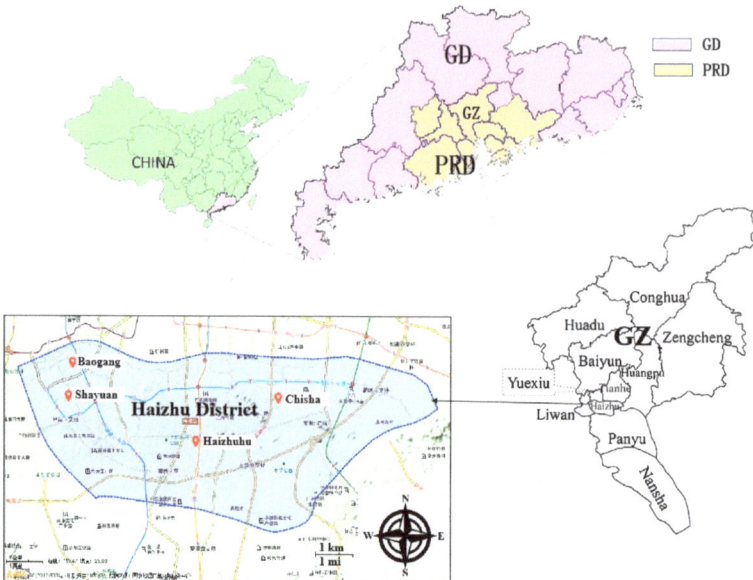

Figure 1. The Haizhu region and its location (GD: Guangdong province; PRD: Pearl River Delta; GZ: Guangzhou) [14,41,42].

2. Data and Model Experiment

2.1. Observation Data

Air quality monitoring data, including the hourly concentrations of pollutants (CO, SO_2, $PM_{2.5}$, PM_{10}, NO_2, O_3) and daily AQI (Air Quality Index), were collected from Guangzhou Environmental Monitoring Centre data network (http://210.72.1.216:8080/gzaqi_new/RealTimeDate.html). Hourly and daily meteorological data [wind speed, wind direction, relative humidity (RH), temperature, satellite cloud images, etc.] were downloaded from the Guangzhou meteorological data network (http://data.tqyb.com.cn/weather/index.jsp). These websites are regularly maintained by the government, and the data are released to the public. The data mentioned above were used to analyse the air quality status and pollution episodes in 2015, and to evaluate the WRF-Chem model's performance.

2.2. Emission Inventory and Scenarios

2.2.1. Emission Inventory

The anthropogenic emissions from the 2012-based Multiresolution Emission Inventory for China (MEIC) [43] (http://www.meicmodel.org/), biogenic emission parameterisation [44], dust emission parameterisation [45], and sea salt emission parameterisation [46], and the marine emission from EDGAR (Emission Database for Global Atmospheric Research, http://edgar.jrc.ec.europa.eu/) were applied in this study. The 2012-based MEIC was a Chinese national emission inventory with a resolution of 0.25° × 0.25° based on 2012 emission status, developed by Tsinghua University. It included the emissions in major sectors, such as transport, industry and human residential, etc. [13]. Previous studies reported that MEIC was well developed by a technology-based emission model [47] and similar emission datasets have been widely used in the numerical simulation for cities in PRD [27,48].

The emissions in the key research area, Guangzhou, were obtained based on the Guangzhou ambient air quality plan (2016–2025) [49]. The emission of air pollutants from different sources in Haizhu district in 2015 was retrieved from the Guangzhou environmental protection bureau's data network (http://www.gzepb.gov.cn/infoindex.htm). Based on the statistics yearbook (http://210.72.4.52/gzStat1/chaxun/njsj.jsp) and the Guangzhou environmental protection bureau's data network, the emission inventory was generated according to Chinese national technical guidelines [50,51] and relevant studies (Zheng et al., 2009 [52]; Zhao et al., 2015 [53]). Spatial allocation of emission inventory depended on the source characteristics. Haizhu district is an urban area, and the major emission sectors are industry, residential, and (road) transport. Emission sectors, corresponding inventory technical guidelines, and special allocation rules in this research are shown in Table S1.

According to the 2015-based Haizhu district emission inventory, the 2020 Haizhu district emission inventory was predicted with the extrapolation function [13] [EI_2020 = f_x (EI_2015 × activity factor), where EI refers to the emission inventory, and the activity factor refers to the trend of emission sections, gained from emission control policies for the 13th Five-Year period [12,49,54] and the trend of city development in the past five to ten years (e.g., society, economy, vehicle and population, etc.) according to the statistical yearbook network (http://210.72.4.52/gzStat1/chaxun/njsj.jsp)]. We note that the emission inventory in Haizhu district is compiled on an annual resolution. The resolution of these inventories was 0.01° × 0.01°. All of the emission inventories have the same major sectors (transport, industry, and residential, etc.).

2.2.2. Emission Scenarios

Because the major and easily controlled sections had been regulated during the 12th Five-Year Plan period (12th FYP), the 13th FYP emission control policy was set more aggressively: industries generating air pollution will be moved out from the central city area, ultraclean/effective technology will be widely used, the ratio of public transport will rise to 70% of motorised travel [12], electric buses will be applied widely, accounting for 63% of public buses [54], vehicle emission standards and fuel standards will be strengthened (specifically, eliminating high pollution vehicles, implementing the new national emission standard of vehicle and the new national vehicle fuel standard, etc.), and VOC emission will be controlled entirely in particular industries (e.g., chemical industry, paint industry, and printing industry, etc.) [49].

Four scenarios were designed to evaluate the impact of the 13th FYP on the air quality in Haizhu district in 2020. The meteorological conditions were assumed to be unchanged, which means that the meteorological conditions in 2015 were used for all scenarios in 2020. The four 2020 scenarios are as follows (also in Table S2):

Scenario 2020A was proposed such that both Haizhu emission control policy and Guangzhou (except Haizhu) emission control policy would follow the 2015 emission control policy tendency. This scenario is a baseline scenario.

Scenario 2020B was designed such that Haizhu emission control would be implemented based on the 13th FYP emission control plan, while Guangzhou (except Haizhu) would still adhere to the 2015 emission control policy.

Scenario 2020C was a 13th FYP policy scenario in which both Haizhu emission control policy and Guangzhou (except Haizhu) emission control policy would be implemented based on the 13th FYP emission control plan.

Scenario 2020D was a scenario for vehicle control in Haizhu due to the significance of vehicle emission in Haizhu. In this scenario, vehicle control in Haizhu would be implemented based on the 13th FYP emission control plan, while other emission source controls in the Haizhu and Guangzhou (except Haizhu) emission control policy would remain as the 2015 emission control policy tendency. We note that the contribution of vehicle emission in Haizhu district is <10% of that in Guangzhou city.

Table 1 shows the changes in emissions of SO_2, NO_x, CO, PM_{10}, $PM_{2.5}$, and VOCs for the different scenarios in the whole city of Guangzhou. The 2020A, 2020B, and 2020D scenario emissions would be higher than the 2015 emissions (SO_2, NO_x, CO, PM_{10}, $PM_{2.5}$, and VOCs would increase 57.2–58.9%, 47.4–48.6%, 60.1–60.2%, 14.6–15.1%, 25.8–26.4%, and 24.9–25.0%, respectively), whereas the 2020C scenario emission would be lower than or roughly equal to the 2015 emission. The difference of emissions among each scenarios indicates that the emission control of the whole Guangzhou city would significantly affect the 2020 emission. Additionally, the emission ratio (VOCs/NO_x) would be around 0.72 in scenarios 2020A, 2020B, and 2020D and 0.60 in scenario 2020C, lower than 0.86 in 2015.

Table 1. Changes in pollutant emissions of the different scenarios in the whole city (Guangzhou) (A, B, C, and D are scenarios 2020A, 2020B, 2020C, and 2020D, respectively).

	SO_2	NO_x	CO	PM_{10}	$PM_{2.5}$	VOCs
(A-2015)/2015	58.9%	48.6%	60.2%	15.1%	26.4%	25.0%
(B-2015)/2015	57.2%	47.4%	60.1%	14.6%	25.8%	24.9%
(C-2015)/2015	−72.9%	−14.2%	1.0%	−36.7%	−41.8%	−33.2%
(D-2015)/2015	58.9%	48.1%	60.1%	15.0%	26.2%	25.0%

The changes in emissions for each sector are shown in Figure 2. SO_2 emission from the sectors residential, vehicle, and others, and NO_x emission from the residential sector in the four 2020 scenarios would be lower than those in the 2015 emission. $PM_{2.5}$ emission and PM_{10} emission from the residential sector in scenarios 2020B and 2020C would also be lower than those in the 2015 emission. In scenarios 2020C, NO_x emission from the vehicle and others sectors, CO emission from the industry, residential, and vehicle sectors, and VOCs emission from the residential sector would be higher than those in the 2015 emission. Overall pollutants emissions in scenarios 2020A, 2020B and 2020D would be higher than those in 2015, pollutants emissions in scenarios 2020C would be lower than in 2015, except that CO emission in scenarios 2020C would be slightly higher than that in 2015.

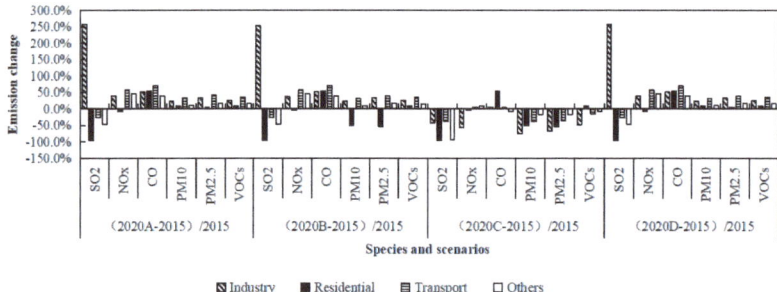

Figure 2. Emission changes in sectors in different scenarios in the whole city (Guangzhou).

2.3. WRF-Chem Model Setup

2.3.1. Description of WRF-Chem Model

WRF-Chem is a chemical transport model developed by the community led by NOAA/ESRL (Earth System Research Laboratory, The United States National Oceanic and Atmospheric Administration) [55,56], and it is widely used for analysing heavy pollution processes and the effectiveness of emission control measures [22,27,57].

2.3.2. Configuration of WRF-Chem

In this study, a three-nested domain was applied for model set-up (Figure S2), with grid cell areas of 9 × 9 km, 3 × 3 km, and 1 × 1 km, respectively. The biggest domain covered the southern China area, the middle domain covered PRD, and the smallest one covered the Guangzhou central area. There were 27 sigma levels for all domains, and in this study, the data of the ground level was mainly used. The NCEP (the United States' National Centers for Environmental Prediction) 6-h FNL (Final Operational Global Analysis data) meteorological data and the emission inventory mentioned above were input for model setup. The simulation data from MOZART (Model for Ozone and Related

Chemical Tracers, https://www.acom.ucar.edu/wrf-chem/mozart.shtml) was used as the initial and boundary chemistry data [58]. The related parameterisation schemes in the simulation were the Regional Acid Deposition Model, version 2 (RADM2) gas-phase chemical mechanism [59] and the MADE/SORGAM (Modal Aerosol Dynamics Model for Europe/Secondary Organic Aerosol Model) aerosol chemical mechanism [60,61]. We have not activated the feedback from chemistry on meteorology. The first 120 h in the simulation were used as the model spin-up.

2.3.3. Cases Setup

(1) Evaluation of overall air quality in 2020

The month of October was used as a proxy of the entire year of 2015 since the pollutant concentrations in October 2015 were very similar to those in the whole year (details are shown in Section 3.1). WRF-Chem simulations were conducted with the emissions in four 2020 scenarios and the emissions in 2015. The simulated concentrations of $PM_{2.5}$, PM_{10}, SO_2, NO_2, CO, and O_3 were compared between results of the four 2020 scenarios and the result of 2015, in order to evaluate the effectiveness of the 13th FYP policy on air quality. The meteorological conditions for 2020 scenarios were assumed to be the same as the meteorological conditions in 2015. The results are shown in Section 3.3.

(2) Evaluation of air quality during pollution episodes

To identify the air quality status, we analysed the annual air quality situation and air pollution events based on the data observed in four urban stations (Baogang, Chisha, Shayuan, and Haizhuhu) in Guangzhou Haizhu district (Figure 1). A pollution episode was defined as a short period consisting of subsequent days (at least one day) with the AQI ≥ 101 [62]. A heavy pollution episode was a pollution episode with AQI ≥ 201 [62].

The pollution episodes in 2015 are shown in Table 2 (more information is shown in Table S3). The table shows that the number and the pollution level of pollution episodes had seasonal characters. The pollution episodes occurred more often in winter, and heavy pollution was observed in the summer and winter. The Shayuan station and Haizhuhu station suffered pollution episodes more than other stations in 2015, and Shayuan was the only station where the heavy pollution episodes were observed in both winter and summer (14–28 January 2015, and 3–8 August 2015, respectively). The two heavy pollution episodes were typical pollution ($PM_{2.5}$ and NO_2 pollution in winter, O_3 pollution in summer [63]). Therefore, the two heavy pollution episodes in Shayuan were analysed with pollution and meteorology progress (results shown in Section 3.1.2). Also, the meteorology conditions of these two episodes were applied in the 2020 scenarios numerical simulations with WRF-Chem to evaluate the effectiveness of the 13th FYP policy on air quality of pollution episodes (specifically, $PM_{2.5}$ and NO_2 pollution for the winter episode, O_3 pollution for the summer episode). The simulation results are shown in Section 3.4.

Table 2. Pollution episodes observed at Haizhu monitoring stations in 2015.

Station	Season	Event Frequency (Units)	key Pollutant	Max Pollution Level
Baogang	Spring	5	$PM_{2.5}$, NO_2, O_3	Moderate pollution
	Summer	4	O_3	**Heavy pollution**
	Autumn	3	$PM_{2.5}$, NO_2, O_3	Moderate pollution
	Winter	9	$PM_{2.5}$, NO_2	Moderate pollution
Chisha	Spring	4	$PM_{2.5}$, NO_2, O_3	Moderate pollution
	Summer	4	O_3	**Heavy pollution**
	Autumn	3	NO_2, O_3	Moderate pollution
	Winter	7	$PM_{2.5}$, NO_2	Moderate pollution
Shayuan	Spring	7	$PM_{2.5}$, O_3	Moderate pollution
	Summer	5	O_3	**Heavy pollution**
	Autumn	12	$PM_{2.5}$, NO_2, O_3	Moderate pollution
	Winter	7	$PM_{2.5}$, NO_2	**Heavy pollution**
Haizhuhu	Spring	6	$PM_{2.5}$, NO_2, O_3	Moderate pollution
	Summer	3	O_3	Light pollution
	Autumn	10	NO_2, O_3	Moderate pollution
	Winter	12	$PM_{2.5}$, NO_2, O_3	Moderate pollution

3. Results and Discussion

3.1. Air Quality in Haizhu in 2015

3.1.1. Overview

The annually-averaged concentrations of $PM_{2.5}$, PM_{10}, SO_2, and NO_2 were 40 µg/m^3, 61 µg/m^3, 14 µg/m^3, and 49 µg/m^3, respectively, in Haizhu district in 2015, which were higher than those in the whole of Guangzhou city (39 µg/m^3, 59 µg/m^3, 13 µg/m^3, and 47 µg/m^3, respectively [64]). The 95th percentile of the CO daily concentration was 0.9 mg/m^3, in Haizhu district in 2015, which was lower than that in the whole of Guangzhou city (1.5 mg/m^3, [64]), The routine pollutants (except for O_3) peaked in January (winter), among which $PM_{2.5}$ and NO_2 were the major pollutants. The 90th percentile of the daily 8 h maximum O_3 concentration (O_3-8h) in 2015 was 139 µg/m^3, slightly lower than that in Guangzhou (145 µg/m^3; [64]). O_3-8h peaked in September (summer to autumn, or hot season). O_3 was also the key pollutant in summer and autumn in Haizhu district [64]. These observational results suggested that air quality in Haizhu was still harmful to human health (the thresholds: $PM_{2.5} \leq 10$ µg/m^3 annual mean, $PM_{10} \leq 20$ µg/m^3 annual mean, $NO_2 \leq 40$ µg/m^3 annual mean, O_3-8h ≤ 100 µg/m^3 8 h mean) [65] and it was worse than the Guangzhou average level in 2015, which highlights the necessity of district/town regulation in air quality control. Please note that this study does not intend to estimate the influence of air quality on human health which is sensitive to the exposure levels (the levels of air pollutants). The pollutant concentrations in October 2015 were very similar to those in the year of 2015 (a-SO_2, a-NO_2, a-PM_{10}, a-$PM_{2.5}$, a-O_3-8h, a-CO, in Figure 3a).

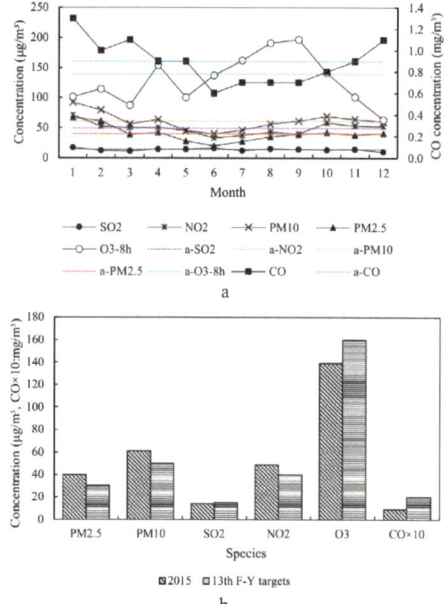

Figure 3. (a) Air pollutant monthly concentrations in 2015 averaged in four stations (Baogang, Chisha, Shayuan, and Haizhuhu) in Haizhu district, (b) comparison of air quality in 2015 in Haizhu district and the 13th Five-Year targets. (In (a), a-SO_2, a-NO_2, a-PM_{10}, a-$PM_{2.5}$ means 2015 annual value for SO_2, NO_2, PM_{10}, $PM_{2.5}$, respectively; a-O_3-8h means the 90th percentile of O_3-8h in 2015; a-CO means the 95th percentile of the CO daily concentration in 2015; other pollutants: monthly-averaged concentrations. In (b), O_3: the 90th percentile of O_3-8h in a year; CO: the 95th percentile of the CO daily concentration in a year; other pollutants: annually-averaged concentrations. These parameters were adopted according to the national standards in China [66].).

Compared with the 13th FYP targets in 2020 ($PM_{2.5} \leq 30$ μg/m^3, $PM_{10} \leq 50$ μg/m^3, $SO_2 \leq 15$ μg/m^3, $NO_2 \leq 40$ μg/m^3, O_3-8h ≤ 160 μg/m^3, and CO ≤ 2 mg/m^3) in Guangzhou [49] (Figure 3b), $PM_{2.5}$, PM_{10}, and NO_2 concentrations in 2015 exceeded the targets, and the air pollutant emission could even rise in the coming years (Table 1). Haizhu district could face worsening air quality, and more emphasis should be placed on the control measures and effectiveness assessment for $PM_{2.5}$ and NO_2 during the 13th FYP so that the measures can be duly adjusted to achieve the 13th FYP targets.

3.1.2. Air Quality during Pollution Episodes

Table 2 summarizes the pollution episodes in Haizhu district in 2015. Shayuan station was the only station where heavy pollution episodes were observed in both winter and summer, and the two pollution episodes were also the typical pollution ($PM_{2.5}$ and NO_2 pollution in winter, O_3 pollution in summer). Therefore, these two typical pollution episodes in Shayuan will be analysed in detail below.

(1) Heavy pollution episode in winter

A winter heavy pollution episode in Haizhu district occurred during 14–28 January 2015 (Table S4 shows the statistical data). In this pollution episode, the air quality was light pollution in the beginning (15–19 January), then the pollution level peaked during 20–21 January. It finally decreased to light pollution again in the third period (22–27 January). On 28 January, the air quality was good again. This episode was a process in which the air quality decreased gradually and then increased gradually. During this episode, the wind speed was in the range of 1.1–2.9 m/s, the RH was in the range of 52–85%, and there was no precipitation. The temperature was in the range of 10.3–18.0 °C, and the surface pressure was in the range of 1009.1–1016.1 hPa. The low wind speed during this winter episode (~1.5 m/s) indicated that the atmosphere was stable, and as reported in previous studies, low or calm wind speed generally leads to high pollution [67,68], because stable atmospheric condition favours the accumulation of pollutants. There is another possibility that the existence of temperature inversion layer could also lead to the accumulation of the air pollutants (e.g., Malek et al., 2006 [69]).

(2) Heavy pollution episode in summer

A summer heavy pollution episode in Haizhu district happened during 3–8 August 2015 (Table S5). The whole episode covered only six days, including the two days of light pollution (4 and 7 August), one day of moderate pollution (5 August), and one day of heavy pollution (6 August). The pollution occurred and disappeared within a short time, but it exhibited high intensity. In this episode, the wind speed was in the range of 1.1–4.4 m/s, the RH was between 65 and 76%, and there was no precipitation. In the first period (3–6 August 2015), the wind speed was in the range of 1.1–1.4 m/s. There was no precipitation nor cloud. The temperature was in the range of 27.6–30.3 °C, and the surface pressure was in the range of 991.7–1002.4 hPa. The hourly temperatures were between 23.4 and 35.6 °C. Sunny weather, low wind, and high temperatures would create favourable conditions for O_3 production and accumulation [10]. Then, the wind speed increased to 3.9–4.4 m/s during 7–8 August 2015. Strong wind transported the air pollutants out of Haizhu, improving the air quality in Haizhu to light pollution. The temperature inversion layer might also enhance the accumulation of the air pollutants (e.g., Malek et al., 2006 [69]).

3.2. Evaluation of WRF-Chem Model Performance

October was the month when the concentrations of $PM_{2.5}$, NO_2 and O_3 (the major pollutants in Haizhu) were all at elevated levels. Therefore, we used the observed data in October to evaluate the performance of the WRF-Chem simulation.

(1) Meteorological simulation evaluation

The simulated surface meteorological parameters (wind speed, wind direction, temperature, and RH) in Guangzhou were validated with the observed data (Table S6 and Figure 4). The temperature, RH, and wind speed in the simulation were generally in line with the observations (Table S6), and the simulated wind directions covered the observed wind directions (Figure 4, wind direction). Similar model behaviour was reported in some previous studies [9,70,71]. The over-prediction of the wind speed in the present study is probably due to the underestimation of the roughness in the cities. For example, the constructions in the urban area are not adequately represented in our model simulation although a very fine resolution of 1 km is already used. Overall, the simulation generally reproduced the meteorological condition in 2015.

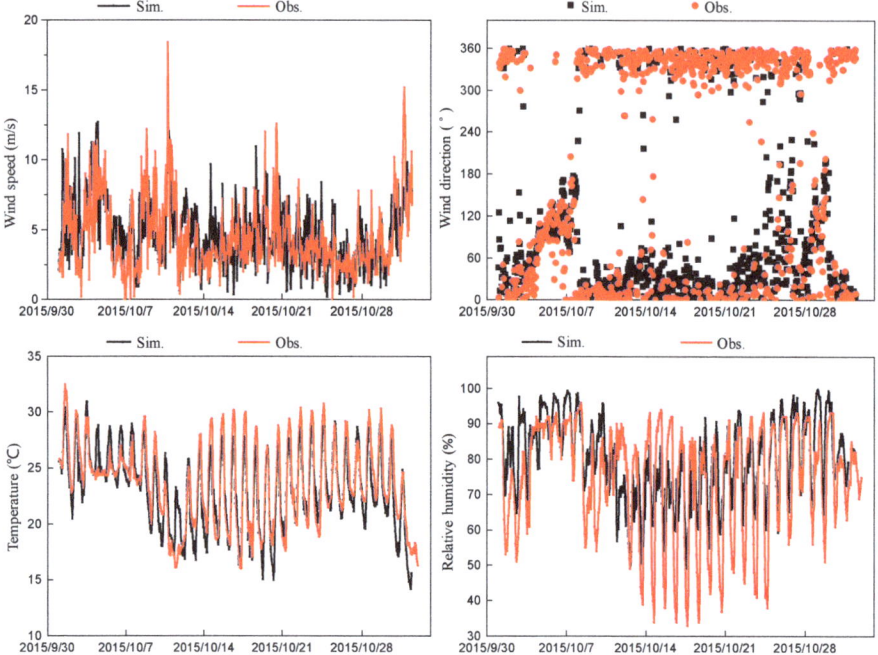

Figure 4. Comparison of simulated and observed meteorological parameters in Guangzhou, 2015. (Hourly statistical data in October 2015).

(2) Chemical simulation evaluation.

Simulated pollutant concentrations were also compared to observed data, including O_3 daily 8 h maximum concentration and other pollutants' daily-averaged concentrations (CO, $PM_{2.5}$, PM_{10}, SO_2, and NO_2). Metrics of evaluation on model performance are shown in Table S7. The ranges of the mean-bias (MB) and root-mean-square-error (RMSE) on pollutant concentrations were −8.07 μg/m³ to 24.34 μg/m³, 9.81 μg/m³ to 48.35 μg/m³, respectively, except CO. The MB and RMSE of CO were −0.32 mg/m³ and 0.43 mg/m³, respectively. Compared to previous studies [13,70–72], the model in this paper showed similar behaviour. Figure 5 shows that the general characteristics of the routine pollutants were captured by the simulation. Monthly averaged simulated concentrations of NO_2, PM_{10}, SO_2, CO, O_3-8h, and $PM_{2.5}$ at the locations of the monitoring sites were 147.6%, 101.1%, 112.0%, 65.1%, 91.4%, and 80.9% of the observations, respectively. Concentrations of NO_2, PM_{10}, and SO_2 were overpredicted in the simulation, and concentrations of CO, O_3-8h, and $PM_{2.5}$ were underestimated. It is noticeable that there was an observed peak in pollution on 14 October which was underestimated in the simulation. Concentrations of NO_2, PM_{10}, SO_2, CO, O_3-8h, and $PM_{2.5}$ in the simulation were 39%, 49%, 42%, 52%, 49%, and 65% less than that of the observation on 14 October, respectively. The most probable reason for the discrepancy is the meteorological condition was not well captured. Other possible causes include the uncertainty of the emission inventory, the uncertainty of the chemical scheme in simulating the formation of secondary air pollutants, etc. The uncertainty of emission inventories results from the uncertainty in the activity level, the emission factor, and the different grid size [73–75]. The simulation of secondary air pollutants (O_3, secondary inorganic and organic aerosol) is a hot research topic and still contains some uncertainty as reported in Ahmadov et al. (2012) [76], Li et al., (2018) [77], etc. Besides, the planetary boundary (PBL) scheme also plays an important role in predicting the level of air pollutants at the surface and the reader is referred to relevant studies (e.g., Banks and Baldasano, 2016 [78]; Shin and Hong, 2011 [79]; Hu et al., 2010 [80]) for the evaluation of various PBL schemes including the one used in the present study, i.e., the YSU (Yonsei University) scheme [81]. There is also the possibility that the representative area of the measurement site differs from the grid cell that contains the site and such difference might lead to the difficulty of directly comparing the simulation with observation (e.g., Schutgens et al. 2016 [82]).

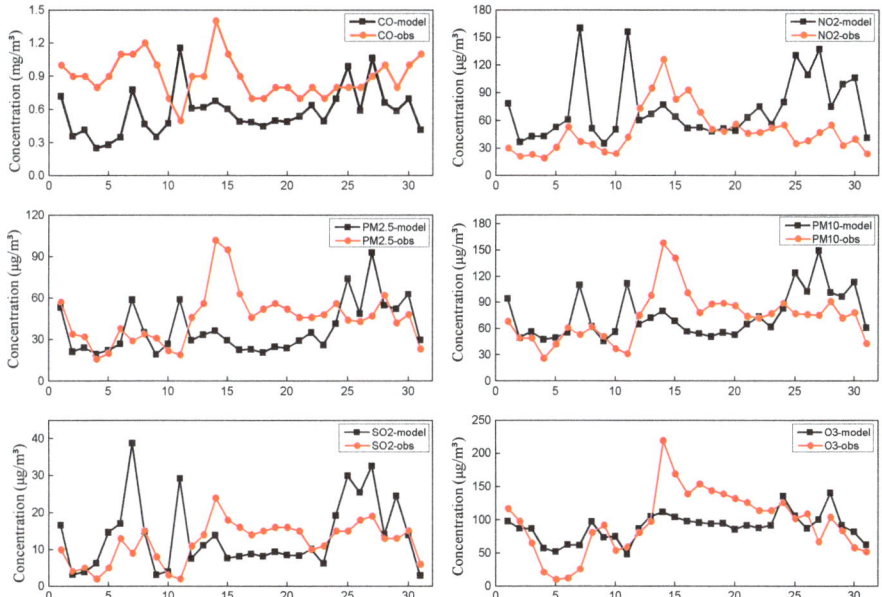

Figure 5. Comparisons of simulated and observed CO, NO_2, $PM_{2.5}$, PM_{10}, SO_2, and O_3. (Daily statistical data in October 2015 in Haizhu district; O_3: daily 8-h maximum concentration; other pollutants: daily-averaged concentration).

In the present study, we follow the relevant air quality modelling studies, e.g., those studies included in Table S7, to use the statistic tools to evaluate the WRF-Chem model performance. While the statistic method can provide some insights into the model abilities, it would help if some advanced diagnostic tools, e.g., the simulation error apportioning techniques proposed by Solazzo et al. (2017a) [83] and Solazzo et al. (2017b) [84], are adopted in future study to identify the critical processes that demand most urgent attention.

3.3. Effect of 13th FYP on Overall Air Quality in Haizhu in 2020

Since the pollutant concentrations in October 2015 were very similar to those in the whole year, the simulated results in the four 2020 scenarios in October are used for evaluating the effect of the 13th FYP on the air quality in Haizhu by comparing to the simulation for October 2015 (Figure 6).

Figure 6 shows that simulated concentrations of five pollutant species (CO, NO_2, $PM_{2.5}$, PM_{10}, and SO_2) in scenario 2020C are lower than pollutants concentrations simulated in 2015. For O_3, 2020 scenarios' simulation results are higher than 2015 result. CO, $PM_{2.5}$, PM_{10}, and SO_2 concentrations in 2020 scenarios A, B, D are higher than the pollutants concentration in 2015 simulation. NO_2 concentrations in the four 2020 scenarios are slightly lower than the concentration in 2015 simulation.

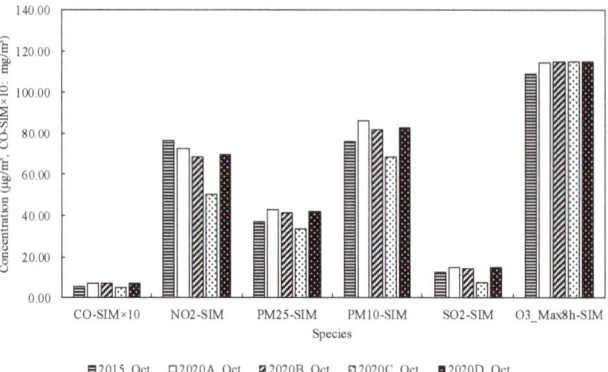

Figure 6. Comparison between simulation on October 2015 and 2020 scenarios simulation on October in Haizhu district (monthly-averaged surface concentrations).

It is noticeable that O_3 concentrations in all of the four 2020 scenarios are higher than those in 2015 simulation, which means that ozone pollution will probably get worse by the end of 13th FYP. Although scenario 2020A had the slightly less pollution than other scenarios had, the level of O_x ($O_3 + NO_2$) in scenario 2020A was higher than that of other scenarios (Figure 6), which means that atmospheric oxidation in scenario 2020A was more elevated [13,85]. It would cause the more intensive formation of secondary aerosols, which would lead to regional pollution [86]. Meanwhile, the level of O_x in scenario 2020C was the lowest.

In Section 3.1, the observed concentrations of $PM_{2.5}$, PM_{10}, and NO_2 in 2015 exceeded the targets. Meanwhile, the WRF-Chem simulations suggested that from 2015 to 2020, the concentrations of $PM_{2.5}$ and PM_{10} would increase in scenarios 2020A, 2020B, and 2020D. Therefore, the real concentrations of aerosols in 2020 could be higher than those in 2015 and further exceed the targets. The observed O_3-8h concentration in 2015 did not exceed the target (Section 3.1), but with the increasing trend from 2015 to 2020 as simulated in all scenarios, the real O_3-8h concentration in 2020 should be considered as a cause for concern.

According to the simulations, by the end of 13th FYP, if the emission control policy just follows 12th FYP policy (scenario 2020A), the air pollution would increase and the control on all emission sectors in Haizhu (scenario 2020B) and control just on vehicle emissions in Haizhu (scenario 2020D) would have a similar impact on air quality, and could not improve air quality very much. The scenario 2020C (the comprehensive emission control measures taken in the whole Guangzhou city) would have the better effect on the emission control than other scenarios would do. Among the four 2020 scenarios, the scenario 2020C is the better scenario.

According to Figure 6, pollutant concentrations between scenario 2020B and 2020D are similar. The effect of controlling every emission sectors in Haizhu district would be almost equal with the effect of only controlling the traffic emission in Haizhu district, suggesting that traffic source would be a key source in Haizhu district.

Wang et al. (2016) [13] used WRF-CMAQ model (a base case in 2010, two cases in 2020) and evaluated the impact of emission control measures on the air quality in the PRD region, and their results showed that reducing NO_x emissions would cause rising $PM_{2.5}$ levels in certain areas, although it would benefit with reduction of regional $PM_{2.5}$. They also noted that O_3 formation in PRD was generally VOCs-limited and cutting VOCs emission could benefit the reduction of overall O_3. Liu et al. (2017) [14] assessed the influence of vehicle emission control measures for 2015–2020 in the PRD region on air quality and climate and their results suggested that, if vehicle emission wasn't controlled, most air pollutants and GHG would increase by 20–64% by 2020.

3.4. Effect of 13th FYP on Air Quality during Heavy Pollution Episodes in Haizhu in 2020

To evaluate the effect of 13th FYP on NO_2 and $PM_{2.5}$ in the winter heavy pollution episode and O_3 in the summer heavy pollution episode, we conducted the WRF-Chem simulations with the four 2020 pollution control scenarios (2020A, 2020B, 2020C, and 2020D). Table 3 shows the changes in pollutant concentrations of the four scenarios. Since Table 3 has summarized the overall impacts (in numbers) of each emission reduction scenarios on the concentration of $PM_{2.5}$, NO_2, and O_3, we will present the spatial variation of the impact in Figures 7–9 in the following text. According to the results of Section 3.2, the simulation during the observational peak was underestimated compared to the measurement, implying that the actual concentrations of pollutants in 2020 could be higher than our simulations.

Table 3. Changes in pollutant concentrations in Haizhu district of scenarios (A, B, C, and D are scenarios 2020A, 2020B, 2020C, and 2020D, respectively; winter episode: 14–28 January; summer episode: 3–8 August).

	Species	Changes in Pollutant Concentrations		
		(B−A)/A	(C−A)/A	(D−A)/A
Winter episode	$PM_{2.5}$	−7.15%	−23.42%	−3.18%
	NO_2	−6.46%	−28.29%	−4.59%
Summer episode	O_3	0.45%	0.35%	0.14%

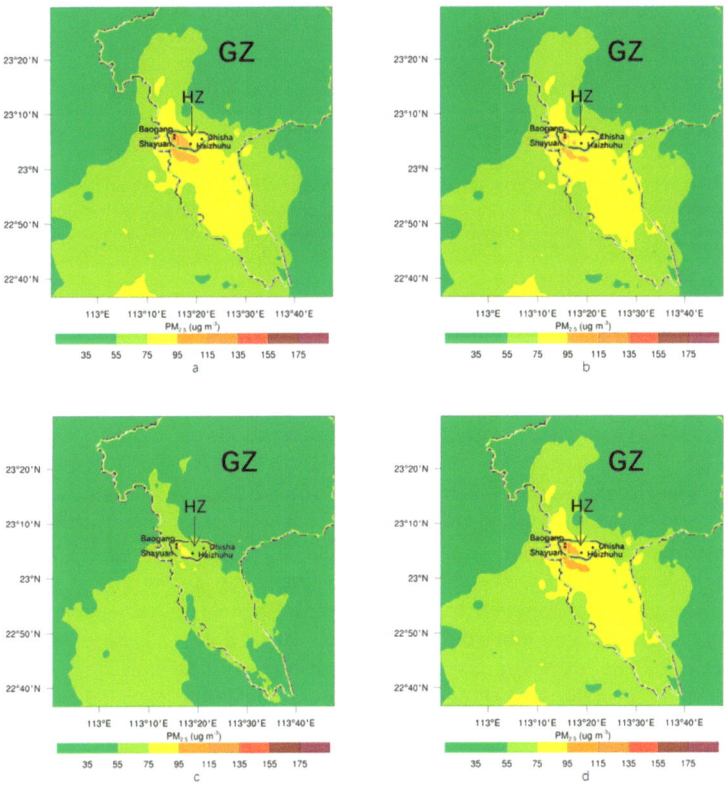

Figure 7. Maps of the simulated mean $PM_{2.5}$ concentration distribution during the period 0000 BJT, 20 January, to 2300 BJT, 21 January, over Haizhu for 2020 scenarios (a: 2020A; b: 2020B; c: 2020C; d: 2020D; HZ: Haizhu; GZ: Guangzhou; BJT: Beijing Time).

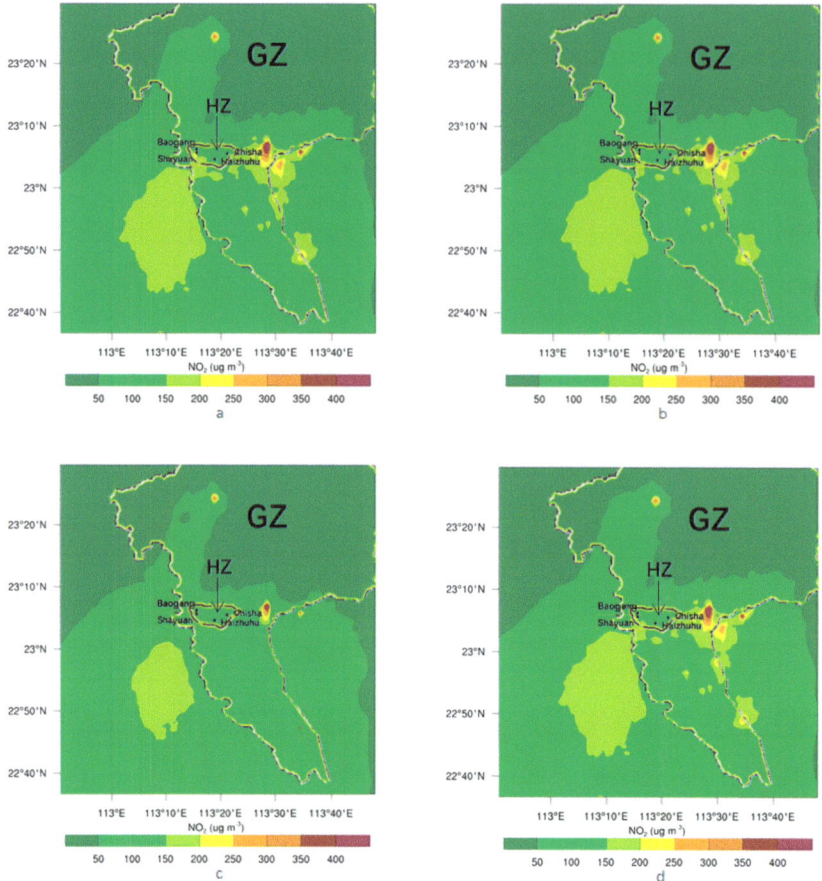

Figure 8. Maps of the simulated mean NO_2 concentration distribution during the period 0000 BJT, 20 January, to 2300 BJT, 21 January, over Haizhu for 2020 scenarios (a: 2020A; b: 2020B; c: 2020C; d: 2020D; HZ: Haizhu; GZ: Guangzhou).

(1) Effect on $PM_{2.5}$ and NO_2 in winter heavy pollution episode

For the simulated concentrations of $PM_{2.5}$ and NO_2 in Shayuan station, the concentrations of $PM_{2.5}$ and NO_2 were in the order of 2020A > 2020D > 2020B > 2020C. Comprehensive emission control measures taken in the whole of Guangzhou (2020C) would be able to reduce concentrations of $PM_{2.5}$ and NO_2 two times more than those just in Haizhu (2020B). In previous research, emission control from outside of Beijing contributed 8–14% to the improvement of the air quality in Beijing during APEC [57]. Therefore, regional control policies are needed. Regional control measures in a whole city could affect the air quality of the urban district, even if the district is just an isolated island. Referring to the measures in Haizhu, the effect of vehicle emission control (2020D) on $PM_{2.5}$ (NO_2) reduction would be ~44% (~71%) of the effect of comprehensive measures (2020B), meaning that vehicles are the major $PM_{2.5}$ (NO_2) emission source in Haizhu. Vehicle emission is a typical problem in urban areas of China [87]. Previous studies showed that vehicle emission accounted for 21%, 10%, and 19.3% of the total $PM_{2.5}$ in Dongguan (2014) [68], Guangzhou (2014) [88], and PRD (2012) [89], respectively.

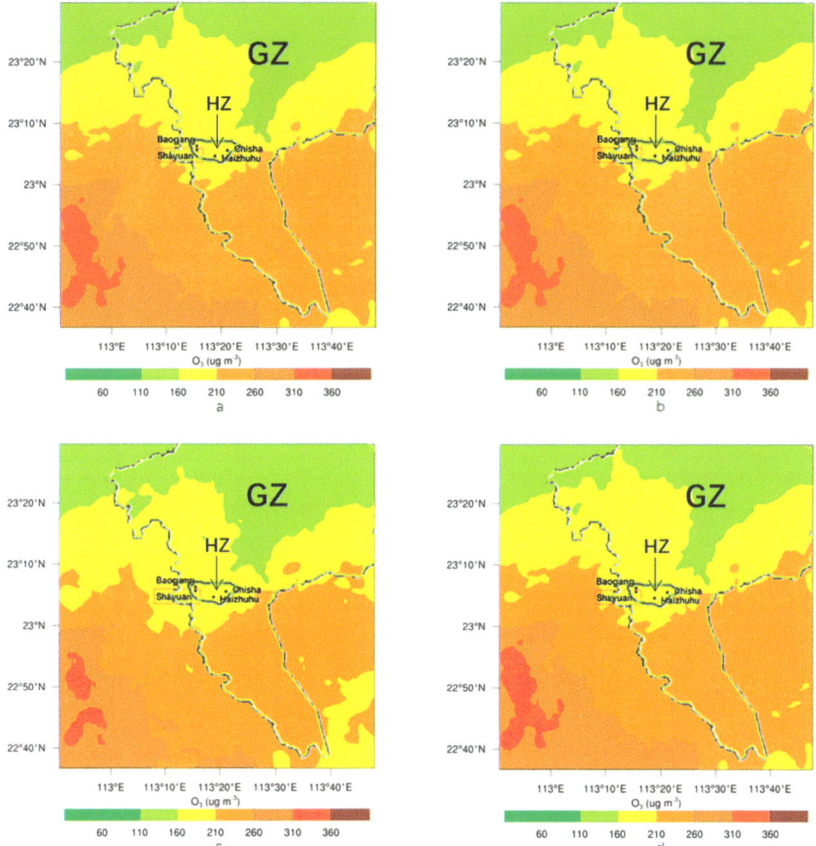

Figure 9. Maps of the simulated O_3 mean daily 8 h maximum concentration distribution over Haizhu for 2020 scenarios on 6 August (a: 2020A; b: 2020B; c: 2020C; d: 2020D; HZ: Haizhu; GZ: Guangzhou).

In the worst meteorological condition, $PM_{2.5}$ pollution on Haizhu district would be improved, as shown in Figures 7c and 7b, which means that the effects on regional air quality of scenarios 2020C and 2020B were relatively significant. In particular, compared with 2020D (Figure 7d), the improvement on regional air quality in 2020B was more obvious, although the $PM_{2.5}$ pollution in the particular station (Shayuan) seemed to have no significant difference. In Figure 8, NO_2 pollution shows a similar situation. It indicates that regional and comprehensive measures could improve regional air quality during heavy pollution days.

(2) Effect on O_3 in summer heavy pollution episode

For the Shayuan station, the O_3 pollution in scenario 2020B would be higher than in other scenarios, and the O_3 pollution in the other scenarios were in the order 2020C > 2020D > 2020A (Table 3). The simulation results suggested that the O_3 level increases from 2015 to 2020 in all emission scenarios even in 2020C in which the rest of air pollutants decrease. Such changes of O_3 concentration is probably because the Haizhu district is a VOC-limited region in which the O_3 level increases (decreases) with the reduction of NO_x (VOCs) emission (e.g., Sillman, 1999 [90]). There are generally two pathways that can be adopted to relieve the O_3 pollution in VOC-limited region like Haizhu district. (1) To significantly reduce the NO_x level and push the Haizhu district into NO_x-limited region in which O_3 level decreases with the reduction of NO_x and VOCs. However, in this pathway, there will be a period in which O_3 level significantly increases before it starts to decrease. (2) To adopt a proper reduction ratio of

VOCs/NO$_x$ and flatten the O$_3$ increase to the largest extent during the transient from VOC-limited region to NO$_x$-limited region. But this pathway requires a great deal of effort to determine a favourable reduction ratio of VOCs/NO$_x$ and cut the emission of VOCs from various sectors.

The findings in the present study highlight the complexity and nonlinear chemistry of O$_3$ formation and call for further investigations. For example, the reduction ratio of VOCs/NO$_x$ for anthropogenic sources was generally suggested to be 1:2 in PRD [10]. However, considering the change of precursor emissions and meteorological conditions, the ratio of VOCs/NO$_x$ might need to be further studied. Besides, the NO$_2$ is overpredicted in the present study implying that the VOC-limited nature in Haizhu district might be overpredicted, although there is very low chance that Haizhu district is a NOx-limited region in 2015 because the NO$_x$ is still at an elevated level and this region (PRD) has been repeatedly diagnosed as a VOC-limited region (e.g., Xue et al., 2014 [91]). Moreover, the formation of O$_3$ depends on the local mixture of NO$_x$ and VOCs. Therefore, the location/sector of the emission of NO$_x$ and VOC, apart from the overall reduction of NO$_x$ and VOC in an area, might have an impact on the changes of O$_3$. Although we have made some effort on such topic by evaluating the effect of only controlling traffic emission on air quality (scenario 2020D), it would be of interest to adopt more sophisticated tools, e.g., a tagging technique (Grewe, 2013 [92]), to evaluate the effect of various combinations of emission reduction in sectors in future study.

The simulated O$_3$-8h distribution on 6 August is shown in Figure 9 to demonstrate the difference of regional O$_3$ pollution in Haizhu district. Ozone pollution would be relieved in scenario 2020C (Figure 9c) compared with that in other scenarios (Figure 9a,b,d). It is noticeable that ozone pollution in the southwest area (red area, suburban and rural area) in 2020C (Figure 9c) could be smaller than those in other scenarios (Figure 9a,b,d), which may be because of the suitable ratio of precursors in that area of scenario 2020C, although the pollution in Haizhu (urban area) seemed to have no significant difference. The difference of regional O$_3$ pollution shows that in the southwest area (suburban and rural area), the policy in scenario 2020C could decrease the O$_3$ significantly (less red area), and in other regions, e.g., in Haizhu (urban area), the effect is less noticeable, which indicated that the control of ozone pollution should be taken according to local conditions, i.e., measures in urban areas should be different from those in rural areas. We note that transboundary transport of O$_3$ and its precursors (CO, VOCs, and NO$_x$) due to mesoscale dynamics complicates O$_3$ pollution control and requires strong and efficient cooperation among the adjacent regions.

4. Conclusions

In this study, the air quality in the year of 2015 and during pollution episodes in Haizhu district were analysed, and the impacts of emission control scenarios by the year of 2020 on air quality were evaluated using the WRF-Chem numerical simulation.

For the air quality in Haizhu 2015, the annually-averaged concentrations of PM$_{2.5}$, PM$_{10}$, SO$_2$, and NO$_2$ were higher than those in the entire Guangzhou city. O$_3$-8h was slightly lower than that in Guangzhou. Pollution episodes in Haizhu in 2015 primarily occurred in summer and winter. The typical winter pollution episode (14 to 28 January) was a process of gradual accumulation of pollution and dissipation, with NO$_2$ or PM$_{2.5}$ as the key pollutants, associated with the wind. The heavy pollution episode in the hot season (3 to 8 August) was a process of pollution that occurred and disappeared quickly, with O$_3$ as the key pollutant, due to suitable local pollution and strong sunshine.

The WRF-Chem simulation generally captured the observed chemical characteristics, suggesting that the WRF-Chem model could be used to simulate air quality in the research case.

Emission control scenario 2020C (comprehensive measures taken in the whole of Guangzhou city) would improve air quality more significantly than other scenarios (measures taken in Haizhu) under all conditions (heavy pollution conditions and annual level). For urban areas, scenario 2020D (vehicle emission control) would account for more than half of the influence of 2020B (all source emission controls) on air quality. By the end of the 13th FYP, it is noticeable that O$_3$ pollution would increase, which indicates that the control ratio of VOCs and NO$_x$ may be unfavourable and requires further assessment. It would be of interest to perform simulations with the meteorology/chemistry interactions to investigate the influence of policies on O_3 and other air pollutants in future research.

Our study suggested that control measures should be strengthened for NO_2, $PM_{2.5}$, and PM_{10}, and control ratio of VOCs and NO_x should be adjusted for controlling O_3. The urban area should focus on vehicle emission control, strengthen regional cooperation on pollution control, and establish short-term measures for heavy pollution conditions.

Supplementary Materials: The following are available online at http://www.mdpi.com/2076-3417/10/15/5276/s1, Figure S1: Research framework, Figure S2: WRF-Chem simulation domain, Table S1: Emission sectors, corresponding inventory technical guidelines and special allocation rules, Table S2: Description of emission control scenarios, Table S3: Detailed information for pollution episodes observed in Haizhu monitoring stations, 2015, Table S4: Statistical data for heavy pollution episode in Haizhu district during 14–28 January 2015, Table S5: Statistical data for heavy pollution episode in Haizhu district during 3–8 August 2015, Table S6: Performance statistics for meteorological simulation (hourly statistical data in October 2015), Table S7: Statistical comparison of model evaluation in this paper and previous study.

Author Contributions: J.Z. and H.D. designed the study. Y.L. and C.F. collected and analysed data. T.G. and L.L. did emission data preparation and generated the figures. R.O. and M.W. did the policy investigation. J.Z. contributed to literature search, numerical simulation and writing the manuscript. All authors have read and agreed to the published version of the manuscript.

Funding: This research was funded by the National Key Research Program (No. 2016YFC0202005), the Natural Science Foundation of China (No. 41975165; No. 11802347), the Fundamental Research Funds for the Central Universities (No. 20lgpy180), and the social science and technology development (key) project of Dongguan city (No. 2019507101161). The APC was funded by the National Key Research Program (No. 2016YFC0202005).

Acknowledgments: The author greatly appreciated Guangzhou Haizhu District Environmental Monitoring Station for providing the data in this work and Tsinghua University for providing MEIC. Thanks also go to Qinyi Li in Spanish National Research Council and Jinpu Zhang in Guangzhou Environmental Monitoring Center for valuable suggestions and fruitful discussion.

Conflicts of Interest: The authors declare no conflict of interest.

References

1. Gulia, S.; Shiva, N.S.M.; Khare, M.; Khanna, I. Urban air quality management—A review. *Atmos. Pollut. Res.* **2015**, *6*, 286–304. [CrossRef]
2. SC (State Council of People's Republic of China. Air Pollution Control Action Plan. 2013. Available online: http://www.gov.cn/zwgk/2013-09/12/content_2486773.htm (accessed on 23 November 2018).
3. CAA (Clean Air Asia). China Air 2015—Air Pollution Prevention and Control Progress in Chinese Cities. 2015. Available online: http://cleanairasia.org/wp-content/uploads/2016/03/ChinaAir2015-report.pdf (accessed on 23 November 2018).
4. CAA (Clean Air Asia). China Air 2016—Air Pollution Prevention and Control Progress in Chinese Cities. 2016. Available online: http://www.allaboutair.cn/a/cbw/bg/2016/0822/472.html (accessed on 23 November 2018).
5. BJMBEE (Beijing Municipal Bureau of Ecoloy and Environment). Beijing Environmental Statement 2015. 2015. Available online: http://www.bjepb.gov.cn/bjhrb/xxgk/ywdt/hjzlzk/hjzkgb65/index.html (accessed on 23 November 2018).
6. SHEP (Shanghai Environmental Protection). Shanghai Air Quality Report. 2015. Available online: http://www.sepb.gov.cn/hb/fa/cms/shhj/list_login.jsp?channelId=5157 (accessed on 23 November 2018).
7. Zhan, J.; Liu, Y.; Lin, L.; Ding, H.; Xu, W. Investigation of spatial and temporal association of $PM_{2.5}$ pollution during the winter of 2014 in typical cities of Pearl River Delta. *Res. Environ. Sci.* **2017**, *30*, 110–120. [CrossRef]
8. Ma, X.; Jia, H. Particulate matter and gaseous pollutions in three megacities over China: Situation and implication. *Atmos. Environ.* **2016**, *140*, 476–494. [CrossRef]
9. Li, M.; Song, Y.; Mao, Z.; Liu, M.; Huang, X. Impacts of thermal circulations induced by urbanization on ozone formation in the Pearl River Delta region, China. *Atmos. Environ.* **2016**, *127*, 382–392. [CrossRef]
10. Wang, T.; Xue, L.; Brimblecombe, P.; Lam, Y.F.; Li, L.; Zhang, L. Ozone pollution in China: A review of concentrations, meteorological influences, chemical precursors, and effects. *Sci. Total Environ.* **2017**, *575*, 1582–1596. [CrossRef]

11. SC (State Council of People's Republic of China). Environmental Protection 13th Five-Year Plan. 2016. Available online: http://www.gov.cn/zhengce/content/2016-12/05/content_5143290.htm (accessed on 23 November 2018).
12. PGGZ (People's Government of Guangzhou). Guangzhou Environmental Protection 13th Five-Year Plan. 2016. Available online: http://www.gz.gov.cn/gzgov/s2812/201612/467e946dfcb048c0a7bff56655808c60.shtml (accessed on 23 November 2018).
13. Wang, N.; Lyu, X.P.; Deng, X.J.; Guo, H.; Deng, T.; Li, Y.; Yin, C.Q.; Li, F.; Wang, S.Q. Assessment of regional air quality resulting from emission control in the Pearl River Delta region, southern China. *Sci. Total Environ.* **2016**, *573*, 1554–1565. [CrossRef]
14. Liu, Y.H.; Liao, W.Y.; Lin, X.F.; Li, L.; Zeng, X.L. Assessment of Co-benefits of vehicle emission reduction measures for 2015–2020 in the Pearl River Delta region, China. *Environ. Pollut.* **2017**, *223*, 62–72. [CrossRef]
15. Maji, K.J.; Dikshit, A.K.; Arora, M.; Deshpande, A. Estimating premature mortality attributable to $PM_{2.5}$ exposure and benefit of air pollution control policies in China for 2020. *Sci. Total Environ.* **2018**, *612*, 683–693. [CrossRef]
16. Yang, X.; Teng, F. The air quality co-benefit of coal control strategy in China. *Resour. Conserv. Recycl.* **2016**, *129*, 373–382. [CrossRef]
17. Wang, Z.; Pan, L.; Li, Y.; Zhang, D.; Ma, J.; Sun, F.; Xu, W.; Wang, X. Assessment of air quality benefits from the national pollution control policy of thermal power plants in China: A numerical simulation. *Atmos. Environ.* **2015**, *106*, 288–304. [CrossRef]
18. Qiu, X.; Duan, L.; Cai, S.; Yu, Q.; Wang, S.; Chai, F.; Gao, J.; Li, Y.; Xu, Z. Effect of current emission abatement strategies on air quality improvement in China: A case study of Baotou, a typical industrial city in Inner Mongolia. *J. Environ. Sci.* **2017**, *57*, 383–390. [CrossRef] [PubMed]
19. Cai, S.; Wang, Y.; Zhao, B.; Wang, S.; Chang, X.; Hao, J. The impact of the "air pollution prevention and control action plan" on PM2. 5 concentrations in jing-jin-ji region during 2012–2020. *Sci. Total Environ.* **2017**, *580*, 197–209. [CrossRef] [PubMed]
20. Li, M.; Patiño-Echeverri, D. Estimating benefits and costs of policies proposed in the 13th FYP to improve energy efficiency and reduce air emissions of China's electric power sector. *Energy Policy* **2017**, *111*, 222–234. [CrossRef]
21. Wei, W.; Li, P.; Wang, H.; Song, M. Quantifying the effects of air pollution control policies: A case of Shanxi province in China. *Atmos. Pollut. Res.* **2017**, *9*, 429–438. [CrossRef]
22. Guo, J.; He, J.; Liu, H.; Miao, Y.; Liu, H.; Zhai, P. Impact of various emission control schemes on air quality using WRF-Chem during APEC China 2014. *Atmos. Environ.* **2016**, *140*, 311–319. [CrossRef]
23. Xu, H.M.; Tao, J.; Ho, S.S.H.; Ho, K.F.; Cao, J.J.; Li, N.; Chow, J.C.; Wang, G.H.; Han, Y.M.; Zhang, R.J.; et al. Characteristics of fine particulate non-polar organic compounds in Guangzhou during the 16th Asian Games: Effectiveness of air pollution controls. *Atmos. Environ.* **2013**, *76*, 94–101. [CrossRef]
24. Shen, X.J.; Sun, J.Y.; Zhang, X.Y.; Zhang, Y.M.; Zhang, L.; Fan, R.X.; Zhang, Z.X.; Zhang, X.L.; Zhou, H.G.; Zhou, L.Y.; et al. The influence of emission control on particle number size distribution and new particle formation during China's V-Day parade in 2015. *Sci. Total Environ.* **2016**, *573*, 409–419. [CrossRef]
25. Tan, J.H.; Duan, J.C.; Chen, D.H.; Wang, X.H.; Guo, S.J.; Bi, X.H.; Sheng, G.Y.; He, K.B.; Fu, J.M. Chemical characteristics of haze during summer and winter in Guangzhou. *Atmos. Res.* **2009**, *94*, 238–245. [CrossRef]
26. Wang, M.; Cao, C.; Li, G.; Singh, R.P. Analysis of a severe prolonged regional haze episode in the Yangtze River Delta, China. *Atmos. Environ.* **2015**, *102*, 112–121. [CrossRef]
27. Zhang, L.; Wang, T.; Lv, M.; Zhang, Q. On the severe haze in Beijing during January 2013: Unraveling the effects of meteorological anomalies with WRF-Chem. *Atmos. Environ.* **2015**, *104*, 11–21. [CrossRef]
28. Ding, A.; Wang, T.; Zhao, M.; Wang, T.; Li, Z. Simulation of sea-land breezes and a discussion of their implications on the transport of air pollution during a multi-day ozone episode in the Pearl River Delta of China. *Atmos. Environ.* **2004**, *38*, 6737–6750. [CrossRef]
29. Shen, J.; Zhang, Y.; Wang, X.; Li, J.; Chen, H.; Liu, R.; Zhong, L.; Jiang, M.; Yue, D.; Chen, D.; et al. An ozone episode over the Pearl River Delta in October 2008. *Atmos. Environ.* **2015**, *122*, 852–863. [CrossRef]
30. Zhao, H.; Wang, S.; Wang, W.; Liu, R.; Zhou, B. Investigation of ground-level ozone and high-pollution episodes in a megacity of Eastern China. *PLoS ONE.* **2015**, *10*, e0131878. [CrossRef] [PubMed]

31. Xu, J.; Zhang, Y.; Fu, J.S.; Zheng, S.; Wang, W. Process analysis of typical summertime ozone episodes over the Beijing area. *Sci. Total Environ.* **2008**, *399*, 147–157. [CrossRef]
32. Qu, Y.; An, J.; Li, J.; Chen, Y.; Li, Y.; Liu, X.; Hu, M. Effects of NOx and VOCs from five emission sources on summer surface O_3 over the Beijing-Tianjin-Hebei region. *Adv. Atmos. Sci.* **2014**, *31*, 787–800. [CrossRef]
33. MEE (Ministry of Ecology and Environment of People's Republic of China). China Vehicle Environmental Management Annual Report 2017. 2017. Available online: http://dqhj.mee.gov.cn/jdchjgl/zhgldt/201706/P020170605550637870889.pdf (accessed on 23 November 2018).
34. EPGD (Environmental Protection of Guangdong Province). Guangzhou Environmental Protection's Implementation of Grid Control. 2016. Available online: http://www.gdep.gov.cn/zwxx_1/hbxx/201607/t20160725_213142.html (accessed on 23 November 2018).
35. EPGD (Environmental Protection of Guangdong Province). 2017 Guangdong Cities Air Quality Ranking Released, Guangdong Achieved Targets for Three Years. 2018. Available online: http://mp.weixin.qq.com/s/YKKBj9pncWaq67lFq0l6mQ (accessed on 23 November 2018).
36. Wang, J.; Kwan, M.P.; Ma, L. Delimiting service area using adaptive crystal-growth Voronoi diagrams based on weighted planes: A case study in Haizhu District of Guangzhou in China. *Appl. Geogr.* **2014**, *50*, 108–119. [CrossRef]
37. Zhang, Z.; Li, H.; Liu, H.; Ni, R.; Li, J.; Deng, L.; Lu, D.; Cheng, X.; Duan, P.; Li, W. A preliminary analysis of the surface chemistry of atmospheric aerosol particles in a typical urban area of Beijing. *J. Environ. Sci.* **2016**, *47*, 71–81. [CrossRef]
38. Ling, Z.H.; Zhao, J.; Fan, S.J.; Wang, X.M. Sources of formaldehyde and their contributions to photochemical O_3 formation at an urban site in the Pearl River Delta, southern China. *Chemosphere* **2017**, *168*, 1293–1301. [CrossRef]
39. HZEPB (Haizhu Environmental Protection Bureau). Haizhu Environmental Protection Bureau 2015 Work Report and 2016 Key Work. 2016. Available online: http://zwgk.haizhu.gov.cn/HZ12/201605/t20160526_346735.html (accessed on 23 November 2018).
40. HZEPB (Haizhu Environmental Protection Bureau). Haizhu Environmental Protection Bureau 2016 Work Report and 2017 Key Work. 2017. Available online: http://zwgk.haizhu.gov.cn/HZ12/201703/t20170324_392921.html (accessed on 23 November 2018).
41. CNGIPSP (China National Geographic Information Public Service Platform). GS(2017)508-1100471 Map Data. 2017. Available online: http://map.tianditu.com/ (accessed on 23 November 2018).
42. Gong, J.; Hu, Z.; Chen, W.; Liu, Y.; Wang, J. Urban expansion dynamics and modes in metropolitan Guangzhou, China. *Land Use Policy.* **2018**, *72*, 100–109. [CrossRef]
43. He, K. Multi-resolution Emission Inventory for China (MEIC): Model framework and 1990–2010 anthropogenic emissions. In *AGU Fall Meeting Abstracts*; Fall Meeting 2012 of American Geophysical Union; American Geophysical Union: San Francisco, CA, USA, 3–7 December 2012; abstract id. A32B-05; Available online: http://adsabs.harvard.edu/abs/2012AGUFM.A32B..05H (accessed on 23 November 2018).
44. Guenther, A.; Zimmerman, P.; Wildermuth, M. Natural volatile organic compound emission rate estimates for US woodland landscapes. *Atmos. Environ.* **1994**, *28*, 1197–1210. [CrossRef]
45. Shaw, W.J.; Allwine, K.J.; Fritz, B.G.; Rutz, F.C.; Rishel, J.P.; Chapman, E.G. An evaluation of the wind erosion module in DUSTRAN. *Atmos. Environ.* **2008**, *42*, 1907–1921. [CrossRef]
46. Gong, S.L.; Barrie, L.A.; Lazare, M. Canadian Aerosol Module (CAM): A size-segregated simulation of atmospheric aerosol processes for climate and air quality models: 2. Global sea-salt aerosol and its budgets. *J. Geophys. Res. Atmos.* **2002**, *107*, 4779. [CrossRef]
47. Lei, Y.; Zhang, Q.; He, K.B.; Streets, D.G. Primary anthropogenic aerosol emission trends for China, 1990–2005. *Atmos. Chem. Phys.* **2011**, *11*, 931–954. [CrossRef]
48. Li, Q.; Zhang, L.; Wang, T.; Tham, Y.J.; Ahmadov, R.; Xue, L.; Zhang, Q.; Zheng, J. Impacts of heterogeneous uptake of dinitrogen pentoxide and chlorine activation on ozone and reactive nitrogen partitioning: Improvement and application of the WRF-Chem model in southern China. *Atmos. Chem. Phys.* **2016**, *16*, 14875. [CrossRef]
49. PGGZ (People's Government of Guangzhou). Guangzhou Air Quality Targets Plan (2016–2025). 2017. Available online: http://www.gz.gov.cn/gzgov/s2811/201712/57727a1d77354f5dbc22bb5831aa7d93.shtml (accessed on 23 November 2018).

50. MEE (Ministry of Ecology and Environment of People's Republic of China). Announcement on the publication of four technical guidelines (Technical Guidelines for the Primary Source Emission Inventory of Atmospheric Fine Particles (Trial), Technical Guidelines for the Emission Inventory of Atmospheric Volatile Organic Compounds (Trial), Technical Guidelines for the Emission Inventory of Atmospheric Ammonia Source (Trial), and Technical Guidelines for Priority Control of Atmospheric Pollution Sources (Trial)). 2014. Available online: http://www.mee.gov.cn/gkml/hbb/bgg/201408/t20140828_288364.htm (accessed on 23 November 2018).
51. MEE (Ministry of Ecology and Environment of People's Republic of China). Announcement on the Publication of Five Technical Guidelines (Technical Guidelines for the Primary Source Emission Inventory of Inhalable Particulate Matter (Trial), Technical Guidelines for the Air Pollutant Emission Inventory for Road Vehicles (Trial), Technical Guidelines for the Air Pollutant Emission Inventory of Non-Road Mobile Source (Trial), Technical Guidelines for the Air Pollutant Emission Inventory for Biomass Combustion Sources (Trial), Technical Guidelines for the Emission Inventory of Dust Particles Discharge (Trial)). 2014. Available online: http://www.mee.gov.cn/gkml/hbb/bgg/201501/t20150107_293955.htm (accessed on 23 November 2018).
52. Zheng, J.Y.; Zhang, L.J.; Che, W.W.; Zheng, Z.Y.; Yin, S.S. A highly resolved temporal and spatial air pollutant emission inventory for the Pearl River Delta region, China and its uncertainty assessment. *Atmos. Environ.* **2009**, *43*, 5112–5122. [CrossRef]
53. Zhao, X.Y.; Hu, Q.H.; Wang, X.M.; Ding, X.; He, Q.F.; Zhang, Z.; Shen, R.Q.; Lü, S.J.; Liu, T.Y.; Fu, X.X.; et al. Composition profiles of organic aerosols from Chinese residential cooking: Case study in urban Guangzhou, south China. *J. Atmos. Chem.* **2015**, *72*, 1–18. [CrossRef]
54. PGGZ (People's Government of Guangzhou). Guangzhou Transportation Development 13th Five-Year Plan. 2016. Available online: http://www.gz.gov.cn/gzgov/s2812/201611/a8ee48d726b649caaba3e4f12572feae.shtml (accessed on 23 November 2018).
55. NOAA (National Oceanic and Atmospheric Administration). Weather Research and Forecasting Model Coupled to Chemistry (WRF-Chem). 2017. Available online: https://ruc.noaa.gov/wrf/wrf-chem/ (accessed on 23 November 2018).
56. NCAR (National Center for Atmospheric Research). Atmospheric Chemistry Observation & Modeling: WRF-CHEM. 2018. Available online: https://www2.acom.ucar.edu/wrf-chem (accessed on 23 November 2018).
57. He, H.; Tie, X.; Zhang, Q.; Liu, X.; Gao, Q.; Li, X.; Gao, Y. Analysis of the causes of heavy aerosol pollution in Beijing, China: A case study with the WRF-Chem model. *Particuology* **2015**, *20*, 32–40. [CrossRef]
58. Emmons, L.K.; Walters, S.; Hess, P.G.; Lamarque, J.F.; Pfister, G.G.; Fillmore, D.; Granier, C.; Guenther, A.; Kinnison, D.; Laepple, T.; et al. Description and evaluation of the Model for Ozone and Related chemical Tracers, version 4 (MOZART-4). *Geosci. Model Dev.* **2010**, *3*, 43–67. [CrossRef]
59. Stockwell, W.R.; Middleton, P.; Chang, J.S.; Tang, X. The second generation regional acid deposition model chemical mechanism for regional air quality modeling. *J. Geophys. Res. Atmos.* **1990**, *95*, 16343–16367. [CrossRef]
60. Ackermann, I.J.; Hass, H.; Memmesheimer, M.; Ebel, A.; Binkowski, F.S.; Shankar, U.M.A. Modal aerosol dynamics model for Europe: Development and first applications. *Atmos. Environ.* **1998**, *32*, 2981–2999. [CrossRef]
61. Schell, B.; Ackermann, I.J.; Hass, H.; Binkowski, F.S.; Ebel, A. Modeling the formation of secondary organic aerosol within a comprehensive air quality model system. *J. Geophys. Res. Atmos.* **2001**, *106*, 28275–28293. [CrossRef]
62. MEE (Ministry of Ecology and Environment of People's Republic of China). Technical Regulation on Ambient Air Quality Index (on trial) (HJ 633-2012). 2016. Available online: http://www.gov.cn/zwgk/2012-03/02/content_2081374.htm (accessed on 23 November 2018).
63. EPGD (Environmental Protection of Guangdong Province). Guangdong-Hong Kong-Macao Pearl River Delta Regional Air Quality Monitoring Network: Report of Monitoring Results 2015. 2016. Available online: http://www.gdep.gov.cn/hjjce/kqjc/index_1.html (accessed on 23 November 2018).
64. GZEP (Guangzhou Environmental Protection Bureau). Guangzhou Air Quality Report 2015. 2016. Available online: http://www.gzepb.gov.cn/zwgk/hjgb/ (accessed on 23 November 2018).
65. WHO (World Health Organization). WHO Air Quality Guidelines (Global Update 2005). 2005. Available online: http://apps.who.int/iris/bitstream/handle/10665/69477/WHO_SDE_PHE_OEH_06.02_eng.pdf;jsessionid=0757FBBE71232DE20E0BB5E7C64E45A1?sequence=1 (accessed on 23 November 2018).

66. MEE (Ministry of Ecology and Environment of People's Republic of China) Technical Regulations for Urban Environmental Air Quality Ranking. 2018. Available online: http://www.mee.gov.cn/gkml/sthjbgw/bgtwj/201808/W020180815576019704248.pdf (accessed on 23 November 2018).
67. Tong, C.H.M.; Yim, S.H.L.; Rothenberg, D.; Wang, C.; Lin, C.Y.; Chen, Y.D.; Lau, N.C. Assessing the impacts of seasonal and vertical atmospheric conditions on air quality over the Pearl River Delta region. *Atmos. Environ.* **2018**, *180*, 69–78. [CrossRef]
68. Zou, B.B.; Huang, X.F.; Zhang, B.; Dai, J.; Zeng, L.W.; Feng, N.; He, L.Y. Source apportionment of $PM_{2.5}$ pollution in an industrial city in southern China. *Atmos. Pollut. Res.* **2017**, *8*, 1193–1202. [CrossRef]
69. Malek, E.; Davis, T.; Martin, R.S.; Silva, P.J. Meteorological and environmental aspects of one of the worst national air pollution episodes (January, 2004) in Logan, Cache Valley, Utah, USA. *Atmos. Res.* **2006**, *79*, 108–122. [CrossRef]
70. Cui, H.; Chen, W.; Dai, W.; Liu, H.; Wang, X.; He, K. Source apportionment of $PM_{2.5}$ in Guangzhou combining observation data analysis and chemical transport model simulation. *Atmos. Environ.* **2015**, *116*, 262–271. [CrossRef]
71. Song, H.; Wang, K.; Zhang, Y.; Hong, C.; Zhou, S. Simulation and evaluation of dust emissions with WRF-Chem (v3. 7.1) and its relationship to the changing climate over East Asia from 1980 to 2015. *Atmos. Environ.* **2017**, *167*, 511–522. [CrossRef]
72. Gao, J.; Zhu, B.; Xiao, H.; Kang, H.; Hou, X.; Shao, P. A case study of surface ozone source apportionment during a high concentration episode, under frequent shifting wind conditions over the Yangtze River Delta, China. *Sci. Total Environ.* **2016**, *544*, 853–863. [CrossRef]
73. Hong, C.; Zhang, Q.; He, K.; Guan, D.; Li, M.; Liu, F.; Zheng, B. Variations of China's emission estimates: Response to uncertainties in energy statistics. *Atmos. Chem. Phys.* **2017**, *17*, 1227–1239. [CrossRef]
74. Huo, H.; Yao, Z.; Zhang, Y.; Shen, X.; Zhang, Q.; He, K. On-board measurements of emissions from diesel trucks in five cities in China. *Atmos. Environ.* **2012**, *54*, 159–167. [CrossRef]
75. Zheng, B.; Zhang, Q.; Tong, D.; Chen, C.; Hong, C.; Li, M.; Geng, G.; Lei, Y.; Huo, H.; He, K. Resolution dependence of uncertainties in gridded emission inventories: A case study in Hebei, China. *Atmos. Chem. Phys.* **2017**, *17*, 921–933. [CrossRef]
76. Ahmadov, R.; McKeen, S.A.; Robinson, A.L.; Bahreini, R.; Middlebrook, A.M.; De Gouw, J.A.; Meagher, J.; Hsie, E.Y.; Edgerton, E.; Shaw, S.; et al. A volatility basis set model for summertime secondary organic aerosols over the eastern United States in 2006. *J. Geophys. Res. Atmos.* **2012**, *117*, D6. [CrossRef]
77. Li, Q.; Zhang, L.; Wang, T.; Wang, Z.; Fu, X.; Zhang, Q. "New" Reactive Nitrogen Chemistry Reshapes the Relationship of Ozone to Its Precursors. *Environ. Sci. Technol.* **2018**, *52*, 2810–2818. [CrossRef] [PubMed]
78. Banks, R.F.; Baldasano, J.M. Impact of WRF model PBL schemes on air quality simulations over Catalonia, Spain. *Sci. Total Environ.* **2016**, *572*, 98–113. [CrossRef] [PubMed]
79. Shin, H.H.; Hong, S.Y. Intercomparison of planetary boundary-layer parametrizations in the WRF model for a single day from CASES-99. *Bound. Layer Meteorol.* **2011**, *139*, 261–281. [CrossRef]
80. Hu, X.M.; Nielsen-Gammon, J.W.; Zhang, F. Evaluation of three planetary boundary layer schemes in the WRF model. *J. Appl. Meteorol. Climatol.* **2010**, *49*, 1831–1844. [CrossRef]
81. Hong, S.Y.; Noh, Y.; Dudhia, J. A new vertical diffusion package with an explicit treatment of entrainment processes. *Mon. Weather Rev.* **2006**, *134*, 2318–2341. [CrossRef]
82. Schutgens, N.A.; Gryspeerdt, E.; Weigum, N.; Tsyro, S.; Goto, D.; Schulz, M.; Stier, P. Will a perfect model agree with perfect observations? The impact of spatial sampling. *Atmos. Chem. Phys.* **2016**, *16*, 6335–6353. [CrossRef]
83. Solazzo, E.; Bianconi, R.; Hogrefe, C.; Curci, G.; Tuccella, P.; Alyuz, U.; Balzarini, A.; Baró, R.; Bellasio, R.; Bieser, J.; et al. Evaluation and error apportionment of an ensemble of atmospheric chemistry transport modeling systems: Multivariable temporal and spatial breakdown. *Atmos. Chem. Phys.* **2017**, *17*, 3001–3054. [CrossRef]
84. Solazzo, E.; Hogrefe, C.; Colette, A.; Garcia-Vivanco, M.; Galmarini, S. Advanced error diagnostics of the CMAQ and Chimere modelling systems within the AQMEII3 model evaluation framework. *Atmos. Chem. Phys.* **2017**, *17*, 10435–10465. [CrossRef]
85. Koch, R.; Knispel, R.; Elend, M.; Siese, M.; Zetzsch, C. Consecutive reactions of aromatic-OH adducts with NO, NO_2 and O_2: Benzene, naphthalene, toluene, m-and p-xylene, hexamethylbenzene, phenol, m-cresol and aniline. *Atmos. Chem. Phys.* **2007**, *7*, 2057–2071. [CrossRef]

86. Barthelmie, R.J.; Pryor, S.C. Secondary organic aerosols: Formation potential and ambient data. *Sci. Total Environ.* **1997**, *205*, 167–178. [CrossRef]
87. Li, Y.J.; Sun, Y.; Zhang, Q.; Li, X.; Li, M.; Zhou, Z.; Chan, C.K. Real-time chemical characterization of atmospheric particulate matter in China: A review. *Atmos. Environ.* **2017**, *158*, 270–304. [CrossRef]
88. Tao, J.; Zhang, L.; Cao, J.; Zhong, L.; Chen, D.; Yang, Y.; Chen, D.; Chen, L.; Zhang, Z.; Wu, Y.; et al. Source apportionment of PM2.5 at urban and suburban areas of the Pearl River Delta region, south China-With emphasis on ship emissions. *Sci. Total Environ.* **2017**, *574*, 1559–1570. [CrossRef] [PubMed]
89. Yin, X.; Huang, Z.; Zheng, J.; Yuan, Z.; Zhu, W.; Huang, X.; Chen, D. Source contributions to $PM_{2.5}$ in Guangdong province, China by numerical modeling: Results and implications. *Atmos. Res.* **2017**, *186*, 63–71. [CrossRef]
90. Sillman, S. The relation between ozone, NO_x and hydrocarbons in urban and polluted rural environments. *Atmos. Environ.* **1999**, *33*, 1821–1845. [CrossRef]
91. Xue, L.K.; Wang, T.; Gao, J.; Ding, A.J.; Zhou, X.H.; Blake, D.R.; Wang, X.F.; Saunders, S.M.; Fan, S.J.; Zuo, H.C.; et al. Ground-level ozone in four Chinese cities: Precursors, regional transport and heterogeneous processes. *Atmos. Chem. Phys.* **2014**, *14*, 13175–13188. [CrossRef]
92. Grewe, V. A generalized tagging method. *Geosci. Model Dev.* **2013**, *6*, 247–253. [CrossRef]

© 2020 by the authors. Licensee MDPI, Basel, Switzerland. This article is an open access article distributed under the terms and conditions of the Creative Commons Attribution (CC BY) license (http://creativecommons.org/licenses/by/4.0/).

Article

Transport Pathways and Potential Source Region Contributions of PM$_{2.5}$ in Weifang: Seasonal Variations

Chengming Li, Zhaoxin Dai *, Xiaoli Liu and Pengda Wu

Chinese Academy of Surveying and Mapping, Beijing 100830, China; cmli@casm.ac.cn (C.L.); liuxl@casm.ac.cn (X.L.); wupd@casm.ac.cn (P.W.)
* Correspondence: daizx@lreis.ac.cn

Received: 3 March 2020; Accepted: 14 April 2020; Published: 20 April 2020

Abstract: As air pollution becomes progressively more serious, accurate identification of urban air pollution characteristics and associated pollutant transport mechanisms helps to effectively control and alleviate air pollution. This paper investigates the pollution characteristics, transport pathways, and potential sources of PM$_{2.5}$ in Weifang based on PM$_{2.5}$ monitoring data from 2015 to 2016 using three methods: Hybrid Single-Particle Lagrangian Integrated Trajectory (HYSPLIT), the potential source contribution function (PSCF), and concentration weighted trajectory (CWT). The results show the following: (1) Air pollution in Weifang was severe from 2015 to 2016, and the annual average PM$_{2.5}$ concentration was more than twice the national air quality second-level standard (35 µg/m^3). (2) Seasonal transport pathways of PM$_{2.5}$ vary significantly: in winter, spring and autumn, airflow from the northwest and north directions accounts for a large proportion; in contrast, in summer, warm-humid airflows from the ocean in the southeastern direction dominate with scattered characteristics. (3) The PSCF and CWT results share generally similar characteristics in the seasonal distributions of source areas, which demonstrate the credibility and accuracy of the analysis results. (4) More attention should be paid to short-distance transport from the surrounding areas of Weifang, and a joint pollution prevention and control mechanism is critical for controlling regional pollution.

Keywords: PM$_{2.5}$; spatiotemporal characteristics; back-trajectory clustering; potential source contribution; concentration weighted trajectory

1. Introduction

With rapid socioeconomic development, accelerated industrialization, and continuously increasing energy consumption, particulates have become major urban air pollutants in China [1–3]. In particular, fine particulates PM$_{2.5}$ can not only reduce atmospheric visibility but also increase mortality and incidence of diseases such as respiratory diseases [4–6], and these particulates have aroused great public concern and attention. Studies have indicated that urban air pollution levels and their spatiotemporal distributions are not only associated with local emissions but also influenced by cross-regional transport from sources in surrounding areas [7–9]. Accurate identification of urban atmospheric pollution characteristics and transport mechanisms is critical to control and mitigate air pollution [10].

Existing studies on the spatiotemporal characteristics of PM$_{2.5}$ and transport mechanisms have mostly focused on geo-statistics analysis [11], air quality models (e.g., large-scale Weather Research and Forecasting (WRF) model) [12], and back-trajectory clustering-based mechanism analysis. For back-trajectory clustering-based transport analysis, previous research has mostly focused on hotspots [13,14], while there is limited research on small but seriously polluted cities in China.

Li et al. (2018, 2019) indicated that $PM_{2.5}$ concentration and its influencing factors varied in different seasons, and they also preliminarily showed that the air pollution in Weifang City was affected by meteorological conditions and pollution of surrounding cities [6,15]. The above-mentioned studies are based on qualitative analysis. To quantitatively analyze the cross-regional influences of Weifang in different seasons, this paper describes an in-depth study of the seasonal variations in internal and external potential sources and provides targeted pollution control measures for government over different seasons. In addition, to improve the credibility and accuracy of the results, different trajectory analysis methods are used to analyze the potential sources of Weifang $PM_{2.5}$ in this paper, which can be used to evaluate the results through mutual verification.

Therefore, based on hourly high-frequency $PM_{2.5}$ data from 38 provincial monitoring stations in Weifang from 2015 to 2016, this paper examines the transport characteristics and potential sources of $PM_{2.5}$ in Weifang City in different seasons by using a back-trajectory clustering method. This research helps to better understand the causes of $PM_{2.5}$ pollution and the sources in Weifang as well as provides important scientific references for joint atmospheric pollution control in Weifang City.

This paper includes five sections. Section 2 summarizes relevant studies on pollutant transport mechanisms and potential sources. Section 3 describes the data sources and methods used. The analytical results of the $PM_{2.5}$ pollution characteristics and potential source contributions in different seasons are presented in Section 4, followed by the conclusions in Section 5.

2. Literature Review

The causes and sources of urban air pollution are closely related to the characteristics of airflow transport trajectories. Methods such as the back-trajectory model, clustering analysis, and the potential source contribution function (PSCF) method have become important ways to address these characteristics [16–18]. For example, Yan et al. (2015) conducted a trajectory analysis on a smog process in Beijing in February 2014 and discovered that Baoding, Hengshui, and Handan are important potential source areas in the region [19]. Donnelly et al. (2015) analyzed the effects of various long-range transport pathways of the concentrations of particulate matter with diameter less than 10 μm (PM_{10}) in Ireland by using the HYSPLIT4 model and showed that air quality in Ireland is heavily dependent on air mass origin and the inherent characteristics of the air mass [20]. Lee et al. (2011) examined the pathways of PM_{10} by back-trajectory analysis and showed that the transboundary pollutants of high-PM_{10} levels are more than twice as high as those from internal sources, especially in winter and spring; in addition, the local pollutants contributing to high-PM_{10} concentrations have decreased through the efforts to reduce emissions, but the transboundary pollutants have not decreased [21]. By analyzing the transport pathways of particulate matter over Guwahati, located in the Brahmaputra River Valley (BRV), Tiwari et al. (2017) found that the turbid air masses transported over Guwahati mostly from the western and southwestern directions contribute to higher PM concentrations, either carrying anthropogenic pollution from the Indo-Gangetic Plains or locally and LRT (long-range transported) dust from BRV and western India, respectively [22]. Li et al. (2017) employed the PSCF and concentration weighted trajectory (CWT) methods to analyze the transport trajectories and potential sources of $PM_{2.5}$ and PM_{10} in Beijing and demonstrated that, in summer and autumn, the impacts of air pollution are mostly from the south and southeast, while those in spring and winter are influenced from the southeast and north [23]. By analyzing the transport pathways and potential sources of PM_{10} pollution in Shanghai, Li et al. (2014) reported that there are two potential sources of PM_{10} pollution in Shanghai: one is located in the northwest (for example, Hebei and Shandong), while the other is in the southwest (for example, Zhejiang and Jiangxi) [24]. Yang et al. (2017) examined the spatiotemporal characteristics and trajectories of a smog process in Beijing in December 2015 and discovered that the $PM_{2.5}$ concentration in Beijing was high in the south but low in the north [25]. The major potential pollutant sources are the deserts in the northwest and the built-up areas in the Beijing-Tianjin-Hebei region.

Based on eight monitoring locations in Chengdu and meteorological data over three months, Liao et al. (2017) analyzed the spatiotemporal characteristics and sources of $PM_{2.5}$ in Chengdu. The results reveal that the major potential sources of $PM_{2.5}$ in Chengdu are located along the western margin of the Sichuan Basin and in the southeastern cities [26]. Xin et al. (2016) adopted daily average PM_{10} concentration data and Global Data Assimilation System (GDAS) data to study the transport trajectories that significantly influence PM_{10} in Xining and found that atmospheric pollution is easily affected by inland trajectories [27]. Based on nine air quality monitoring locations in Qingdao in winter, Li et al. (2017) analyzed the characteristics of atmospheric pollution and pollutant sources, the results revealed that $PM_{2.5}$ is a major urban atmospheric pollutant in Qingdao, and the greatest contributions are from Shanxi, the southern part of Hebei, and the western part of Shandong [23]. Additionally, pollutants such as aerosols from the deserts in Inner Mongolia and the Yellow Sea are also a cause of atmospheric pollution. Lv et al. (2015) discovered that the $PM_{2.5}$ concentration in Guangzhou is more sensitive to the velocities of air mass movements than their directions. The regional $PM_{2.5}$ contributions in spring, summer, autumn, and winter are 15%, 28%, 16%, and 22%, respectively [28].

However, previous studies have shortcomings with respect to air pollution transport characteristics. Most studies: (1) have focused on typical large-scale regions, while limited research has focused on small yet heavily polluted cities, such as Weifang City [24]; and (2) have adopted low temporal resolution data (with 6- or 24- h resolution); however, high time resolution data have been shown to contribute to improved resolutions of source areas in PSCF calculations [23]. Moreover, in previous studies, several air quality monitoring data have mostly been used only to obtain average pollutant concentrations in PSCF models, whereas sparse and unevenly distributed monitoring data cannot truly represent the concentrations over the study area.

3. Data and Methodology

3.1. Study Area and Data Sources

Weifang is located in the middle of the Shandong Peninsula (Figure 1), with Zibo to the west, Linyi to the south, Qingdao to the east and Laizhou Bay and the Bohai Sea to the north, covering a total area of 16,000 square kilometers. Weifang City is high in the south (with ground elevation 100–1032 m) and low in the north (with ground elevation under 7 m). The south is mainly covered with hills and low mountains, while the northeast is mainly characterized by plains and lakes. Weifang City has been experiencing industrial and economic advancements and was one of the most rapidly developing cities in Shandong Province. As of 2019, the number of registered vehicles had exceeded 2 million. This growth has resulted in damage to the Weifang City environment. Existing studies on the characteristics of air pollution in Weifang have mainly focused on spatiotemporal patterns, while few reports have described the transport trajectories and mechanisms as well as the potential sources of air pollution. This paper utilizes hourly $PM_{2.5}$ monitoring data and a back-trajectory model to analyze the transport pathways and potential source contributions of $PM_{2.5}$ pollution in Weifang and can provide important scientific support for joint atmospheric pollution control and management.

The $PM_{2.5}$ monitoring data used in this paper were acquired from 5 national monitoring stations, 4 provincial monitoring stations, and 29 city monitoring stations in Weifang (Table A1 in Appendix A). The data were obtained by automatic air quality monitors through 24-h continuous monitoring and cover the period from 1 March 2015 to 29 February 2016. The data were acquired from the Data Center of the Ministry of Environmental Protection of the People's Republic of China (http://datacenter.mep.gov.cn/index). Thermo Fisher 1405F monitoring devices were used to measure the $PM_{2.5}$ concentrations, and this instrument operates on the principle of measuring $PM_{2.5}$ concentrations by a filter dynamic measurement system (FDMS) with the tapered element oscillating microbalance (TEOM) and the beta-attenuation method [6].

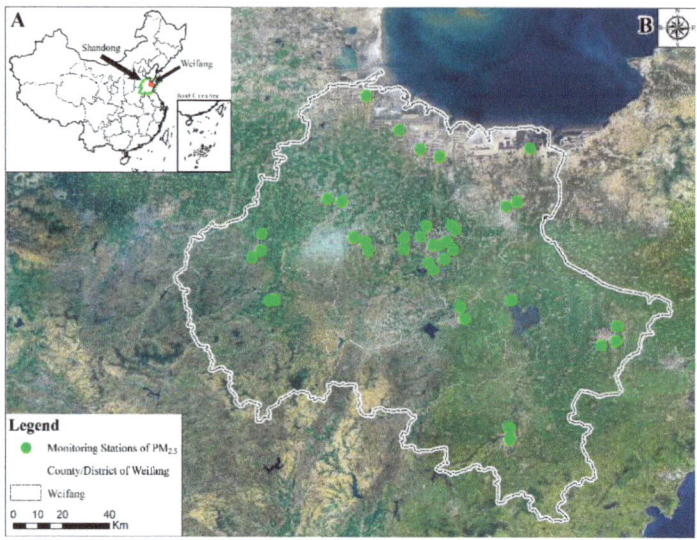

Figure 1. Study area and spatial distribution of monitoring sites.

The major data structure is illustrated in Figure 2a. The meteorological data used in the back-trajectory model were obtained from GDAS data provided by the National Centers for Environmental Prediction (NCEP). The 1.0° resolution global reanalysis data are adopted. These data are recorded every 6 h, namely, at 00:00, 06:00, 12:00 and 18:00 (UTC). Their structure is shown in Figure 2b.

StationID	Date	Longitude	Latitude	PM2.5
W001	2015-03-01 00:00	119.078	36.7336	85.29
W001	2015-03-01 01:00	119.078	36.7336	86.48
W001	2015-03-01 02:00	119.078	36.7336	85.55
………	………	………	………	
W206	2016-02-29 08:00	118.520	36.500	101.09
W206	2016-02-29 09:00	118.520	36.500	103.17
………	………	………	………	

(**a**) Structure of $PM_{2.5}$ monitoring data

Date	Press	TMP (K)	Relative humidity (%)	PWAT (kg/m2)	………	u wind (m/s)	………
2015-03-01 00:00	1019.6	305.69	62%	0	………	6.9	………
2015-03-01 01:00	1019.5	305.51	62%	0	………	6.9	………
2015-03-01 02:00	1019.2	306.05	61%	0	………	6.9	………
………	………	………	………	………	………	………	………
2016-02-29 00:00	1023.3	301.37	63%	0	………	5.4	………
2016-02-29 01:00	1023.3	301.73	63%	0	………	5.4	………
………	………	………	………	………	………	………	………

(**b**) Structure of GDAS data

Figure 2. Structures of monitoring station data and GDAS data.

3.2. Methodology

(1) HYSPLIT model

The Hybrid Single-Particle Lagrangian Integrated Trajectory (HYSPLIT) model established by the Air Resources Laboratory of the National Oceanic and Atmospheric Administration (NOAA) is used in this paper to analyze sources and transport trajectories of air pollutants [29,30]. This model gives highly accurate and temporally continuous simulation results and has been widely used in research on the transport and diffusion of various pollutants in different areas. The HYSPLIT model is divided into two parts: the backward transport model and the forward diffusion model, which solve problems concerning sources and sinks, respectively. In this paper, the backward transport model of the HYSPLIT model is employed to simulate 72-h backward airflow transport trajectories near the ground surface in Weifang during the period of interest. The characteristics of the airflow movements in the study area are thereby reflected.

(2) Trajectory clustering analysis

Trajectory clustering analysis, a multivariate statistical analyses technique, was used to divide the trajectory data into several classes or clusters. Data in the same class or cluster share a higher degree of similarity, whereas those in different classes or clusters vary more significantly [31]. This paper uses TrajStat, a plugin of MeteoInfo; this plugin can view, query, and cluster trajectories and includes two clustering methods: Euclidean distance and angle distance. Because this paper aims to determine the direction from which the air masses that reach the site have originated, the angle distance clustering method is utilized to cluster airflow trajectories. The angle distance is often used to define the mean angle between the two trajectories, which varies between 0 and π. The details of the angle distance clustering method can be found in the work of Sirois and Bottenheim (1995) [32].

(3) Potential source contribution function (PSCF)

The PSCF model is a simple method that links residence time in upwind areas with high concentrations through a conditional probability field [33]. This method can identify pollutant sources by analyzing airflow trajectories and a given threshold [34,35]. It can be calculated as follows:

$$PSCF_{ij} = \frac{n_{ij}}{N_{ij}} \qquad (1)$$

where N_{ij} is the total number of airflow trajectories' endpoints that fall in the ijth grid and n_{ij} is the total number of airflow trajectories' endpoints for which the measured $PM_{2.5}$ concentration exceeds a given threshold in the same grid. In this study, the 24-h average Grade II standard $PM_{2.5}$ concentration (75 μg/m^3) in ambient air quality standards of China (GB3095-2012) was selected as the threshold value [26]. The trajectories were calculated hourly. Studies have demonstrated that great uncertainty exists in the calculation result when N_{ij} is extremely small. To eliminate this uncertainty, an arbitrary weight function, W_{ij}, was applied when the number of the endpoints in a particular cell was less than three times the average number of endpoints for each cell [36,37].

$$WPSCF_{ij} = W_{ij} \times PSCF_{ij} \qquad (2)$$

$$W_{ij} = \begin{cases} 1.00 & N_{ij} > 80 \\ 0.70 & 20 < N_{ij} \leq 80 \\ 0.42 & 10 < N_{ij} \leq 20 \\ 0.05 & 0 < N_{ij} \leq 10 \end{cases} \qquad (3)$$

(4) Concentration weighted trajectory (CWT)

The CWT method first computes the weighted concentrations of trajectories and then obtains the weighted concentrations of grids [38]. The calculation formula of the CWT method is given as follows:

$$CWT_{ij} = \frac{\sum_{k=1}^{N} C_k \tau_{ijk}}{\sum_{k=1}^{N} \tau_{ijk}} \quad (4)$$

where CWT_{ij} is the weighted average concentration of grid ij; N is the total number of trajectories; k denotes a trajectory; C_k is the $PM_{2.5}$ concentration of trajectory k when it passes through grid ij, which can be calculated by the HYSPLIT model; and τ_{ijk} is the duration in which trajectory k stays in grid ij [39,40]. In addition, the CWT method gives rise to great uncertainties, thus the weight coefficient W_{ij} is needed to reduce these uncertainties. Similarly, W_{ij} is determined using Equation (3), and the introduction of the coefficient is as follows:

$$WCWT_{ij} = W_{ij} \times CWT_{ij} \quad (5)$$

For PSCF and CWT methods, the input data and applicable resolution of grid are the same. The difference between CWT and PSCF is that PSCF usually uses a concentration threshold to evaluate the potential sources of $PM_{2.5}$. It means that it may have the same PSCF value when sample concentrations are either only slightly higher or much higher than the criterion. As a result, it may not distinguish moderate sources from strong ones. For CWT method, the limitation of PSCF can be overcome by assigning a weighted concentration by averaging the sample concentrations that have associated trajectories that cross the grid cell.

4. Results and Analyses

4.1. $PM_{2.5}$ Pollution Characteristics

Hourly $PM_{2.5}$ concentration data for Weifang from 2015 to 2016 are examined to analyze the annual, seasonal, and monthly characteristics of the $PM_{2.5}$ concentration. The annual average $PM_{2.5}$ concentration in Weifang is 73.03 μg/m^3, which is more than twice the national second-level standard (35 μg/m^3) and nearly five-fold the national first-level standard (15 μg/m^3). Figure 3 presents the seasonal and monthly variations in the $PM_{2.5}$ concentration. In general, the monthly average $PM_{2.5}$ concentration presents a U-shaped curve. The $PM_{2.5}$ concentration is higher in winter (December, January, and February), approximately 101.64 μg/m^3, which significantly exceeds the national 24-h atmospheric quality second-level standard (75 μg/m^3). The concentrations in autumn (September to November) and spring (March to May) are 74.81 and 66.53 μg/m^3, respectively. The lowest concentration occurs in summer (June to August), with only 49.12 μg/m^3. The $PM_{2.5}$ concentrations in the four seasons all exceed the national 24-h atmospheric quality first-level standard (35 μg/m^3). The highest concentration occurs in January (122.86 μg/m^3) and the lowest in July (48.02 μg/m^3). The monthly average considerably decreases in March but rapidly increases in October.

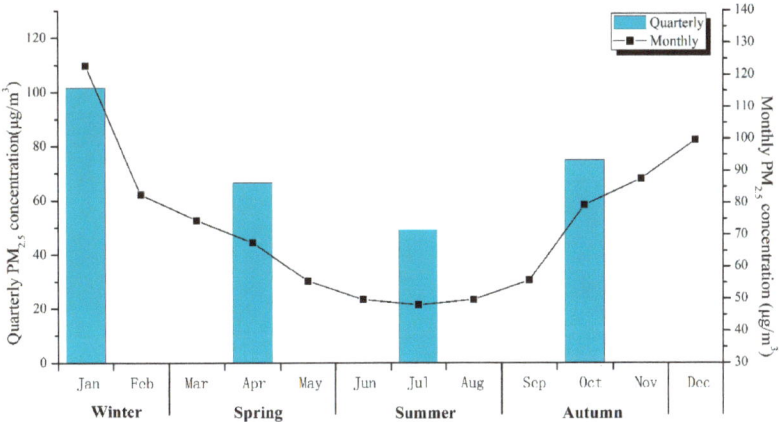

Figure 3. Quarterly and monthly characteristics of $PM_{2.5}$ concentration in Weifang.

4.2. Backward Trajectory Clustering

TrajStat was used to process airflow data in Weifang in different seasons to obtain transport trajectories in different seasons. However, from these airflow trajectories, it is not possible to determine the exact number of trajectories from different directions. Therefore, according to the consistency in the spatial distributions of various airflow trajectory types, trajectories in spring, summer, autumn, and winter are integrated into four, five, two, and two clusters, respectively (Figure 4).

Figure 4 shows that, in spring, the airflow trajectories originating from northern inland areas (Trajectories 1–3) dominate. More specifically, airflows originating from Russia, Mongolia, central Inner Mongolia, and Hebei account for 36.91% of the total trajectories, and those from the Yellow Sea and central Shandong account for 29.38%. In summer, the airflow trajectories are shorter and follow a star-shaped distribution. Influenced by warm and humid airflows from the ocean, these trajectories are dominated by southerly and southeasterly winds and account for 67.1% of the total trajectories. In autumn, as cold air masses move southward, the airflow trajectories from the southeast weaken. As a result, the airflow trajectories originating from southeastern Mongolia, central Inner Mongolia, and northeastern Hebei account for 59.07%. The airflow trajectories from northeastern Hebei are shorter, accounting for 40.30%. They pass over the waters of the Bohai Sea and finally return to Weifang via the inland areas of Shandong. In winter, under the influence of the Siberian cold current, the airflow trajectories originating from the northwest dominate (68.91%). These trajectories are longer, and the air masses move more quickly. Moreover, in winter, the airflow trajectories from southwestern Shandong account for 31.06% and are shorter and slower.

4.3. Transport Trajectories in Different Seasons

Based on the airflow backward trajectory clustering results in different seasons, combined with the hourly $PM_{2.5}$ concentration data, this paper analyzes the pollution characteristics of different transport trajectories. According to the second-level $PM_{2.5}$ concentration (75 µg/m^3) in Ambient Air Quality Standards (GB 3905-2012), back-trajectories are classified into "low polluted" trajectories (<75 µg/m^3) and "polluted" trajectories (75 µg/m^3). The statistics for different types of trajectories are described in Table 1.

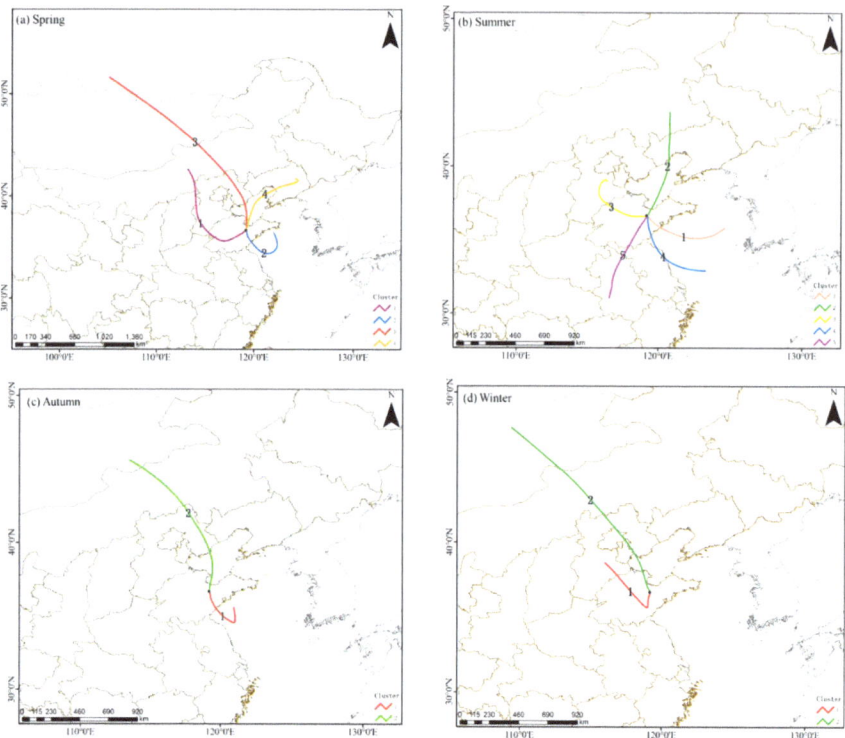

Figure 4. Cluster mean back-trajectories in different seasons in Weifang from March 2015 to February 2016.

Table 1. Trajectory proportions and $PM_{2.5}$ concentrations based on all trajectories and pollution trajectories.

Season	Trajectory Cluster	All Trajectories		Polluted Trajectories	
		Proportion (%)	Average ($\mu g/m^3$)	Proportion of Seasonal Total Polluted Trajectories (%)	Average ($\mu g/m^3$)
Spring	1	12.08	83.35	29.43	109.71
	2	45.36	69.63	31.55	99.46
	3	27.15	53.72	18.83	98.21
	4	15.41	76.64	20.20	111.62
	All		70.83		104.87
Summer	1	34.06	41.50	12.04	89.41
	2	22.71	42.93	4.45	88.29
	3	6.63	62.95	20.68	99.94
	4	21.16	68.16	43.19	94.13
	5	15.44	47.59	19.63	96.24
	All		52.63		94.92
Autumn	1	25.80	94.52	52.60	124.22
	2	74.20	66.53	47.40	116.84
	All		80.53		120.72
Winter	1	37.11	148.08	55.72	158.08
	2	62.89	82.99	44.28	114.11
	All		115.54		138.61

154

(1) Pollution characteristics in spring

The average $PM_{2.5}$ concentration of spring trajectories is 70.83 µg/m^3. The trajectory clusters in descending order are: 2 > 3 > 4 > 1. The airflow trajectories originating from the north (Inner Mongolia) have the highest average $PM_{2.5}$ concentrations, reaching 83.35 µg/m^3. This may be due to two reasons: one is that these airflows mainly come from the western arid regions of China, and, in spring, northwesterly winds easily blow loose topsoil and fine sand from dry surfaces to form sandstorms; the other reason is that these trajectories pass over the industrial areas in southern Hebei and easily transport local pollutants to Weifang City to form accumulation. Type 3 trajectories from the northwest are longer and have the lowest $PM_{2.5}$ concentration of 53.72 µg/m^3. This may due to two reasons: (1) these trajectories travel large distances at high wind speeds; and (2) they subsequently pass over regions with high vegetation cover and relatively low population density [41], such as Xilingol league in Inner Mongolia, Chengde City in northern Hebei, and waters of the Bohai Sea. Both the abovementioned cases may lead to low pollution transport and effective adsorption, diffusion, and dilution of pollutants.

The average $PM_{2.5}$ concentration of the polluted trajectories is 104.87 µg/m^3. In particular, Type 1 and 2 trajectories account for 29.43% and 31.55% of the total polluted trajectories, respectively. These types together account for 60.98% of the seasonal total polluted trajectories, suggesting that $PM_{2.5}$ is mainly transported from the north and southeast directions to Weifang in spring.

(2) Pollution characteristics in summer

The average $PM_{2.5}$ concentration of summer trajectories is the lowest, only 49.40 µg/m^3. In summer, many plants flourish, and the total leaf area of surface vegetation significantly increases, which is conducive to the adsorption of atmospheric particulates. Moreover, precipitation in summer is more concentrated, and the increased precipitation greatly facilitates wet deposition and dilution of atmospheric pollutants. However, under the control of the subtropical high and typhoon, the trajectories are diverse. The trajectory clusters in descending order of average $PM_{2.5}$ concentration are: 4 > 3 > 5 > 2 > 1. The airflow trajectories from southeast have the highest value (68.16 µg/m^3), followed by those from northwest (62.95 µg/m^3). This pattern indicates that external pollutant emissions from southeast and northwest have a substantial effect on $PM_{2.5}$ pollution in Weifang. Although Type 4 originated from the Yellow Sea, affected by subtropical auxiliary high pressure and wheat straw burning in south China, pollutants and suspended particles from eastern Zhejiang and southern Shandong are easily carried by the trajectory and accumulate Weifang. Type 1 trajectories originating from the Bohai Sea have the lowest average $PM_{2.5}$ concentration (41.50 µg/m^3) and make up the largest proportion of the total trajectories. These trajectories mainly pass over waters and coastal areas in eastern Shandong. These areas are characterized by strong air masses, cleanliness, and intensive atmospheric wet deposition, which facilitate wet deposition and dilution of pollutants [1,6,42].

The average $PM_{2.5}$ concentration of polluted trajectories is 94.92 µg/m^3. Types 3–5 trajectories account for 20.68%, 43.19%, and 19.63% of the total polluted summer trajectories, respectively. They together account for 83.50% of the seasonal total polluted trajectories, indicating that $PM_{2.5}$ is mainly transported in easterly directions to Weifang in summer.

(3) Pollution characteristics in autumn

Autumn trajectories have an average $PM_{2.5}$ concentration of 80.53 µg/m^3. The autumn trajectory clusters in descending order of $PM_{2.5}$ concentration are: 1 > 2. In particular, Type 2 trajectories are longer and have lower $PM_{2.5}$ concentrations. These trajectories originate mostly from northwestern Mongolia, central Inner Mongolia, and northern Hebei and travel over extensive areas at high wind speeds. In addition, they pass over the waters of the Bohai Sea and enter Weifang directly. The air masses are relatively clean, which is conducive to the diffusion of pollutants. In contrast, the $PM_{2.5}$ concentrations of Type 1 trajectories are higher, with an average of 94.52 µg/m^3. This is because short and slow Type 1 trajectories do not favor the diffusion of pollutants. Furthermore, these airflow trajectories pass over the northeastern part of Shandong before they reach Weifang and often carry pollutants from Shandong, leading to higher $PM_{2.5}$ concentrations.

The polluted trajectories have an average $PM_{2.5}$ concentration of 120.72 $\mu g/m^3$. Type 1 trajectories account for 52.60% of the total polluted trajectories. It follows that the northeasterly direction is the major $PM_{2.5}$ transport direction to Weifang in autumn.

(4) Pollution characteristics in winter

Compared to the other three seasons, the average $PM_{2.5}$ concentration of winter trajectories is the highest, reaching 115.54 $\mu g/m^3$. More specifically, Type 2 trajectories have the highest $PM_{2.5}$ concentrations, with an average of 148.08 $\mu g/m^3$. This is mostly because these airflow trajectories originate from south-central Hebei and pass over inland cities in northern Shandong. In winter, these areas are in the heating period, leading to increases in anthropogenic pollutant emissions, such as from coal burning [8,43]. As air masses pass over these areas, they usually carry soil, dust, and pollutants from the inland areas to Weifang. In addition, these airflow trajectories are relatively short and travel slowly, making pollutants not able to diffuse easily. Hence, the average $PM_{2.5}$ concentration of these trajectories is relatively high. In contrast, the average concentration of Type 2 trajectories is the lowest, but they are also significantly higher than those in the other seasons, at 82.99 $\mu g/m^3$. These trajectories originate from Siberia and arrive in Weifang through the Mongolian Plateau, Inner Mongolia, Hebei, Beijing, Tianjin, and the Bohai Sea and are relatively long and travel at higher wind speeds. These trajectories pass over the waters of the Bohai Sea, which may slightly facilitate the diffusion and elimination of pollutants.

The average $PM_{2.5}$ concentration of polluted winter trajectories is 138.61 $\mu g/m^3$. The average $PM_{2.5}$ concentration of Type 1 trajectories reaches 158.08 $\mu g/m^3$, which includes not only the external source from southern Hebei Province but also local emissions and is the main pollution source of Weifang City in winter.

Overall, in autumn and winter, Weifang is mostly influenced by pollutants from inland areas of Hebei and Shandong. Airflows from these areas often carry particulates emitted from the passed areas. This reflects the knock-on effects of $PM_{2.5}$ pollution within Shandong Province. In summer, due to the influences from coastal cities in the east and clean airflows from the ocean, the concentrations of pollutants are lower. In spring, severe pollution in Weifang is closely associated with sandstorms in the arid areas in the west, where loose topsoil and fine sand are blown from dry surfaces.

4.4. Potential Source Regions

To further investigate the sources of atmospheric pollutant transport in Weifang, this paper analyzes the potential source regions of $PM_{2.5}$ pollution. First, $PM_{2.5}$ concentration data are added to airflow trajectories, and areas covering all airflow trajectories are uniformly divided into 0.5° × 0.5° grids. Then, the weighted potential source contribution function (WPSCF) value of each grid is computed. WPSCF reveals the spatial distribution of $PM_{2.5}$ potential sources obtained by combining back-trajectories and measurements of $PM_{2.5}$ concentration. A high PSCF value signifies a potential source location. The greater the WPSCF value of a grid is, the higher the contribution levels of potential source regions to $PM_{2.5}$ pollution in Weifang, given that other factors remain stable.

Figure 5 shows the PSCF results for $PM_{2.5}$ in Weifang from 2015 to 2016. The colors represent the contribution levels of potential source areas; the black color is associated with high concentrations, while green represents low $PM_{2.5}$ concentrations. Distinct seasonal variations are noted in the distribution of the potential source areas of $PM_{2.5}$. (1) In spring, high PSCF values are mainly found in north-central Jiangsu and southwestern Shandong. Additionally, areas such as Tianjin, Liaoning and western Jilin have key influences on the sources of pollution in Weifang. (2) In summer, PSCF values are generally smaller, indicating that Weifang is less affected by pollutants from the surrounding areas in summer. Compared to those in spring, the potential sources are shifted eastward and are located in Shandong and the waters of the Bohai Sea and Yellow Sea. This is may because the southeasterly monsoon carries pollutants emitted from passed areas when it moves northward. Furthermore, stubble burning is the most intensive in some southern regions in summer, creating large fumes that easily travel northward with the monsoon. (3) In autumn, higher PSCF values are increasing and are mainly found in regions such as northern Anhui, northeastern Henan, and southwestern Shandong. In addition, areas such as Hebei, Beijing, Tianjin, eastern Shanxi, and central Inner Mongolia make certain potential contributions. (4) In winter, potential source areas transit and extend northwestward. High values are found in the whole city of Shandong, northern Jiangsu, northeastern Henan, and southern Hebei; in these regions, their situations are very similar to that of Weifang, with the same climate, industrial emission, population density, and winter heating. In addition, those values in eastern Shanxi and western Inner Mongolia increase.

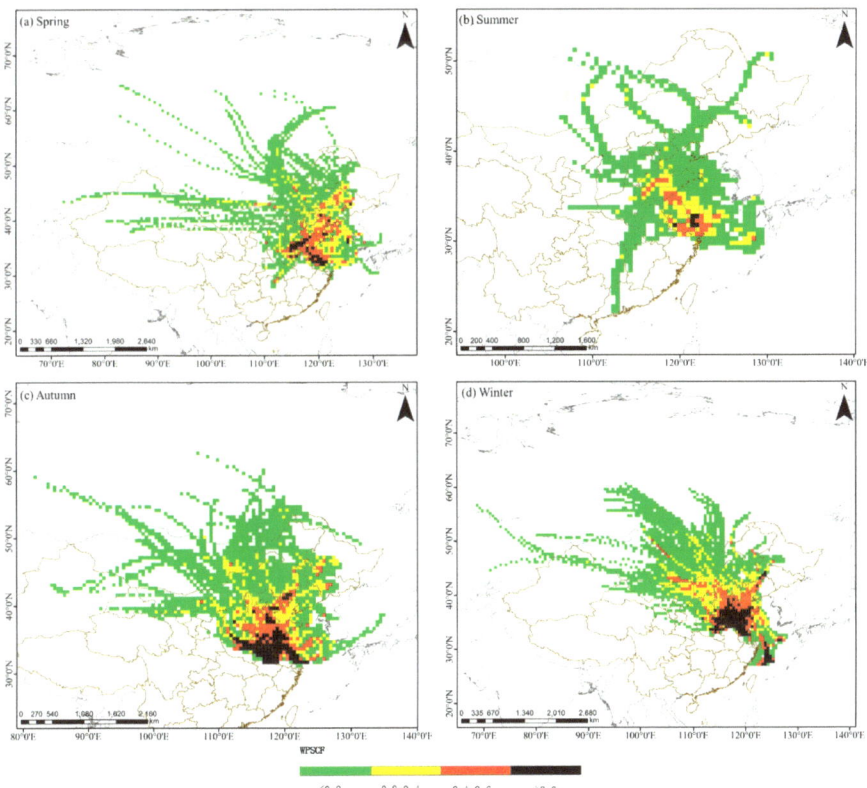

Figure 5. Spatial distribution of WPSCF values of $PM_{2.5}$ in spring, summer, autumn, and winter from March 2015 to February 2016.

In summary, in winter, spring, and autumn, Weifang is more substantially affected by nearby inland cities in Jiangsu, Henan, and Shandong, and these areas are potential source regions of pollutants in Weifang. In contrast, in summer, Weifang is less influenced by pollutants from surrounding areas. Moreover, comparisons reveal that the distributions of high PSCF values in different seasons agree well with those of the major trajectory areas shown in Table 1. This finding indicates that the potential source contribution results obtained by the PSCF method in this paper are reasonably reliable.

4.5. Potential Source Region Contributions

The CWT method is adopted to calculate the weighted concentrations of trajectories originating from various potential source regions. The CWT value represents a weighted specific concentration value by averaging sample concentrations that have associated trajectories that cross the grid cell, under the condition of other factors remaining relatively constant. The larger is the CWT value, the greater is the contribution of the grid cell to the pollutant concentration of Weifang City. The results are illustrated in Figure 6. The regions colored in black correspond to the main contributing sources associated with the highest $PM_{2.5}$ values, while the green color represents regions with low values.

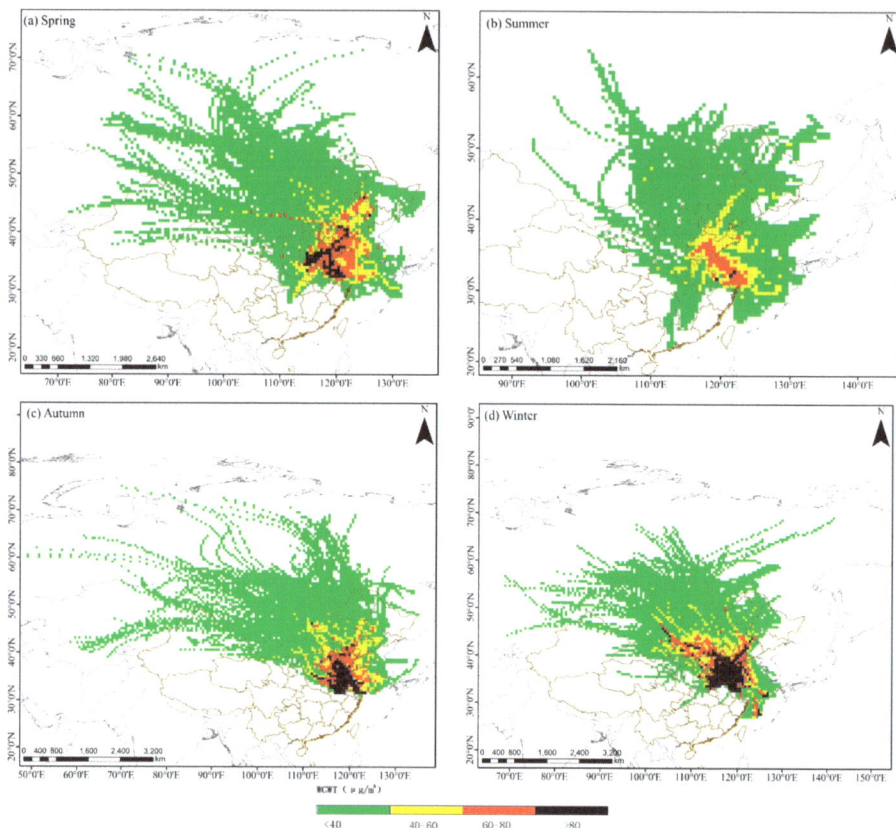

Figure 6. Spatial distribution of WCWT values of $PM_{2.5}$ in spring, summer, autumn, and winter from March 2015 to February 2016.

In spring, higher WCWT values are mostly concentrated in northeastern Henan and southwestern Shandong, which are important sources of $PM_{2.5}$ in Weifang. These areas have daily average $PM_{2.5}$ concentration contributions above 80 μg/m^3. Additionally, other regions in Shandong, western Tianjin, and the central Bohai region influence pollution in Weifang. For these areas, the daily average $PM_{2.5}$ concentration contributions are more than 60 μg/m^3.

In summer, WCWT values are generally smaller. The highest CWT values covering the map were distributed in coastal areas in the east, such as Zhejiang and Jiangsu, with values of 60–80 μg/m^3. This is possibly because southeasterly wind dominates in Weifang in summer. Sea-salt aerosols from the south are easily transported to the Weifang region via the monsoon traveling northward.

Compared to those in summer, higher WCWT values in autumn are found in areas further south and northeast. These values are mainly observed at the intersection of Anhui, Jiangsu, and Shandong, with daily average $PM_{2.5}$ concentration contributions of above 80 μg/m^3. Furthermore, relatively substantial contributions are observed in certain localized areas such as eastern Hebei, southern Beijing, and Tianjin, and the daily average $PM_{2.5}$ concentration contributions of these areas exceed 60 μg/m^3.

The WCWT values in winter are greater than those in other seasons. The greatest contributions to the $PM_{2.5}$ concentration in Weifang are throughout Shandong as well as certain nearby areas such as northeastern Henan, northern Jiangsu, and southwestern Hebei. The daily average contributions are more than 80 μg/m^3. This is probably because many pollutants are emitted when coal is burned for heating in the north in winter.

When CWT results are compared to the WPSCF results, they have basically consistent results for areas contributing to the $PM_{2.5}$ concentration in Weifang, which demonstrates the credibility and accuracy of the analysis results. Nevertheless, the potential source regions obtained via the CWT method for spring, summer, and autumn cover larger areas than those simulated by the PSCF method, while the results for winter are similar. These results are consistent with the findings reported by Yan et al. (2018) in Yinchuan [44]. These observations may arise because the CWT method takes into account all the concentrations rather than a subset of high concentrations in the PSCF method. The findings demonstrate that the seasonal distributions of $PM_{2.5}$ source areas obtained by these two methods share generally similar characteristics, suggesting the credibility and accuracy of the analysis results.

5. Discussion and Conclusions

Based on $PM_{2.5}$ monitoring data from 2015 to 2016 from 38 air quality monitoring stations, this paper describes an in-depth study of the seasonal variations in internal and external potential sources in Weifang by using different trajectory analysis methods.

(1) Seasonal differences in the contributions of potential source regions of $PM_{2.5}$ in Weifang. In winter, spring, and autumn, airflows are mostly from the northwesterly and northerly directions and significantly influence the $PM_{2.5}$ concentration in Weifang. However, in summer, airflow trajectories are scattered, and warm and humid airflows from the ocean in the southeastern direction dominate. In winter, spring, and autumn, Weifang is more greatly affected by pollutant transport from nearby inland cities in Shandong and Henan. These transport pathways are short in general, and the wind speeds are low, leading to accumulation of the carried pollutants in Weifang. In contrast, in summer, Weifang is less influenced by pollutants from the surrounding areas, and the potential source regions are mainly located in coastal areas in the east, such as Jiangsu, the Bohai Sea, and the Yellow Sea. It should be noted that the region covering central Inner Mongolia and southern Liaoning is also a potential source area for Weifang.

(2) Policy implications for $PM_{2.5}$ pollution in Weifang. The results indicate that, in formulating relevant pollution control and prevention measures, the government should focus on the control of pollutant sources and take the migration of regional pollution caused by these sources into account. For example, according to the transport patterns of pollutants in Weifang, the surrounding pollutant source regions can be divided and classified (such as key control zones). This finding suggests that different control and management policies can be implemented, and region-specific pollution control and prevention measures can be formulated. Because Weifang is more significantly affected by short-distance pollutant transport from the surrounding cities and provinces as well as inland areas in Shandong Province, the government should pay more attention to short-distance transport from these regions. For instance, more intensive greening measures can be introduced to reduce the short-distance transport of $PM_{2.5}$. Furthermore, the interactions between the city and its surrounding areas should be considered, and joint control and cooperation between different regions should be enhanced.

(3) Evaluation of the analysis results. The results demonstrate that the seasonal distributions of $PM_{2.5}$ source areas obtained by two methods (PSCF and CWT) share generally similar characteristics, demonstrating the credibility and accuracy of the analysis results. Both methods can effectively reflect the potential source areas of pollution in the region. However, there is a limitation in validating the modeling results with the real transport values because emissions and depositions of air pollutants along the trajectories cannot be captured easily. Nevertheless, our results are able to locate the source direction and areas and identify source contributions to a certain extent. To support this point, we analyzed the notice issued by Shandong Provincial Environmental Protection Office, which indicates that the primary source of external dust transport from Inner Mongolia is the main reason that leads to the rapid increase in pollutants in Weifang City in April [45]. Thus, some references indicated that in northern China (such as Weifang), the main pollution sources come from anthropogenic emissions related to coal burning and their transportation [46]. These cases both support our findings regarding spring and winter. In addition, PSCF and CWT are appropriate tools that help identify source contributions to the concentration variations at the destination, assuming that other elements/factors remain the same, such as the period of the kite festival (in April every year). In the future, additional work that combines emission sources and external monitored $PM_{2.5}$ concentration data is needed to improve the prediction of $PM_{2.5}$ source regions and validate the analysis results quantitatively.

This paper analyzes the external and internal potential source regions of $PM_{2.5}$ pollution and their influences on Weifang City. Our findings can provide scientific support for the design of region-specific measures for atmospheric pollution prevention and the development of a chemical transport model combined with meteorology. However, this work also has limitations. For instance, the resolution of the study grid based on the backward trajectory model is not very high and may not be applied to small-scale regions. Furthermore, the estimation of the sources of $PM_{2.5}$ is not perfect because the contributing sources were calculated only based on meteorological data without information such as the production and deposition of dust. In future research, based on the results in this paper, auxiliary parameters such as the local emission sources and external $PM_{2.5}$ monitoring value will be combined to improve the simulation analysis accuracy and evaluate the accuracy of potential sources contributing to the $PM_{2.5}$ in Weifang quantitatively. Furthermore, analyses for various years will be further conducted to assess the inter-annual variability.

Author Contributions: Conceptualization, C.L.; methodology, C.L. and Z.D.; software, Z.D. and X.L.; validation, Z.D., X.L. and P.W.; formal analysis, Z.D., X.L. and P.W.; investigation, Z.D. and X.L.; resources, C.L.; writing—original draft preparation, Z.D.; writing—review and editing, C.L. and Z.D.; supervision, C.L.; and funding acquisition, C.L. and Z.D. All authors have read and agreed to the published version of the manuscript.

Funding: This research was funded by the National Natural Science Foundation of China, grant numbers 41907389 and 41871375, National Key Research and Development Program of China, grant number 2018YFB2100700, and Basic Foundation of Chinese Academy of Surveying and Mapping (AR2010).

Conflicts of Interest: The authors declare no conflict of interest.

Appendix A

Table A1. Information of Air Monitoring Stations.

Order	Station Name	Latitude	Longitude	Type	Availability
1	Weifang arbitration committee	36.702	119.122	National monitoring stations	
2	Weifang environmental protection bureau	36.702	119.144		
3	Hanting Station	36.774	119.191		
4	Weifang No. 7 High School	36.687	119.017		
5	Weifang Fangzi post	36.652	119.164		
6	Weifang College	36.715	119.176	Provincial monitoring stations	
7	Weifang government	36.728	119.018		
8	Jincheng Middle School	36.772	119.098		
9	Xinhui group	36.637	119.107		
10	Shouguang monitoring station	36.869	118.735	City monitoring stations	Available
11	Zhucheng Safety Supervision Bureau	36.004	119.406		
12	Changle Sports Bureaus	36.730	118.834		
13	Changyi No. 7 High School	36.859	119.431		
14	East side in Binhai district	37.020	119.145		
15	Qingzhou Guangtong group	36.742	118.494		
16	Gaomi college town	36.343	119.748		
17	Anqiu Qingyunhu village	36.479	119.223		
18	Changyi Xiaying school	37.051	119.479		
19	Zhucheng Technology School	36.045	119.404		
20	Xiashan water works	36.503	119.411		
21	Gaomi Ruiguang electronic	36.411	119.807		
22	Changle Wutu street	36.684	118.887		
23	Fangzi Luneng school	36.614	119.124		
24	Linqu qushan	36.503	118.543		
25	Experimental school in Gaoxin district	36.687	119.198		
26	West side in Binhai district	37.116	118.999		
27	Hanting foreign language school	36.758	119.205		
28	Shouguang business district	36.859	118.787		
29	Shouguang Hou village	37.047	119.075		
30	Qingzhou monitoring station	36.681	118.491		
31	Shouguang Yangkou village	37.240	118.879		
32	Qingzhou Shuangbeistadium	36.657	118.456		
33	Normal university of Special education	36.734	119.078		
34	Linqu water works	36.500	118.520		
35	Gaomi Sports Bureaus	36.359	119.802		
36	Changle Zhuliu street	36.715	118.875		
37	Changyi highway bureau	36.842	119.390		
38	Anqiu Qingyunshan scenic spot	36.437	119.240		

References

1. He, J.; Ding, S.; Liu, D.F. Exploring the spatiotemporal pattern of PM2.5 distribution and its determinants in Chinese cities based on a multilevel analysis approach. *Sci. Total Environ.* **2019**, *659*, 1513–1525. [CrossRef] [PubMed]
2. Xie, R.; Zhao, G.; Zhu, B.Z.; Chevallier, J. Examining the Factors Affecting Air Pollution Emission Growth in China. *Environ. Modeling Assess.* **2018**, *23*, 389–400. [CrossRef]
3. Liu, J.; Kiesewetter, G.; Klimont, Z.; Cofala, J.; Heyes, C.; Schöpp, W.; Zhu, T.; Cao, G.; Gómez-Sanabria, A.; Sander, R.; et al. Mitigation pathways of air pollution from residential emissions in the Beijing-Tianjin-Hebei region in China. *Environ. Int.* **2019**, *125*, 236–244. [CrossRef] [PubMed]
4. Burnett, R.; Chen, H.; Szyszkowicz, M.; Fann, N.; Hubbell, B.; Pope, C.A.; Apte, J.S.; Brauer, M.; Cohen, A.; Weichenthal, S.; et al. Global estimates of mortality associated with long-term exposure to outdoor fine particulate matter. *Proc. Natl. Acad. Sci. USA* **2018**, *115*, 9592–9597. [CrossRef] [PubMed]

5. Fang, D.; Wang, Q.G.; Li, H.M.; Yu, Y.; Lu, Y.; Qian, X. Mortality effects assessment of ambient PM2.5 pollution in the 74 leading cities of China. *Sci. Total Environ.* **2016**, *569–570*, 1545–1552. [CrossRef] [PubMed]
6. Li, Y.; Dai, Z.; Liu, X. Analysis of Spatial-Temporal Characteristics of the PM2.5 Concentrations in Weifang City, China. *Sustainability* **2018**, *10*, 2960. [CrossRef]
7. Cheng, Z.H.; Li, L.S.; Liu, J. Identifying the spatial effects and driving factors of urban PM2.5 pollution in China. *Ecol. Indic.* **2017**, *82*, 61–75. [CrossRef]
8. Liu, H.; Fang, C.; Zhang, X.; Wang, Z.; Bao, C.; Li, F. The effect of natural and anthropogenic factors on haze pollution in Chinese cities: A spatial econometrics approach. *J. Clean. Prod.* **2017**, *165*, 323–333. [CrossRef]
9. Ma, Y.; Ji, Q.; Fan, Y. Spatial linkage analysis of the impact of regional economic activities on PM2.5 pollution in China. *J. Clean. Prod.* **2016**, *139*, 1157–1167. [CrossRef]
10. Wei, X.; Liu, M.; Yang, J.; Du, W.-N.; Sun, X.; Huang, Y.-P.; Zhang, X.; Khalil, S.K.; Luo, D.-M.; Zhou, Y.-D. Characterization of PM2.5-bound PAHs and carbonaceous aerosols during three-month severe haze episode in Shanghai, China: Chemical composition, source apportionment and long-range transportation. *Atmos. Environ.* **2019**, *203*, 1–9. [CrossRef]
11. Yan, D.; Lei, Y.; Shi, Y.; Zhu, Q.; Li, L.; Zhang, Z. Evolution of the spatiotemporal pattern of PM2.5 concentrations in China–A case study from the Beijing-Tianjin-Hebei region. *Atmos. Environ.* **2018**, *183*, 225–233. [CrossRef]
12. Zhang, Z.; Wang, W.; Cheng, M.; Liu, S.; Xu, J.; He, Y.; Meng, F. The contribution of residential coal combustion to PM2.5 pollution over China's Beijing-Tianjin-Hebei region in winter. *Atmos. Environ.* **2017**, *159*, 147–161. [CrossRef]
13. Zhang, Y.; Chen, J.; Yang, H.; Li, R.; Yu, Q. Seasonal variation and potential source regions of PM2.5-bound PAHs in the megacity Beijing, China: Impact of regional transport. *Environ. Pollut.* **2017**, *231*, 329–338. [CrossRef]
14. Zhang, Y.; Zhang, H.; Deng, J.; Du, W.; Hong, Y.; Xu, L.; Qiu, Y.; Hong, Z.; Wu, X.; Ma, Q.; et al. Source regions and transport pathways of PM2.5 at a regional background site in East China. *Atmos. Environ.* **2017**, *167*, 202–211. [CrossRef]
15. Li, C.; Dai, Z.; Yang, L.; Ma, Z. Spatiotemporal Characteristics of Air Quality across Weifang from 2014–2018. *Int. J. Environ. Res. Public Health* **2019**, *16*, 3122. [CrossRef]
16. Ghosh, S.; Biswas, J.; Guttikunda, S.; Roychowdhury, S.; Nayak, M. An investigation of potential regional and local source regions affecting fine particulate matter concentrations in Delhi, India. *J. Air Waste Manag. Assoc.* **2015**, *65*, 218–231. [CrossRef] [PubMed]
17. Gogikar, P.; Tyagi, B. Assessment of particulate matter variation during 2011–2015 over a tropical station Agra, India. *Atmos. Environ.* **2016**, *147*, 11–21. [CrossRef]
18. Hao, T.; Cai, Z.; Chen, S.; Han, S.; Yao, Q.; Fan, W. Transport Pathways and Potential Source Regions of PM2.5 on the West Coast of Bohai Bay during 2009–2018. *Atmosphere* **2019**, *10*, 345. [CrossRef]
19. Yan, R.; Yu, S.; Zhang, Q.; Li, P.; Wang, S.; Chen, B.; Liu, W. A heavy haze episode in Beijing in February of 2014: Characteristics, origins and implications. *Atmos. Pollut. Res.* **2015**, *6*, 867–876. [CrossRef]
20. Donnelly, A.; Broderick, B.; Misstear, B. The effect of long-range air mass transport pathways on PM10 and NO$_2$ concentrations at urban and rural background sites in Ireland: Quantification using clustering techniques. *J. Environ. Sci. Health Part A* **2015**, *50*, 647–658. [CrossRef] [PubMed]
21. Lee, S.; Ho, C.-H.; Choi, Y.-S. High-PM10 concentration episodes in Seoul, Korea: Background sources and related meteorological conditions. *Atmos. Environ.* **2011**, *45*, 7240–7247. [CrossRef]
22. Tiwari, S.; Dumka, U.; Gautam, A.; Kaskaoutis, D.; Srivastava, A.; Bisht, D.; Chakrabarty, R.; Sumlin, B.; Solmon, F. Assessment of PM2.5 and PM10 over Guwahati in Brahmaputra River Valley: Temporal evolution, source apportionment and meteorological dependence. *Atmos. Pollut. Res.* **2017**, *8*, 13–28. [CrossRef]
23. Li, D.; Liu, J.; Zhang, J.; Gui, H.; Du, P.; Yu, T.; Wang, J.; Lu, Y.; Liu, W.; Cheng, Y. Identification of long-range transport pathways and potential sources of PM2.5 and PM10 in Beijing from 2014 to 2015. *J. Environ. Sci.* **2017**, *56*, 214–229. [CrossRef] [PubMed]
24. Li, M.; Huang, X.; Zhu, L.; Li, J.; Song, Y.; Cai, X.; Xie, S. Analysis of the transport pathways and potential sources of PM10 in Shanghai based on three methods. *Sci. Total Environ.* **2014**, *414*, 525–534. [CrossRef] [PubMed]
25. Yang, W.; Wang, G.; Bi, C. Analysis of Long-Range Transport Effects on PM2.5 during a Short Severe Haze in Beijing, China. *Aerosol Air Qual. Res.* **2017**, *17*, 1610–1622. [CrossRef]

26. Liao, T.; Wang, S.; Ai, J. Gui, K.; Duan, B.; Zhao, Q.; Zhang, X.; Jiang, W.; Sun, Y. Heavy pollution episodes, transport pathways and potential sources of PM2.5 during the winter of 2013 in Chengdu (China). *Sci. Total Environ.* **2017**, *584–585*, 1056–1065. [CrossRef]
27. Xin, Y.; Wang, G.; Chen, L. Identification of Long-Range Transport Pathways and Potential Sources of PM10 in Tibetan Plateau Uplift Area: Case Study of Xining, China. *Environ. Pollut.* **2014**, *224*, 44–53. [CrossRef]
28. Lv, B.; Liu, Y.; Yu, P.; Zhang, B.; Bai, Y. Characterizations of PM2.5 Pollution Pathways and Sources Analysis in Four Large Cities in China. *Aerosol Air Qual. Res.* **2015**, *15*, 1836–1843. [CrossRef]
29. Draxler, R.; Hess, G. An overview of HYSPLIT-4 modeling system for trajectories dispersion and deposition. *Aust. Met. Mag.* **1998**, *47*, 295–308, ISSN: 0004-9743.
30. Sahu, S.; Zhang, H.; Guo, H.; Hu, J.; Ying, Q.; Kota, S. Health risk associated with potential source regions of PM2.5 in Indian cities. *Air Qual. Atmos. Health* **2019**, *12*, 327–340. [CrossRef]
31. Yang, K.; Li, Q.; Yuan, M.; Guo, M.; Wang, Y.; Li, S.; Tian, C.; Tang, J.; Sun, J.; Li, J.; et al. Temporal variations and potential sources of organophosphate esters in PM2.5 in Xinxiang, North China. *Chemosphere* **2019**, *215*, 500–506. [CrossRef] [PubMed]
32. Sirois, A.; Bottenheim, J. Use of backward trajectories to interpret the 5-year record of pan and O_3 ambient air concentrations at Kejimkujik National Park, Nova Scotia. *J. Geophys. Res.* **1995**, *100*, 2867–2881. [CrossRef]
33. Ashbaugh, L.; Malm, W.; Sadeh, W. A residence time probability analysis of sulfur concentrations at Grand Canyon National Park. *Atmos. Environ.* **1985**, *19*, 1263–1270. [CrossRef]
34. Wang, Y.; Zhang, X.; Draxler, R. TrajStat: GIS-based software that uses various trajectory statistical analysis methods to identify potential sources from long-term air pollution measurement data. *Environ. Model. Softw.* **2009**, *24*, 938–939. [CrossRef]
35. Dimitriou, K.; Kassomenos, P. Combining AOT, Angstrom Exponent and PM concentration data, with PSCF model, to distinguish fine and coarse aerosol intrusions in Southern France. *Atmos. Res.* **2016**, *172–173*, 74–82. [CrossRef]
36. Polissar, A.; Hopke, P.; Poirot, R. Atmospheric aerosol over Vermont: Chemical composition and sources. *Environ. Sci. Technol.* **2001**, *35*, 4604–4621. [CrossRef]
37. Wang, L.; Liu, Z.; Sun, Y.; Ji, D.; Wang, Y. Long-range transport and regional sources of PM2.5 in Beijing based on long-term observations from 2005 to 2010. *Atmos. Res.* **2015**, *157*, 37–48. [CrossRef]
38. Zhang, K.; Shang, X.; Herrmann, H.; Meng, F.; Mo, Z.; Chen, J.; Lv, W. Approaches for identifying PM2.5 source types and source areas at a remote background site of South China in spring. *Sci. Total Environ.* **2019**, *691*, 1320–1327. [CrossRef] [PubMed]
39. Stohl, A. Trajectory statistics-a new method to establish source-receptor relationships of air pollutants and its application to the transport of particulate sulfate in Europe. *Atmos. Environ.* **1996**, *30*, 579–587. [CrossRef]
40. Stohl, H.; Krom-kolb, H. Origin of ozone in Vienna and surroundings, Austria. *Atmos. Environ.* **1994**, *28*, 1255–1266. [CrossRef]
41. Liu, H.; Fang, C.; Huang, J.; Zhu, X.D.; Zhou, Y.; Wang, Z.B.; Zhang, Q. The spatial-temporal characteristics and influencing factors of air pollution in Beijing-Tianjin-Hebei urban agglomeration. *J. Geogr. Sci.* **2018**, *73*, 177–191. [CrossRef]
42. Yang, Q.; Yuan, Q.; Li, T.; Shen, H.; Zhang, L. The Relationships between PM2.5 and Meteorological Factors in China: Seasonal and Regional Variations. *Int. J. Environ. Res. Public Health* **2017**, *14*, 1510. [CrossRef]
43. Zhan, D.; Kwan, M.; Zhang, W.; Yu, X.; Meng, B.; Liu, Q. The driving factors of air quality index in China. *J. Clean. Prod.* **2018**, *197*, 1342–1351. [CrossRef]
44. Yan, X.; Gou, X.; Wu, W.; Huang, F.; Yang, J.; Liu, Y. Analysis of atmospheric particulate transport path and potential source area in Yinchuan. *Acta Sci. Circumstantiae* **2018**, *38*, 1727–1738. [CrossRef]
45. Shandong's Outer Dust in April Became the Main Force of PM10 Pollution. Available online: http://weifang.iqilu.com/rdjj/2016/0512/2785473.shtml (accessed on 12 May 2016).
46. Wang, S.; Zhou, C.; Wang, Z.; Feng, K.; Hubacek, K. The characteristics and drivers of fine particulate matter (PM2.5) distribution in China. *J. Clean. Prod.* **2017**, *142*, 1800–1809. [CrossRef]

© 2020 by the authors. Licensee MDPI, Basel, Switzerland. This article is an open access article distributed under the terms and conditions of the Creative Commons Attribution (CC BY) license (http://creativecommons.org/licenses/by/4.0/).

Article

Improved Interpolation and Anomaly Detection for Personal PM$_{2.5}$ Measurement

JinSoo Park [1] and Sungroul Kim [2],*

[1] Department of Industrial Cooperation, Soonchunhyang University, Asan 31538, Korea; vtjinsoo@gmail.com
[2] Department of Environmental Sciences, Soonchunhyang University, Asan 31538, Korea
* Correspondence: sungroul.kim@gmail.com; Tel.: +82-41-530-1266

Received: 10 November 2019; Accepted: 7 January 2020; Published: 11 January 2020

Abstract: With the development of technology, especially technologies related to artificial intelligence (AI), the fine-dust data acquired by various personal monitoring devices is of great value as training data for predicting future fine-dust concentrations and innovatively alerting people of potential danger. However, most of the fine-dust data obtained from those devices include either missing or abnormal data caused by various factors such as sensor malfunction, transmission errors, or storage errors. This paper presents methods to interpolate the missing data and detect anomalies in PM$_{2.5}$ time-series data. We validated the performance of our method by comparing ours to well-known existing methods using our personal PM$_{2.5}$ monitoring data. Our results showed that the proposed interpolation method achieves more than 25% improved results in root mean square error (RMSE) than do most existing methods, and the proposed anomaly detection method achieves fairly accurate results even for the case of the highly capricious fine-dust data. These proposed methods are expected to contribute greatly to improving the reliability of data.

Keywords: data interpolation; anomaly detection; bootstrap; fine dust; PM$_{2.5}$

1. Introduction

Korea has been experiencing severe environmental health problems caused by exposure to fine dust [1]. Thus, many stakeholders, including government officials, are trying hard to find solutions for the environmental issues. As part of these efforts, various artificial intelligence (AI)-based technologies are drawing attention as a way to predict future exposure levels as well as to reduce real-time exposure to PM$_{2.5}$ in our daily life. PM data is closely related to personalized health-care service and preventive medicine, which are research areas that have attracted much interest from many researchers today. The personalized healthcare service prompted us to develop predictive analytics technology, which requires the acquisition of data related to individual activity patterns [2,3]. Such data can be seen as person-specific data that is different from the population-based data to be used for the existing broadcasting-type environmental information service aimed at a large audience [2,4]. The pico-scale data is usually collected from each individual sensor device [4]. Unfortunately, such data are more likely to be incomplete than data collected from stationary sensors, because sensors attached to human subjects are affected significantly by the person's activity patterns, meteorological conditions, or the malfunction of the installed device or sensor itself. These kinds of incomplete data can lead to the provision of wrong services, because they may invoke wrong algorithmic decisions caused by the data. Such incomplete data typically include missing or abnormal data. In order to provide high-quality environmental information services, it is essential to conduct studies to deal with these two issues.

Much research has been done on the two issues mentioned above. Most of such research tends to focus on detecting missing or anomalous data only as follows. Conventional techniques related to missing data imputation in time-series data include methods based on machine learning, such as random forests [5], maximum likelihood estimation [6], expectation maximization [7], or nearest neighbors [8]. Technologies belonging to anomalous data detection include prediction-based [9], distance-based [10], probability-based [11], and linear models [11]. Researches dealing with both technologies include [12,13] both using machine learning techniques. As we have seen, there are few studies that simultaneously address these two problems.

Thus, in this study, we present our data-mining skills approach that deals with the two problems of missing and abnormal data included in the $PM_{2.5}$ data obtained from a personal portable sensor. Especially, we demonstrate that our kernel regression-based interpolation method and abnormal data detection method can be applied to our real personal $PM_{2.5}$ measurement data. We attempt to extend a well-known interpolation method incorporating a simple linear interpolation method to interpolate the bursty $PM_{2.5}$ data. The performance of the proposed method is provided in comparison to those of existing interpolation methods. In addition, details on the method are presented in the algorithmic method section of this paper.

2. Methods

2.1. Proposed Algorithm

In this paper, we proposed two algorithms: the kernel regression-based interpolation method and the subsequent abnormal-data detection method. The algorithm presented in this paper was done in the order shown in Figure 1. First, we chose the part of the total data that had no missing data and estimated the bandwidth for the chosen data part. Here, the bandwidth was the value to use for the kernel regression-based interpolation (KRBI) (dotted line in Figure 1), which will be explained in the next subsection. Afterwards, we examined missing data for the entire dataset. If there was missing data, we used linear interpolation (LI) to interpolate the missing data. Afterwards, the interpolation was done again by applying the KRBI algorithm [14]. In this case, the optimal bandwidth value, which was previously obtained, was used. If there were no missing data, abnormal-data detection began for the entire dataset.

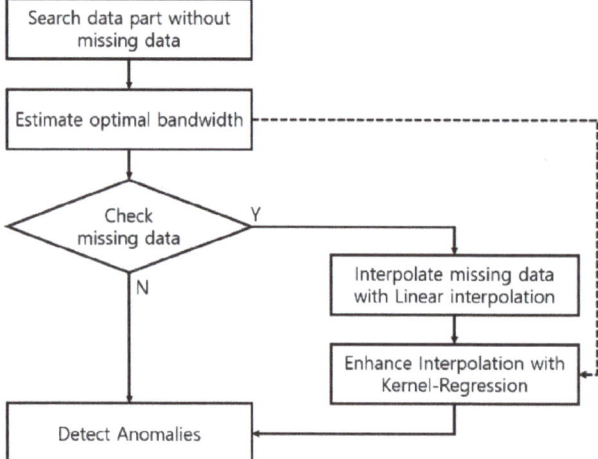

Figure 1. Flowchart of the proposed interpolation and anomaly detection.

2.2. Optimal Bandwidth Selection Based on Leave One Out Cross-Validation (LOOCV)

In order to calculate the appropriate bandwidth, we took all of the data and split it into training and verification data. As the bandwidth was changed from the small value to the large value, the verification data value was predicted using the training data. Finally, we calculated the estimated error in terms of the actual verification data value. Among the bandwidth values that contributed to the error calculation, we programmed our algorithm to take the bandwidth value that provided the lowest error as an appropriate bandwidth for the data used. The corresponding pseudo-code is available in [15]. The bandwidth selection process is explained in the experimental section.

2.3. Kernel Regression-Based Interpolation Using Linear Interpolation

The method proposed in this paper is based on the kernel regression [14], but it took the advantages of the linear interpolation method [16]. The LI method can be used appropriately, especially when the time-series trend is clear. For instance, when the time-series pattern appears to be rising up or decreasing, the LI method can be applied for the interpolation of the corresponding data pattern more appropriately. We used this property of the LI method in order to improve the performance of the KRBI method. In other words, we used the LI method for the bursty missing data a priori and then applied the KRBI method for the final interpolation of the missing data.

The kernel regression algorithm can be summarized as follows. First, we defined the time-series data as (t_i, y_i) where t_i and y_i represented the time and measurement of data at time t_i. Kernel regression was to set the representative value $\hat{y}$ of y_is where $p \leq i \leq q$ and the bandwidth h are defined as $h = p - q$. In this case, the representative value $\hat{y}$ might be calculated as a weighted average value of $W_i y_i$ where the weight W_i could be generated by following well-known statistical models such as Gaussian or uniform distributions. The algorithm can be expressed mathematically as follows.

$$\hat{y} = \frac{\sum_{i=1}^{n} K_h\left(\frac{x-x_i}{h}\right) y_i}{\sum_{j=1}^{n} K_h\left(\frac{x-x_j}{h}\right)} = W_i y_i$$

where the choice of weight $W_i = \frac{\sum_{i=1}^{n} K_h\left(\frac{x-x_i}{h}\right)}{\sum_{j=1}^{n} K_h\left(\frac{x-x_j}{h}\right)}$ and $K_h(\cdot)$ is a kernel chosen [14].

In order to apply the KRBI method, a proper bandwidth calculation must be done. For this purpose, we used a part of the data that had no missing data to estimate the proper bandwidth for the data interpolation. A detailed description is provided in the next section.

2.4. Context-Aware Anomaly Detection

Once we finished the bandwidth selection for interpolation and optimal bandwidth, we developed another algorithm for detecting anomalies in the time-series data. Many techniques have been presented to detect anomalies using various techniques. In this paper, we defined anomalies as data that showed significant changes in values within a very short time. For instance, if $PM_{2.5}$ concentration, measured every 10 seconds, showed significant drops or jumps during the 10-s period (for instance, observations that fall below Q_1-1.5IQR or above Q_3+1.5IQR in the box-and-whisker plot), we considered the value to be abnormal, because the amount of change in $PM_{2.5}$ concentration is assumed to stay stable or similar within a very short time. However, this rule was not an appropriate criterion for detecting outliers, because too much data fell into this category, and it was hard to think that they were all anomalies, given the data context. Thus, in this research, we defined data as anomalous when the following conditions were met. In reality, the $PM_{2.5}$ concentration does not change significantly in most cases. This phenomenon is reflected in the detection of anomalies in our time-series data analysis.

$$d_i = |y_i - y_{i-1}| > th$$

In other words, if d_i (difference of adjacent y_is) exceeds a certain threshold, th, then y_i can be an anomaly. Thresholds are chosen by visual inspection according to the characteristics of the PM data at the moment. Details are available in the experimental section.

3. Experimental Tests

3.1. Bootstrap Simulation on Real Dataset

In this section, we verified the effectiveness of the proposed interpolation method. In order to verify its validity, we (1) randomly removed some of the actual data, (2) interpolated the removed data based on our algorithm, and then (3) compared them with each other in terms of certain performance criteria, including a comparison of applying results using other known methods. We executed our experiments based on a bootstrapping test using the given data in three different scenarios. We assumed that the arc shape of time-series data could be classified into three different patterns in general: up slope, down slope, and flat. Based on this assumption, the validity of the method could be evaluated only for data belonging to each shape pattern. We used the bootstrapping test on the chosen dataset belonging to each shape pattern. Samples from each pattern section of the data were randomly selected and then deleted on purpose. Next, we estimated the interpolating values for the deleted data and compared the error rate between the real values and the estimated interpolation values. The time-series data for these validation tasks are given in Section 3.3.2. Given our thorough examination of the data, the data corresponding to the following time index were specifically chosen because they appeared to match the three typical shape patterns and did not have missing values in the patterns: 3600 to 4600 for the up slope, 3250 to 3500 for the down slope, and 19,000 to 20,000 (Flat 1), 27,000 to 28,000 (Flat 2), and 37,000 to 40,000 (Flat 3) for the flat pattern. Specifically, for the flat pattern case, we chose three sections in order to examine any significant performance differences, because the duration of the flat pattern section is relatively longer than the other part of the time-series data. In our bootstrapping test, 40, 60, 80, and 100 datapoints were randomly deleted to create missing data, and the interpolation results on the deleted data were evaluated compared to those of the original data in terms of RMSE (root mean squared error).

The performance of our proposed method was compared to those of the LOCF (last observation carried forward), Agg (aggregate), and Spline methods [17]. The LOCF method takes the most recent values prior to itself. The Agg method takes the mean of a few previous values. The Spline method is a smoothing technique that comes with base R; it was used for the interpolation of missing data. The results of comparing these four methods are given in Table 1, which shows that our proposed method had lower RMSEs than those of the existing three methods, except for the flat pattern, for which the RMSEs of all four methods were not statistically different. R programming language was used to analyze the performance of the interpolation methods, and R packages stats, zoo, and Metrics were used for the interpolation and calculation of error rates [18]. The simulation was performed on a Microsoft Windows 10 computer with Intel® Core™ i7-6500U CPU at 2.60GHz.

As demonstrated by the real data-set experiments, our proposed method worked better than the existing methods did, although there were very similar results for a few cases, such as the flat pattern. Based on these experimental results, we finally applied our proposed methods to the remaining cases of missing values.

Table 1. Root mean squared errors (RMSEs) of the four interpolation methods applied to our real-time personal monitoring data. LOCF: last observation carried forward.

Data Pattern	Interpolation Method	Number of Missing Data			
		40	60	80	100
Up slope	Proposed	26.919	26.426	25.733	28.599
	Spline	37.052	36.679	35.682	37.908
	LOCF	36.604	36.909	35.707	38.393
	Agg	656.741	657.476	659.759	658.761
Down slope	Proposed	24.687	25.704	27.840	28.945
	Spline	26.741	27.817	30.233	31.834
	LOCF	34.223	35.888	38.719	39.818
	Agg	283.824	278.758	280.763	280.666
Flat 1	Proposed	4.551	3.981	4.341	4.376
	Spline	3.871	3.326	3.555	3.535
	LOCF	5.215	4.803	4.874	5.403
	Agg	4.426	3.918	4.052	4.071
Flat 2	Proposed	3.633	3.751	3.694	3.656
	Spline	3.505	3.610	3.514	3.493
	LOCF	4.554	4.817	4.893	4.783
	Agg	4.031	4.129	4.062	4.062
Flat 3	Proposed	2.335	2.310	2.429	2.325
	Spline	2.229	2.080	2.237	2.187
	LOCF	2.991	3.027	3.226	2.875
	Agg	2.271	2.125	2.293	2.233

3.2. Optimal Bandwidth Selection

Before we carried out our interpolation on the data, we needed to decide on the value of the bandwidth used for our interpolation method. For this task, we randomly chose a complete part of the data that did not have missing data to estimate optimal bandwidth for the data set. Then, the complete data were split into two different sets for training and validation. The training dataset was used to set up a kernel regression model, and then the validation was carried out for the remaining data set. In this bandwidth-selection step, we increased the bandwidth from 1 to 100 to find out which bandwidth provided the smallest error that could be applied to our interpolation procedure. This bandwidth estimation step can in fact depend on the nature of the data. That is, since the results of the estimation may differ significantly depending on the shape of the data pattern, different bandwidths for each of the three data patterns were estimated and used for this study. However, because it was hard to detect the changing point of these different data patterns in our time-series data, we used the bandwidth estimated for the flat data pattern for the following two real-world data experiments. Further research on automatically calculating the appropriate bandwidth depending on the data patterns is necessary in the future.

3.3. Interpolation and Anomaly Detection with Real-World Personal Data

We applied our method described above our separated $PM_{2.5}$ data set containing missing data. These $PM_{2.5}$ data were collected at 10-s intervals from portable personal $PM_{2.5}$ monitors attached to the subjects. The data were measured between 25 January 2019 and 1 February 2019.

3.3.1. Application of Interpolation and Anomaly Detection Method with Real-World Personal Dataset 1

As shown in Figure 2, the level of $PM_{2.5}$ was stable, implying that the subject was relatively calm with minimal abrupt activity changes. The length of the dataset was 59,422 at 10-s intervals, but it had 17,968 missing datapoints in total. Before we went on to detect the anomalies, we first selected an optimal bandwidth for the dataset as we did previously.

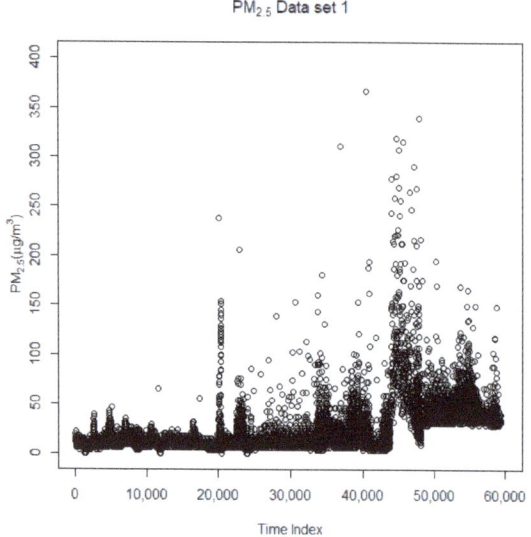

Figure 2. This is a data set that was collected for 6 days between 25 January 2019 and 31 January 2019. The length of the dataset is 59,422, having 17,968 missing datapoints in total.

After deciding on an optimal bandwidth for the data set, as seen in Figure 3, we interpolated the missing values based on the algorithm mentioned before. The corresponding results are given in Figure 4.

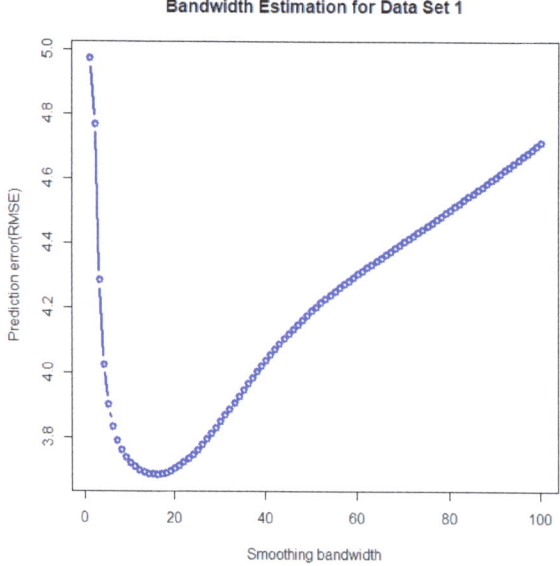

Figure 3. By increasing the bandwidth from small to large, we calculate the RMSE of the estimation and examined at what value of the bandwidth the corresponding RMSE is the minimum.

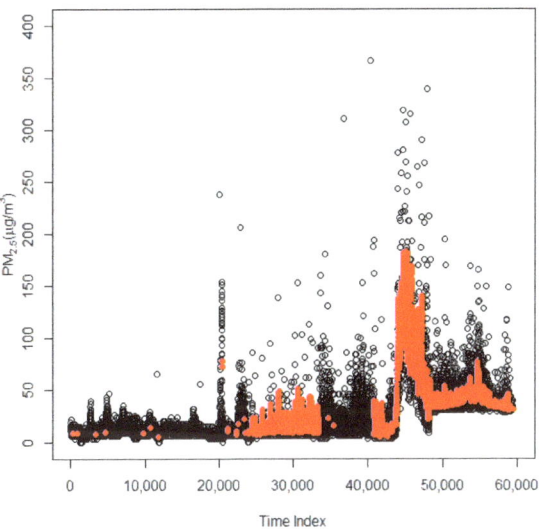

Figure 4. Missing values are interpolated and shown in red.

Figure 4 shows the interpolated data (red dots) superimposed on the actual data. It indicates that the missing data appeared in a small or large aggregated formation. In particular, it can be seen that much of data loss occurred between 40,000 and 60,000 according to the time index. After checking the raw data, we confirmed that this large data loss occurred between 13:00 and 16:00, presumably because of various activities in the afternoon. So much missing data could be very difficult to interpolate by methods other than linear interpolation. In such a case, it is preferable to carry out the initial interpolation using linear interpolation, as in our method, and then to apply other methods to improve the result. As shown in the figure, the interpolation goes well with the overall pattern of the data.

As the next step, for the entire dataset, we detected anomalies in the dataset, based on the method described in the method section. When the difference of adjacent $PM_{2.5}$ values is above a certain threshold (in this experiment, 200), we considered them to be anomalies. The corresponding anomalies are shown in red in Figure 5, which shows there are eight anomalies detected by visual inspection; the four red dots on the top seems to be the true outliers. The other red dots at the bottom do not appear to be true outliers, but they can also be regarded as outliers, because the PM concentration significantly dropped from the previous state by more than 200 µg/m^3 in 10 s, which is not acceptable as a normal degree of change. This result explains that the proposed outlier detection method produces fairly reliable results to some degree, even in a highly capricious environment. In selecting a threshold, we took an empirical approach to decide the threshold. Some data that could be visually regarded as anomaly data were selected as reference data, and then these data were examined to see if they were detected as actual anomaly data. As we performed the experiments with varying thresholds, the values when these reference data were detected were selected as thresholds.

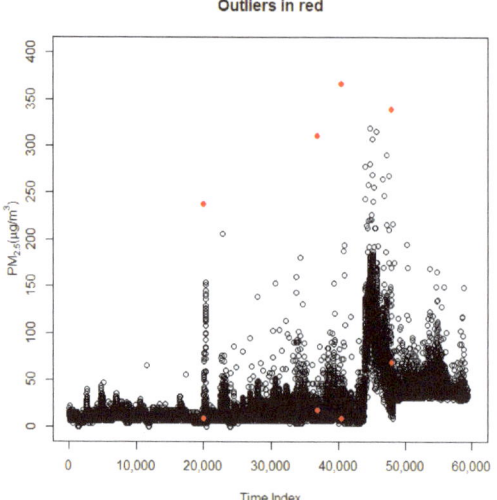

Figure 5. Anomalies are shown in red.

3.3.2. Application of Interpolation and Anomaly Detection with Real-World Personal Dataset Two

This dataset was collected for 8 days between 25 January 2019 and 1 February 2019. The length of the dataset is 62,878, having 74 missing datapoints in total. Dataset 1 used in the previous experimental test was very stable, because the distribution of PM data was mostly less than 100 μg/m^3 during the data acquisition period. However, the PM data in Dataset 2 (Figure 6) reached up to 2000 to 8000 μg/m^3 with 10-s intervals and showed a more dynamic change in the distribution of PM data, implying that the subject had various activities or was exposed to many different environmental conditions containing all the data-distribution patterns of rising, falling, and stable PM$_{2.5}$ concentrations.

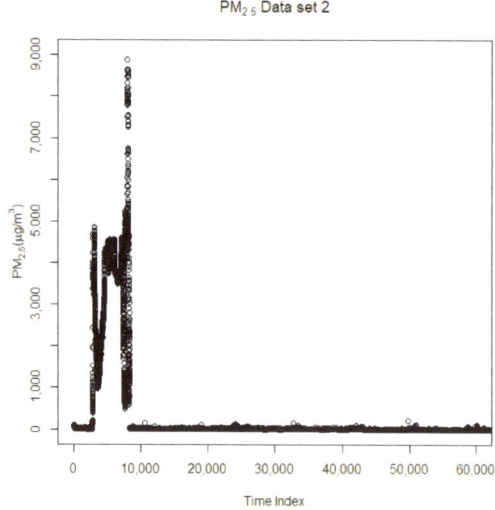

Figure 6. This is Dataset 2, collected for 8 days between 25 January 2019 and 1 February 2019. The length of the dataset is 62,878, having 74 missing datapoints in total.

The same bandwidth estimation steps were also carried out as described previously. Figure 7 shows the chosen bandwidth of 20 estimated for the flat part of the data between 33,000 and 33,600 in the time index. The interpolated missing data are shown in red in Figure 8, overlaid on the entire dataset.

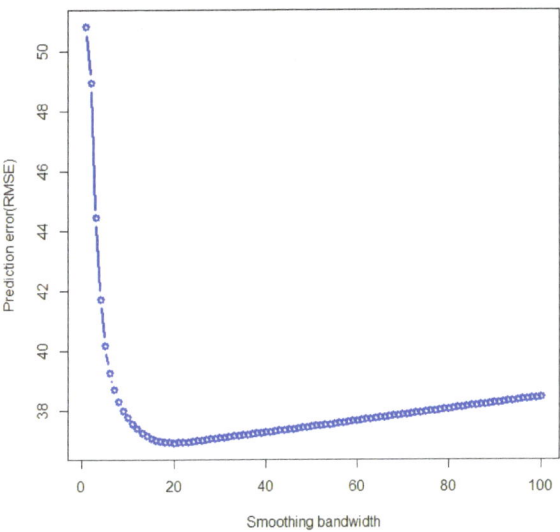

Figure 7. Optimal bandwidth for this dataset is chosen as being 19.

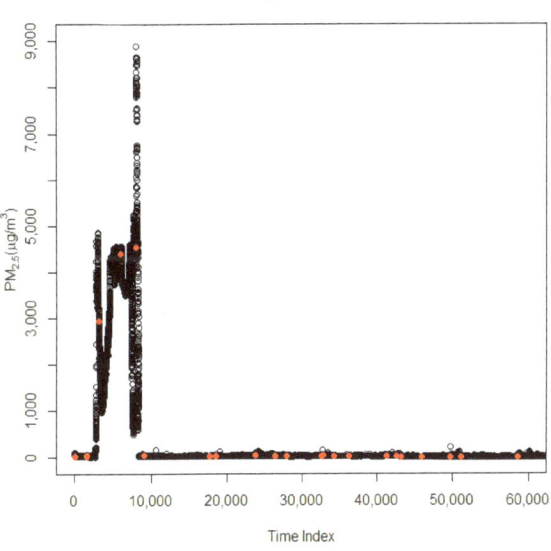

Figure 8. Missing data is interpolated and shown in red.

Figure 8 can be seen as two sets of data with different natures; in the front part, the data fluctuates heavily, but in the back part, the data are relatively stable. In particular, the front data were generated between 14:00 and 18:00, probably by a wide variety of movements. Unlike the previous Dataset 1, the data loss is relatively small, even under the condition of dynamic movements; so, a very stable performance sensor may be used. In the case of the latter part of the data, it appears that more data loss occurs than from the front part, but not much data are actually lost. Overall, a fairly stable interpolation result can be observed.

Finally, anomalies are also shown in red, in Figure 9. Overall, eight anomalies seem to be detected by visual inspection. If you look at the enlarged part of the data (between 2750 and 4000 out of the whole dataset), you can see that the detected outliers can be accepted as real outliers, because the corresponding PM concentration significantly changes within a 10-s period. In this experiment set, the threshold value 1000 was used as the criterion to detect outliers, and the threshold was also empirically chosen, analogously to the approach described in Section 3.3.1.

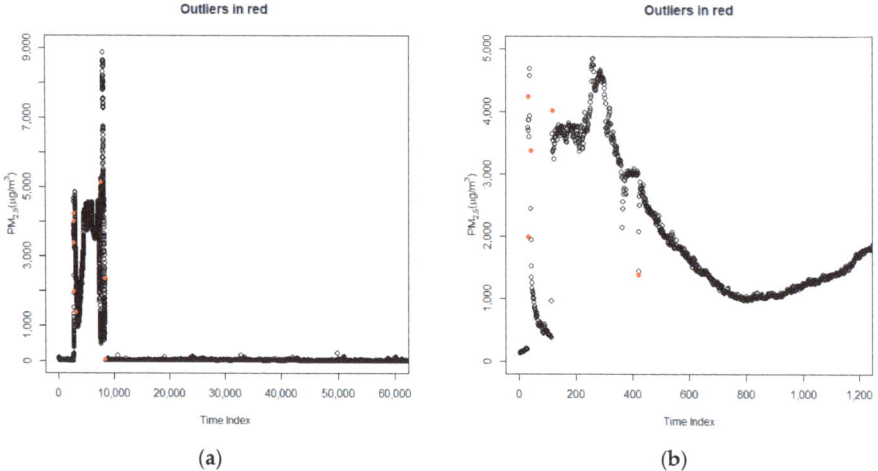

Figure 9. Anomalies are detected and shown in red (**a**) outliers for the entire dataset (**b**) outliers shown enlarged for the part of the dataset between 2750 and 4000.

4. Discussion

This paper presented the results of research on two important technologies related to artificial intelligence environmental information services, such as $PM_{2.5}$ exposure level prediction, and providing alerts based on the prediction. Most environmental data inevitably contain both missing data and anomalies caused by various factors, such as sensor malfunctions, errors in transmission, or errors in storage. Such incomplete data can lead to the miscalculation of data-based analysis and so must be processed appropriately before we use them for data analysis. Most of the related studies have tended to deal with either interpolation or abnormal data problems only. However, it is desirable that interpolation and abnormal data should be handled together for viable data-based services. Therefore, we conducted the research on these two problems simultaneously.

The first technique was interpolation, which can be regarded as a technique to cope with data lost from various sensors. Data tend to be missed mainly in time-series data, which is known to occur in three major forms: missing completely at random (MCAR), missing at random (MAR), and not missing at random (NMAR). In particular, the MAR type is an appropriate form of model for describing the data that are missing in most fine-dust time-series data [17]. This model assumes that the pattern of the time-series data can be described by a certain mathematical generative model and also attempts to take advantages of correlation information in multivariate environments to solve the missing-data problem [17]. In a stationary environment where sensors are fixed to certain objects, the generative model may be used to analyze the data. However, in a univariate case where data are collected from portable wearable devices, it is very challenging to estimate the missing data, because the number of attributes to use is very limited, and a large amount of data is likely to be missing [19]. In these cases, it is very difficult to apply a mathematical generation model to describe the data-distribution pattern or representative values for a specific activity. There are very few articles addressing the interpolation methods for univariate time-series data. Articles by Junninen studied the univariate algorithm in 2004, but do not consider time-series aspects [20]. Authors applied ARIMA (autoregressive integrated moving average) and SARIMA (seasonal autoregressive integrated moving average) models for the univariate model interpolation and provided comparison results [21]. A performance comparison is provided, using built-in interpolation methods in R [19]. Data loss in one instant or a short period of time can be easily interpolated with a simple error-recovery method such as linear interpolation. However, this simple interpolation method may be inadequate for a long bursty loss of sensor-provided data. When such long bursty data loss occurs, the interpolation method with the prediction technique may be more appropriate.

In addition, anomalous-data detection technology should be applied to detect data that deviate from the characteristics of data distribution. The anomalies can be commonly found in various industries including the environmental and finance fields. Anomalous time-series data can be regarded as data that disturb the continuity of data based on temporal flow [11]. Most techniques to detect anomalies can be classified as supervised and unsupervised according to the presence or absence of data labels [22]. The supervised method is a technique for detecting abnormal data by means of a learning algorithm when the data are labeled. The unsupervised method can be used when there is no label in the data, and can be used more flexibly than the supervised method can be, because often data labels are not available. As another category of classification, point-anomalous data detection technology detects one datapoint that has abnormal characteristics among much normal data [22]. In contrast, the statistical method extends the point method, and is a technology that detects data when that fall inside or outside a specific range of values in order to find anomalies. The disadvantage of these technologies is that most of these values are set up by hand. Recently, a lot of context-based methods have been studied. These can be seen as techniques for identifying abnormal data depending on the context of a situation. We proposed a method to detect abnormal data that show significant incremental or decremental changes as measured by time.

Although the excellence of the proposed algorithms has been proved by experiments, we address the limitations in conducting this study. First, the bootstrapping test was used to prove the excellence of the interpolation method, because we did not have any reference data for our interpolation algorithm. However, under such circumstances, we assumed that the bootstrapping test was the best choice for generating a missing value dataset randomly from a secure dataset. In addition, we did three scenario-specific experiments to evaluate the performance of the proposed method, which could be considered a simplified approach. The patterns of data distribution can be more complicated; since data distribution patterns were not considered in this study, some other methods might work better, depending on the type of data. In addition, we did interpolation by applying the same bandwidth estimated for the flat data pattern to the entire dataset. In the future, it is necessary to conduct research to automatically apply different bandwidth values for each data pattern. Finally, because the method proposed in this study was a kind of context-aware based detection method, we acknowledge that other methods previously proposed could get a completely different result. In addition, the threshold value was chosen empirically by visual inspection at the moment, but we need to develop a sophisticated algorithm to choose the value automatically. We may introduce such an automatic method in our near future study.

Despite the limitations of this study, it has a very important academic significance in that it presents a solution to the problem of interpolation and abnormal data of fine-dust data acquired from personal mobile terminals. We believe that these technologies will be a cornerstone for personalized environmental data services in the near future. Future research will include the prediction of PM data concentrations in both indoor and outdoor environments based on machine-learning technologies.

5. Conclusions

We think out findings have contributed greatly to overcoming the incompleteness of environmental data obtained from individual sensors and to providing an academic basis for more reliable data analysis. If the proposed algorithm is further improved, it will contribute a lot to advancing personalized healthcare and preventive medicine research.

Author Contributions: Conceptualization, J.P. and S.K.; Methodology, J.P.; Software, J.P.; Validation, J.P. and S.K.; Data Analysis, J.P. and S.K.; Resources, J.P.; Data Curation, J.P. and S.K.; Writing—Review and Editing, J.P. and S.K.; Visualization, J.P.; Project Administration, S.K.; Funding Acquisition, S.K. All authors have read and agreed to the published version of the manuscript.

Funding: This study was funded by The Environmental Health Research Center Project (2016001360003) by The Korea Environmental Industry and Technology Institute, Ministry of Environment, South Korea. This research was also supported by the Soonchunhyang University Research Fund.

Acknowledgments: Authors thank the study participants and their parents.

Conflicts of Interest: The authors declare no conflict of interest.

References

1. Nakao, M.; Ishihara, Y.; Kim, C.H.; Hyun, I.G. The Impact of Air Pollution, Including Asian Sand Dust, on Respiratory Symptoms and Health-related Quality of Life in Outpatients with Chronic Respiratory Disease in Korea: A Panel Study. *J. Prev. Med. Public Health* **2018**, *51*, 130–139. [CrossRef]
2. Bae, W.D.; Alkobaisi, S.; Horak, M.; Narayanappa, S.; AbuKhousa, E.; Park, C.-S. Predictive and Exposure Analytics: A Case Study of Asthma Exacerbation Management. *J. Ambient Intell. Smart Environ.* **2019**, *11*, 527–552.
3. McAullay, D.; Williams, G.; Chen, J.; Jin, H.; He, H.; Sparks, R.; Kelman, C. A delivery framework for health data mining and analytics. In Proceedings of the Twenty-eighth Australasian conference on Computer Science (ACSC '05), Newcastle, Australia, January 2005; pp. 381–387.
4. Ashana, S.; Strong, R.; Megahed, A. Health Advisor: Recommendation System for Wearable Technologies enabling Proactive Health Monitoring. *arXiv* **2016**, arXiv:1612.00800.
5. Stekhoven, D.J.; Bühlmann, P. Missforest—Non-parametric missing value imputation for mixed type data. *Bioinformatics* **2012**, *28*, 112–118. [CrossRef] [PubMed]
6. mvnmle: ML Estimation for Multivariate Normal Data with Missing Values. Available online: https://cran.r-project.org/web/packages/mvnmle/index.html (accessed on 13 October 2019).
7. mtsdi: Multivariate Time Series Data Imputation. Available online: https://cran.r-project.org/web/packages/mtsdi/index.html (accessed on 13 October 2019).
8. Crookston, N.L.; FinleyNakao, A.O. An r package for knn imputation. *J. Stat. Softw.* **2013**, *23*, 1–16.
9. Anomaly Detection with Time Series Forecasting. Available online: https://towardsdatascience.com/anomaly-detection-with-time-series-forecasting-c34c6d04b24a (accessed on 13 October 2019).
10. Keogh, E.; Lin, J.; Lee, S.H.; Herle, H.V. Finding the most unusual time series subsequence: Algorithms and applications. *Knowl. Inf. Syst.* **2006**, *11*, 1–27. [CrossRef]
11. Aggarwal, C. Time Series and Multidimensional Streaming Outlier Detection. In *Outlier Analysis*, 2nd ed.; Springer: Grewerbestrasse, Switzerland, 2017; pp. 273–310.
12. Akouemo, H.N.; Povinelli, R.J. Time series outlier detection and imputation. In Proceedings of the 2014 IEEE PES General Meeting, National Harbor, MD, USA, 27–31 July 2014; pp. 1–5.
13. Diettetich, T.; Zemicheal, T. Anomaly Detection in the Presence of Missing Values for weather data quality control. In Proceedings of the 2nd ACM SIGCAS Conference on Computing and Sustainable Societies (COMPASS '19), Accra, Ghana, 3–5 July 2019; pp. 65–73.
14. Nonparametric Regression. Available online: http://faculty.washington.edu/yenchic/17Sp_403/Lec8-NPreg.pdf (accessed on 14 October 2019).
15. Cross-Validation (Statistics). Available online: https://en.wikipedia.org/wiki/Cross-validation_(statistics) (accessed on 14 October 2019).
16. Linear Interpolation. Available online: https://en.wikipedia.org/wiki/Linear_interpolation (accessed on 14 October 2019).

17. Moritz, S.; Sardá, A.; Bartz-Beielstein, T.; Zaefferer, M.; Stork, J. Comparison of different Methods for Univariate Time Series Imputation in R. *arXiv* **2015**, arXiv:1510.03924.
18. CRAN. Packages by Name. Available online: https://cran.r-project.org/web/packages/available_packages_by_name.html (accessed on 13 December 2019).
19. Moritz, S.; Bartz-Beielstein, T. imputeTS: Time Series Missing Value Imputation in R. *R J.* **2017**, *9*, 207–218. [CrossRef]
20. Junninen, H.; Niska, H.; Tuppurainen, K.; Ruuskanen, J.; Kolehmainen, M. Methods for imputation of missing values in air quality data sets. *Atmos. Environ.* **2004**, *38*, 2895–2907. [CrossRef]
21. Walter, Y.O.; Kihoro, J.M.; Athiany, K.H.O.; Kibunja, H.W. Imputation of incomplete non-stationary seasonal time series data. *Math. Theory Model.* **2013**, *3*, 142–154.
22. Numenta. *The Science of Anomaly Detection*; Numenta: Redwood City, CA, USA, 2015.

© 2020 by the authors. Licensee MDPI, Basel, Switzerland. This article is an open access article distributed under the terms and conditions of the Creative Commons Attribution (CC BY) license (http://creativecommons.org/licenses/by/4.0/).

Contamination and Potential Risk Assessment of Polycyclic Aromatic Hydrocarbons (PAHs) and Heavy Metals in House Settled Dust Collected from Residences of Young Children

A. Stamatelopoulou [1,*], M. Dasopoulou [1], A. Bairachtari [1], S. Karavoltsos [2], A. Sakellari [2] and T. Maggos [1]

1. Atmospheric Chemistry and Innovative Technologies Laboratory, I.N.RA.S.T.E.S., NCSR "DEMOKRITOS", 15310 Athens, Greece; mdasopoulou@ipta.demokritos.gr (M.D.); kyriaki@ipta.demokritos.gr (K.B.); tmaggos@ipta.demokritos.gr (T.M.)
2. Laboratory of Environmental Chemistry, Department of Chemistry, National and Kapodistrian University of Athens, Panepistimiopolis, 15784 Athens, Greece; skarav@chem.uoa.gr (S.K.); esakel@chem.uoa.gr (A.S.)
* Correspondence: mina.stam@ipta.demokritos.gr; Tel.: +30-2106-503-719

Abstract: House settled dust (HSD) contains various hazardous materials, including polycyclic aromatic hydrocarbons (PAHs) and metals. Exposure to toxicants contained in HSD is of paramount concern especially in the case of young children, due to their particular behavioral characteristics. In this context, extracts of sieved vacuum cleaner dust from 20 residences with young children were examined for the presence of PAHs and trace metals, in Athens, Greece. The results indicated that PAHs and metals were ubiquitous in the studied residences. The calculated enrichment factors (EF) of trace metals indicated that Cu, Se, Zn, Hg, Cd, and Pb were mainly of anthropogenic. According to the PCA analysis, the main sources of household dust were: smoking inside the houses, combustion processes, resuspension of soil dust, and vehicle traffic. In general, the cancer risk due to PAHs exposure was found lower than the threshold value. The ingestion of house dust was the most important route of exposure to metals. The dose of almost all elements for the children was found 1–2 orders of magnitude lower than the corresponding reference values. Both the carcinogenic and noncarcinogenic risks of exposure were within the safety limits.

Keywords: settled house dust; PAHs; metals; enrichment factor; sources; health risk assessment

1. Introduction

Indoor air pollution plays a key role in human health since people spend the largest part of their time in indoor environments and 70–80% of their day in residential environments [1]. Indoor air and dust are the two main pathways ways of exposure to environmental contaminants. Depending on the nature of the contaminant itself, other routes like dietary exposure could also be of great importance. Compared to indoor air, indoor dust is more suitable for estimating human exposure to various chemical substances [2]. Polycyclic aromatic hydrocarbons (PAHs) and trace metals are both ubiquitous in house settled dust (HSD). Young children are more seriously affected by toxicants in HSD due to their particular behavioral characteristics, such as hand-to-mouth behavior, crawling, frequent mouth breathing, and sucking or chewing dirty toys [3–5]. Considering that children spend almost all of their day at home and their breathing zone is very close to the floor, where residential dust tends to accumulate, makes them more susceptible to environmental stressors [6].

Indoor sources of PAHs include smoking, cooking, gas-fired appliances, and the penetration of polluted outdoor air, since the incomplete combustion processes, traffic, industrial emissions, and heating with fossil fuels constitute some of the major outdoor sources of PAHs [2,7–10]. Several PAHs are known or suspected carcinogens, most prevalent among them being benzo[a]pyrene (B[a]P) and benz[a]anthracene (B[a]A) [7]. Whilst PAHs have not been directly associated with asthma and allergies so far, results from a number of studies suggest that traffic-derived PAHs and those emitted from smoking, cooking, and space heating may increase the risk for asthma-related symptoms and seroatopy in children [11].

Trace metals in indoor settled dust originated from both indoor and outdoor sources. External sources include soil, road dust, industrial and vehicular emissions, while decorative paints, consumer and cosmetic products, appliances, and combustion products of cooking, heating, and smoking constitute important indoor sources [12,13]. Inhalation, dust ingestion, and dermal contact are the main routes of human exposure to toxic metals, while it has been proved that incidental oral ingestion constitutes the most important exposure pathway, especially in the case of young children [12,14]. Taking into account that metals accumulated in the tissues and internal organs of the human body, affecting the central nervous system and acting as cofactors or promoters of other disorders, makes infants and young children more vulnerable to metal exposure when compared to adults [15,16]. In particular, trace metals such as Cd, Cr, Ni, and Pb can have cumulative effects, causing growth retardation in children, kidney disease, cancer, and several other adverse health effects [17].

To the best of our knowledge, there has been very limited research in Europe to examine the concentrations and sources of PAHs and trace metals in indoor dust [18–20] and even scarce focusing on residences with infants and young children [7]. Additionally, the present work is the first of its kind to be conducted in Greece. The objectives of this study were: (a) to identify the concentrations and profiles of PAHs and trace metals in indoor dust collected from Greek households, (b) to examine the relationship between the household characteristics and occupant's activities with PAH and trace metals concentrations in indoor dust, (c) to determine the sources of PAHs and trace metals in indoor dust of Greece, (d) to conduct a health risk assessment and to evaluate the cancer risks of exposure to PAHs and metals in residential environments.

2. Material and Methods

2.1. Study Design and Dust Sample Collection

In the present study, twenty residences located in various areas across the Athens region were investigated. Participants with a child below three years of age were eligible and were recruited both through advertising via flyers in several facilities of interest and word of mouth. The field campaigns took place during the warm season of 2015 (June–October). Before starting the sampling procedures, an informed consent form was signed by each participant. Vacuum cleaner bags were collected from all the 20 residences. The residents were asked to vacuum floors only but they were not limited to certain rooms. The idea behind that is that the residents carry the dust through movement from one room to another. The bags were removed from the vacuum cleaner, placed in zip-seal plastic bags, and transported to the laboratory where they were stored at $-20\,°C$, as described by Zhu et al. [21]. The samples were then analyzed for 25 PAHs and 15 trace metals.

In addition, detailed data regarding the building characteristics, such as location and type of the residence, potential nearby pollution sources and presence of attachedgarage, were recorded. Participants were asked to fill out an additional questionnaire concerning the characteristics of the residences (e.g., age of the house, previous water damage, presence of pets, smoking status).

2.2. Samples Clean Up and Preparation

One day before the beginning of the analysis, the samples were placed in a fume hood to reach room temperature. Subsequently, each dust sample was sieved to achieve homogenization. At the end of the process, the sieved dust was weighed using a high precision electronic scale (GIBERTINI, E42S-B, Novate Milanese (MI), Italy) and stored at $-20\,°C$ until analysis.

2.3. Laboratory Analysis of PAHs

Out of the 20 dust samples that were collected, 3 were destroyed during the analytical process. Therefore, the results that were obtained from 17 residences are subsequently presented. The dust extracts were analyzed for 25 PAHs using gas chromatography/mass spectrometry (GC/MS), including naphthalene (Nap), 2-methylnaphthalene (2M-Nap), 1-methylnaphthalene (1M-Nap), 1,2-dimethylnaphthalene (1,2 DM-Nap), 2,6- dimethyl-naphthalene (2,6 DM-Nap), acenaphthene (Ace), 2,3,5-trimethylnaphthalene (2,3,5-TM-Nap), acenaphthylene (Acy), anthracene (Ant), benzo[a]anthracene (B[a]A), benzo[a]pyrene (B[a]P), benzo[e]pyrene (B[e]P), benzo[b]fluoranthene (B[b]F), benzo[k]fluoranthene (B[k]F),benzo[g,h,i]perylene (B[ghi]Per), chrysene (Chry), dibenzo[a,h]anthracene (dBaAnt), fluoranthene (Fla), fluorene (Fl), perylene (Per), indeno (1,2,3-c,d) pyrene (IndP), phenanthrene (Phe), dibenzothiophene (dbt) pyrene (Pyr), 3,6-dimethylphenanthrene (3,6-dMePhe), 1-methylphenanthrene (1-MePhe). The analysis of the polyaromatic hydrocarbons was performed according to ISO12884 with the use of the certified material NIST Urban Dust (1649b). The main steps of the analytical procedure are the following. Samples were extracted in a Soxhlet extractor for 24 h at a reflux rate of about 4 cycles per hour. Before the extraction, deuterared PAHs (d8-Nap, d10-A, d10-Phe, d10-Chr, d10-Pyr, d12-B[ghi]P and d12-Perylene) were added as internal standards to monitor recovery. Subsequently, the extracts were concentrated in a rotary evaporator, loaded onto activated silica gel column chromatography, and eluted with n-hexane and n-hexane/dichloromethane (3:2). PAHs fraction was concentrated under a gentle stream of nitrogen, and an aliquot was analyzed by GC/MS (Agilent Technologies 7890A GC). Twenty-five PAHs were detected, including a group of suspected carcinogens PAHs.

2.4. Laboratory Analysis of Trace Metals

All materials that came into contact with the samples were previously washed thoroughly, soaked in dilute HNO_3 (Merck, Darmstadt, Germany), and rinsed with ultrapure water of 18.2 MΩ cm (Millipore, Bedford, MA, USA). For the preparation of all required solutions, class A volumetric glassware was used. For the determination of As, Ba, Cd, Cr, Cu, Mn, Ni, Pb, Sr, V, the samples were digested with HNO_3 65% supra pure (Merck) with the subsequent addition of H_2O_2 30% (Merck), according to the procedure described by [22]. For Hg and Se, samples were acid digested with HNO_3 in Teflon vials closed and left overnight at room temperature. The following day, the vials were thermostated at 80 °C for 3 h. The samples were allowed to cool at room temperature before their dilution with Milli-Q water. The digested samples were analyzed through inductively coupled plasma mass spectrometry (ICP-MS) by a Thermo Scientific ICAP Qc (Waltham, MA, USA). Measurements were carried out in a single collision cell mode, with kinetic energy discrimination (KED) using pure He. Matrix induced signal suppressions and instrumental drift were corrected by internal standardization (^{45}Sc, ^{103}Rh). Analyses of Al, Fe, and Zn were carried out by flame atomic absorption spectrometry (FAAS) (SpectrAA 200; Varian, Mulgrave, Vic, Australia), following digestion of the samples with a mixture of HNO_3 (65%), HCl (30%) and HF (40%). The calibration curves matched the matrix (acidity) of the samples.

2.5. QA and QC

2.5.1. PAHs

The quantification of PAHs was carried out through a linear calibration curve, which was built on matrix-matched standards (dust on filters). For the determination of the curve a standard polyaromatic solution containing secondary PAHs derivatives was used as internal standards. The solutions were of known concentration (0.05–10 ng/μL). The expanded uncertainty (Uexp k = 2) ranged from 6.3% to 23%, while the LOD of the method ranged from 0.6 pg/μL to 12.8 pg/μL.

2.5.2. Trace Metals

The method detection limits (MDLs) were within the range 0.500–5.00 ng g^{-1} for ICP-MS determined elements and 25.0–50.0 µg g^{-1} for FAAS determined elements. For statistical calculations, values below the MDLs were assigned the method detection limits divided by $\sqrt{2}$. The quality assurance was provided by analyzing the certified reference material (CRM) NIST 1649a (urban dust), which includes reference values for selected trace elements. The recoveries for As, Ba, Cr, Cu, Fe, Mn, Ni, Pb, V, and Zn in the CRM were in the range ±20%. Further, recovery efficiency for spiking sample analysis was ±25% for all elements.

2.6. Statistical Analysis

Statistical analysis was performed with SPSS 22 software (SPSS Inc., Chicago, IL, USA) for Microsoft Windows®. The nonparametric Kruskal–Wallis test was used to examine whether there were significant differences in chemical concentrations among residences with different household characteristics and different occupant's activities. Statistical significance was set at two stages $p < 0.05$ (95% confidence interval) and $p < 0.1$ (90% confidence interval). Pearson's correlation coefficient analysis and principal component analysis (PCA) identified the relationship between contaminants and possible sources. PCA constitutes one of the most common multivariate statistical methods widely used in dust contamination studies [2,9,17,23].

2.7. Enrichment Factors

In order for the metal enrichment in house dust and possible natural or anthropogenic sources to be determined, enrichment factors (EF) were calculated using the Equation (1).

$$EF = \frac{\left(C_x/C_{ref}\right)_{sample}}{\left(C_x/C_{ref}\right)_{background}} \quad (1)$$

where C_x/C_{ref}: the ratio of concentrations of trace metal to the corresponding concentration of a reference metal in the sample and background. In the present study, Al was selected as reference metal, as it is considered that anthropogenic activity contributes the least to its presence. In general, as long as the enrichment factor is less than 10, the natural origin of the elements dominates, while for values greater than 10, their enrichment due to anthropogenic activity is important due to anthropogenic activity [15,23], especially for metals for which the factor is estimated to be greater than 40.

2.8. Cancer Risk Assessment of PAHs in Settled House Dust

The estimation of the excess lifetime cancer risks for young children associated with nondietary ingestion of PAHs in settled house dust was conducted, according to the methodology of Maerterns et al. [18], using the following equation:

$$\text{Lifetime cancer risk} = \sum_{i=1}^{n}\left(\frac{(C_i \times PEF_i \times IR \times EF \times SF \times AF)}{BW \times 1000}\right) \quad (2)$$

where C: the concentration (µg/g) of each of the carcinogenic PAHs (B[a]A, B[a]P, B[b]F, B[k]F, Chr, D[ah]A and Ind), PEF (Potency Equivalency Factor): the factor that expresses the potency of each PAH in terms of B[a]P which were as follows: B[a]A = 0.1, B[b]F = 0.1, B[k]F = 0.1, Chr = 0.001, D[ah]A = 5 and Ind = 0.1, IR (Ingestion Rate): Daily ingestion rate of dust which was considered as 0.05 g/day and 0.1 g/day, EF (Exposure Factor): the rate of exposure, which was calculated taking into account that a preschool-aged children is active for 12 h (while the remaining 12 asleep) and assuming that is exposed to these concentrations until the age of 5, BW (Body Weight): the average body weight, which was considered equal to 13 Kg, SF (Slope Factor): the estimate of the probability of a response occurring per unit intake of the PAH over a lifetime and for the present study an oral slope factor for B[a]P equal to 7.3 (mg·Kg/day) was used, AF (Adjustment Factor): the factor accounts for children's exposure during early life stages. For exposure to carcinogens, U.S.EPA recommends: AF = 0 for children up to 2 years of age and AF = 3 for children between 2 and 15 years of age. It has been estimated that preschool-aged children (1–5 years old) ingest 0.05 to 0.1 g of dust per day, depending on the season and the amount of time spent indoors [18]. For this reason, two different scenarios for the calculation of the risk of carcinogenesis were included: a moderate exposure scenario (0.05 g/day) and a high exposure scenario (0.1 g/day). For the calculations, the average value of B2PAHs that was obtained taking into account the concentrations of all the studied residences, was used.

2.9. Health Risk Assessment of Metals in Settled House Dust

The present study aims to calculate the exposure of children and their parents to metals in house dust based on the assessment method of human exposure risk developed by US Environmental Protection Agency [24]. Residents are exposed to dust through three main pathways: ingestion, inhalation, and dermal contact. The dose received through each of the pathways was calculated by the following equations:

$$D_{ingestion} = C \times \frac{IngR \times EF \times ED}{BW \times AT} \times 10^{-6} \quad (3)$$

$$D_{dermal} = C \times \frac{SA \times SL \times ABS \times EF \times ED}{BW \times AT} \times 10^{-6} \quad (4)$$

$$D_{inhalation} = C \times \frac{InhR \times EF \times ED}{PEF \times BW \times AT} \quad (5)$$

The available studies on the carcinogenic risk parameters in the current assessment standards have been conducted, taking into account only the pathway of inhalation [25]. Therefore, the daily average exposure for life through inhalation of a carcinogenic metal ($LD_{inhalation}$) was calculated by:

$$LD_{inhalation} = \frac{C \times EF}{PEF \times AT} \times \left(\frac{InhR_{child} \times ED_{child}}{BW_{child}} + \frac{InhR_{adult} \times ED_{adult}}{BW_{adult}} \right) \quad (6)$$

where C is the concentration (mg/kg) of the metal in the dust. IngR is the ingestion rate, estimated in the present study to be equal to 200 (mg/day) for children and 100 (mg/day) for their parents [26]. InhR is the inhalation rate and was considered as 5.71 for children and 19.02 for adults (m^3/day) [27]. Exposure frequency (EF), taking into account holidays, was assumed as 335 days per year. Exposure duration (ED) was taken as 5 years for children and 25 years for adults. AT is the averaging time (ED × 365 days for noncarcinogens and 70 × 365 days for carcinogens). BW is the average body weight (13 Kg for children and 70 Kg for adults). The surface area of skin exposure (SA) was considered to be 1150 cm^2 for children and 2145 cm^2 for their parents. The skin adhesive capacity (SL) was taken as 0.2 for children and 0.07 for adults. The skin absorption factor (ABS) was considered as 0.001 [15]. PEF is the particulate emission factor, in this study, 1.36×10^9 m^3/kg.

The doses that were calculated for each element and exposure pathway were subsequently divided by the corresponding reference dose (RfD) (mg/kg × day) to obtain the noncancer risk (HQ), while the dose was multiplied by the corresponding slope factor (SF) to calculate the level of cancer risk [15]. The hazard index (HI) is the sum of HQ. It is supported that if HQ or HI < 1, the risk is small, whereas in case HQ or HI >1, there is a noncarcinogenic risk. Risk refers to the risk of cancer, indicating the probability of cancer, which is usually expressed as the proportion of cancer population in unit population. The acceptable risk is within the range of 10^{-6} to 10^{-4} [15,27].

3. Results and Discussion

3.1. PAH and Trace Metals Concentrations

3.1.1. PAHs

Overall, the 25 PAHs were detected in all dust samples, except for perylene, which was not detected in six residences. The concentrations of all the individual PAHs are presented in Table 1. The total PAH concentration in indoor dust ranged between 1.4 and 7.3 μg/g, across the houses, with a median value of 2.2 μg/g. The house with the lowest total concentration was a recently renovated apartment, and its occupants had moved in about a month before the start of the campaign. This observation is consistent with previously conducted studies, where lower ΣPAH concentrations in newly constructed or renovated houses have been reported [18,21]. The house with the highest ΣPAH concentration was 10 years old, and the only distinct feature it had when compared to the other houses of the study was the presence of a large carpet in the living room. This finding is in agreement with the observations of Maertens et al. [18], who reported elevated total PAH concentration in a residence that was 90% carpeted. In general, the ΣPAH concentrations in the present study were lower with respect to previous studies. Specifically, the mean levels of total PAHs in indoor dust in two studies that were conducted in China were much higher as follows: 21.9–329.6 and 30.9 μg/g, respectively [2,9]. Similarly, higher concentrations were reported in USA (29.2 μg/g) [28,29], Canada (12.9 μg/g) [18] and Italy (5.11) [19].

Among the measured PAHs, phenanthrene was the most predominant, followed by fluorene, while perylene was the congener detected in the lowest concentrations in all the studied households. In terms of their molecular weight, PAHs are categorized in three main categories: PAHs with High Molecular Weight (HMW), Medium Molecular Weight (MMW) and Low Molecular Weight (LMW). HMW PAHs are in general more toxic and harmful with respect to the LMW, which are less toxic. In the present study, LMW PAHs were the predominant PAHs in indoor dust, accounting for 87.5 % of the total PAHs.

Table 1. Concentrations of polycyclic aromatic hydrocarbons (PAHs) in indoor dust samples (ng/g).

PAH (ng/g)	Mean	Median	SD	Min	Max
Nap	339	116	609	75.2	2479
2M-Nap	242	172	198	118	965
1M-Nap	126	91.6	107	61.1	518
Acy	15.7	13.2	7.82	8.24	39.4
1,2 DM-Nap	49.4	40.1	30.7	29.5	158
2,6 DM-Nap	11.7	9.08	7.04	6.43	33.2
Ace	94.6	78.0	69.0	60.8	353
2,3,5-TM-Nap	10.5	8.56	7.33	3.12	28.8
Fl	421	424	207	221	1116
DBT	25.4	21.2	20.9	4.96	91.8
Phe	925	905	465	528	2525
1M-Phe	67.9	45.8	79.0	14.9	358
3,6 DM-Phe	42.3	21.5	75.7	7.59	334
Ant	22.6	19.4	12.7	10.3	57.3
Flu	99.0	86.0	61.8	30.2	308
Pyr	100	74.4	117	20.1	542
BaA	11.3	8.25	12.2	1.68	54.9
Chr	45.3	27.2	72.9	4.25	325
BbF	20.7	16.9	19.9	2.01	90.4
BkF	10.7	7.22	12.3	1.01	55.7
Bep	14.0	9.79	14.8	1.49	65.9
Bap	7.06	5.45	5.69	1.05	24.0
Per	1.02	1.08	1.06	0.00	3.39
IndP	11.1	8.39	10.8	1.05	49.2
dBaAnt	2.06	1.69	1.83	0.39	8.39
B[ghi]Per	13.8	10.3	11.2	1.36	49.3

Out of the 25 measured PAHs, 16 (naphthalene, acenapthylene, acenaphtene, fluorene, phenanthrene, anthracene, fluoranthene, pyrene, bez(a)anthracene, chrysene, benzo(b)fluoranthene, benzo(k)fluoranthene, benzo(a)pyrene, indeno(1,2,3-c,d)pyrene, dibenzo(a,h)anthracene) have been designated high priority pollutants by Environmental Protection Agency (EPA), while 7 (bez(a)anthracene, chrysene, benzo(b)fluoranthene, benzo(k)fluoranthene, benzo(a)pyrene, indeno(1,2,3-c,d)pyrene, dibenzo(a,h)anthracene) have been classified as human carcinogens by the U.S.EPA (2003), referred as B2 PAHs. The contribution of B2 PAHs was estimated equal to 5% of the total PAHs. Especially benzo (a) pyrene has been classified as particularly dangerous and has been extensively studied worldwide [7,18,30–32]. The German Federal Environmental Agency's Commission for indoor air quality has established the limit of 10 µg/g for exposure to B[a]P in house dust [18]. B[a]P concentrations in the present study were found 3 orders of magnitude lower than the above limit. Compared to the results of earlier studies, the concentrations of B[a]P found in this study were either lower [18,31,32] or comparable [7].

3.1.2. Trace Metals

The concentrations of the heavy metals identified in the dust samples are summarized in Table 2. Among the elements that were examined, Fe was found to have the highest concentration (4.9 µg/g), followed by Al (4.2 µg/g), Zn (0.4 µg/g), Cu (0.34 µg/g) and Ba (0.25 µg/g). The elevated concentrations of both soil-related (Fe, Al) and anthropogenic (Zn, Cu, Ba) elements in the dust samples implies that the dust accumulation in the residential indoor Environment is mainly affected by the resuspension of natural dust (soil) and road dust. The elements Hg and Cd had the lowest concentrations.

Table 2. Metal concentrations in indoor dust samples (µg/g).

Elements	Mean	SD	Min	Max
Zn	401	211	136	1031
Fe	4913	5650	851	26100
Al	4217	9391	269	44900
Hg	0.4	0.5	0.0	2.1
V	9.0	5.4	2.4	21.3
Cr	65.2	36.4	14.2	147
Mn	128	67.9	28.9	272
Ni	29.9	23.9	5.2	103
Cu	339	711	11.8	3051
As	4.0	4.4	1.3	19.8
Se	1.0	0.6	0.2	2.8
Sr	118	97.6	37.7	396
Cd	0.5	0.6	0.1	2.3
Ba	251	237.0	35.0	1130
Pb	46.1	71.5	6.5	343

The highest concentration of Fe (26.1 mg/g) was observed in a residence located in an urban area on a high circulation road. Indeed, Fe has been linked with traffic as it originates either from road dust resuspension or emissions from brake wear [33]. The highest concentration of Al (45 mg/g) was observed in a residence which was located in a rural area, while the highest levels of Zn, Cu, and Sr were detected in a ground floor house located in the Athens city centre, and thus significantly affected by vehicular emissions [34]. Due to the absence of established legislation, limits for trace metals in indoor dust were compared with concentrations reported in the literature. Compared to the metal concentrations reported in a recent study examining residential dust samples in Toronto, Canada, the levels of the present study were all higher (Cr (65.2 vs. 42 µg/g), Cu (339 vs. 136 µg/g), Ni (29.9 vs. 23 µg/g), Pb (46.1 vs. 36 µg/g), and Zn (401 vs. 386 µg/g), except for Cd (0.5 vs. 1.7 µg/g) [12]. Higher concentrations compared to those of the present work were also presented in an earlier study conducted in Japan [31], where Shakya 2013, reported lower concentrations for Cr, Ni, Pb, and Zn, but higher for Cd (0.5 vs. 8.2 µg/g) in Kathmandu, Nepal [32].

3.2. Relationship between PAH and Trace Metals Concentrations and Household Characteristics

The relationship between the individual PAH and trace metal concentrations and a series of household characteristics was examined.

3.2.1. PAHs

Compared to residences located in suburban areas, houses in urban areas were found to exhibit significantly higher concentrations of Anthracene (26.2 vs. 16.1 ng/g, $p < 0.05$) and other PAHs derived from combustion processes (Flut, Pyr, B[a]A, Chr, B[b]F, B[k]F, B[a]P, Ind και B[ghi]P), referred to hereafter as COMPAHs, such as B[a]P (8.5 vs. 5.8 ng/g). This is in agreement with the literature, as it has been reported that proximity to combustion sources such as vehicle exhaust emissions is probably the most important factor affecting the accumulation of PAHs in indoor dust [2,7,35]. The age of the house was found to affect both the Pyr (75 vs. 114 ng/g, $p < 0.05$) and all the B2PAH (87 vs. 140 ng/g, $p < 0.05$) concentrations, with the houses that have been built in the last 20 years presenting lower concentrations than the older ones. Houses on the ground floor presented higher concentrations of Flut, B[a]A, Chr, B[b]F, Ind, D[ah]A ($p < 0.05$), which are known to originate from combustion processes.

Both the B2PAH and COMPAH concentrations were considerably higher in the houses with an attached garage. Especially for BaP a significant difference was noticed (13.6 vs. 6.2 ng/g, $p < 0.05$). The existence of a metro or tram station nearby (< 200 m) was associated with increased concentrations of COMPAHs (479 vs. 217 ng/g, $p < 0.05$), especially those of B[a]A, B[a]P, Ind and D[ah]A. This is probably due to the heavy traffic and the parking stations that are located close to such stations. The presence of a carpet in the house was associated with increased concentrations of Nap (1319 vs. 209 ng/g, $p < 0.05$), while the habit of residents wearing shoes inside their home was associated with higher concentrations of COMPAHs (475 vs. 221 ng/g, $p < 0.05$), as well as B2PAHs (191 vs. 73 ng/g), especially those of B[a]P, Pyr, Chr, D[ah]A and B[ghi].

3.2.2. Metals

The location of the house seems to be an important factor in the presence of Zn ($p = 0.083$), Fe ($p = 0.071$), and Sr $p = 0.071$) in house dust, with houses located in urban areas having higher concentrations. Figure 1 illustrates the elevated concentrations of the aforementioned elements detected in urban areas, which are lower in suburban areas and especially in rural areas. This observation is explained by the common emission sources of these elements, which are traffic emissions and road dust resuspension. Also, houses located close to a highway (<200 m) indicated higher concentrations of Se (1.3 vs. 0.6 µg/g, $p = 0.073$).

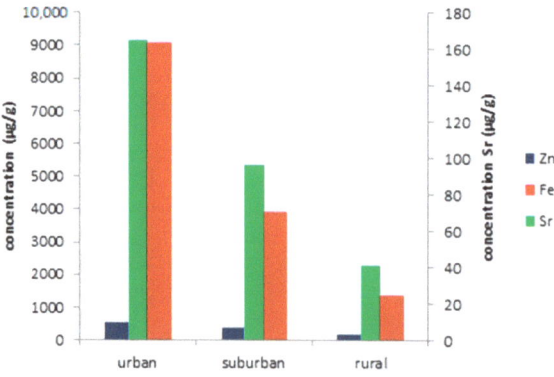

Figure 1. Concentrations of Fe, Zn and Sr in urban, suburban and rural areas.

Residences located either on the ground floor or the first floor indicated higher Hg concentrations with respect to those at higher floors (0.62 vs. 0.16 µg/g, $p < 0.05$). The age of the house was found to affect the levels of Sr, as higher concentrations in homes older than 20 years (128 vs. 77 µg/g, $p < 0.05$) were observed. In houses with a total area of less than 100 m^2, higher concentrations of Fe (6928 vs. 2526 µg/g, $p < 0.05$), Al (2691 vs. 1562 µg/g, $p < 0.05$), V (10.6 vs. 5.7 µg/g, $p < 0.05$) and Ba (327 vs. 133 µg/g, $p < 0.05$) were found. This fact can be related to the lower number of openings (windows and doors) to houses of smaller total area, which consequently leads to lower infiltration rates. Houses that used fireplaces as the main source of heating indicated significantly higher concentrations of Pb (343 vs. 32 µg/g, $p < 0.05$), probably due to the burning of painted or treated wood [33,34], while those that were heated using electricity were associated with significantly higher concentrations of Fe (10729 vs. 3529, $p < 0.05$) and V (12.8 vs. 7.3 µg/g, $p = 0.059$), likely due to dust resuspension occurring during the operation of air conditioning. In houses that remained closed (without the presence of occupants) for several hours during the day, higher values of As (8.7 vs. 2.5 µg/g, $p < 0.05$) and Sr (152 vs. 93, $p < 0.05$) were detected in household dust. The type of glazing appeared to affect Pb levels in household dust, as higher concentrations were observed in homes with single glazing compared to those with double glazing (57.8 vs. 46.2, $p = 0.078$). In terms of glazing on indoor Pb concentrations, we believe that the effect is not directly related to the type of glazing (or to be more accurate not only related to that) but on the type of the windows used. Double glazed windows are of newer technology and have been found in previous studies to significantly reduce the infiltration rates [35]. Since Pb is emitted from outdoor anthropogenic activities and resides in the fine fraction, it is expected that its indoor concentrations will be attributed mainly to infiltration. Finally, as with PAHs, the habit of residents wearing shoes inside their dwelling significantly affected the levels of Al (2684 vs. 1850 µg/g, $p = 0.076$) and Ba (275 vs. 217 µg/g, $p < 0.05$) in house dust. As both elements can be found in urban dust, there are transferred inside the residents by the dust that is found on their shoes.

3.3. Relationship between PAH and Trace Metals Concentrations and Occupant Activities

Apart from the household characteristics, the impact of the occupants' activities on the detection of PAHs and trace metals in dust was also examined.

3.3.1. PAHs

The analysis demonstrated that smoking inside the house significantly contributes to the increase of the concentrations of Phe (1275 versus 780 ng/g, $p = 0.0730.1$) and Ant (32.1 versus 18.7, $p = 0.073$), whereas no statistically significant differences for the other COMPAHs were observed. However, the concentrations of all COMPAHs in smoker's residences were found to be significantly higher than in nonsmoker's dwellings (Figure 2). In particular, Chr indicated 2.5 times higher concentration in smokers' homes (72 vs. 30 ng/g), while Pyr (153 vs. 63 ng/g), B[b]F (30 vs. 14 ng/g), B [k] F (16 vs. 7 ng/g) and B [a] p (10 vs. 5 ng/g) were found twice as high in smokers' houses as in nonsmokers. In a review that was conducted by Maertens et al. [36], a moderate but statistically significant relationship between smoking and PAHs concentration has been reported, while Langer et al. (2010) pointed out in their study that smoking was not a strong determinant in the presence of PAHs in house dust. However, other studies indicated no significant difference between smoking and nonsmoking dust [29,37], with Qi et al. [2] underlying that the effect of smoking on PAH concentration in dust remains uncertain. The use of air conditioning was strongly associated with both COMPAHs (548 vs. 227 ng/g, $p < 0.05$) and B2 PAHs (226 vs. 66 ng/g, $p < 0.05$). Finally, an inverse association between the concentration of ΣPAHs and cleaning activities was found. In particular, houses that were mopped more than once per week indicated significantly lower levels of ΣPAHs (3.4 versus 2.3 µg/g, $p < 0.05$). This finding suggests that frequent cleaning activities can help reduce the concentration of PAHs in household dust. This observation is in agreement with Maertens et al. [18], who reported a weak but statistically significant negative correlation between the frequency of vacuum use and the concentration of PAHs.

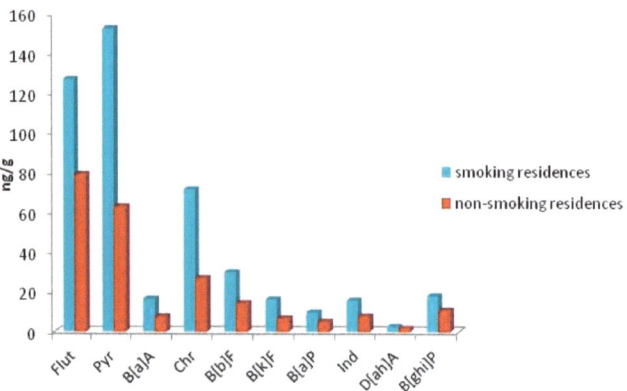

Figure 2. Concentrations of COMPAHs in smoking and nonsmoking houses.

3.3.2. Metals

In terms of residents' activities, smoking indoors was associated with increased concentrations of Cu (593 vs. 216 µg/g, $p < 0.05$), Sr (141 vs. 89 µg/g, $p < 0.05$) and Zn (422 vs. 379 µg/g, $p = 0.058.1$). Frequent cleaning activities (> 5 activities/week) were associated with increased Al concentrations (2801 vs. 1918 µg/g, $p < 0.05$) and the use of cleaning sprays was associated with increased levels of Pb (75.3 vs. 19.5 µg/g, $p < 0.05$). The use of air conditioning appeared to increase the concentrations of Cu (765 vs. 120 µg/g, $p = 0.070.1$), Zn (540 vs. 321 µg/g, $p = 0.092$) and Cd (1.0 vs. 0.3 µg/g, $p = 0.0090.1$). Finally, in terms of different cooking methods, frying was associated with increased levels of Ni (34.0 vs. 25.3 µg/g, $p < 0.05$) and Mn (147 vs. 100 µg/g, $p = 0.076$) in house dust, while frequent boiling (using of pot) was linked with higher levels of As (6.1 vs. 3.2 3 µg/g, $p < 0.05$) and Cd (1.1 vs. 0.3 3 µg/g, $p < 0.05$).

3.4. Enrichment Factors and Principal Component Analysis

Figure 3 presents EF of elements in the house dust samples. Several metals such as Fe, V, Mn, Sr, and Ba were of natural origin (EF < 10), while other metals such as Cu, Se, Zn, Hg, Cd, Pb (EF > 40) and Cr, Ni, As (20 < EF < 40) appeared to derive from anthropogenic processes.

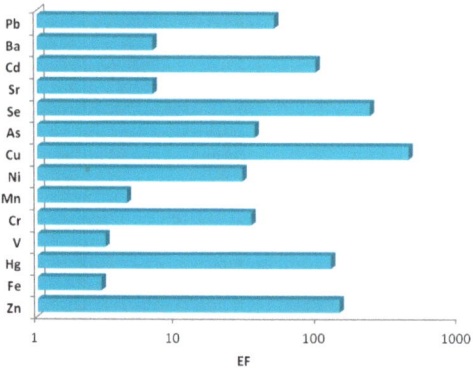

Figure 3. Enrichment factor (EF) of elements in house dust.

Principal Component Analysis (PCA) was performed for the PAHs and the metal contents of all dust samples. Eight PCs (Principal Components) were extracted with eigenvalues >1, by which 91.9% of the total variance was explained. The predominant components of each factor (>0.5) are presented in bold (Table 3).

Table 3. Results of Principal Component Analysis (PCA).

	Factor 1	Factor 2	Factor 3	Factor 4	Factor 5	Factor 6	Factor 7	Factor 8
	Smoking	Combustion Processes	Soil Dust	Vehicle Emissions	Road Dust	Abrasion of Plated Surfaces	Soil	Vehicle Emissions
Eigenvalue	11.522	4.840	3.540	2.127	2.051	1.269	1.160	1.056
Cumulative %	38.41	54.54	66.34	73.43	80.27	84.50	88.36	91.88
Acy	0.187	**0.951**	−0.005	−0.035	0.068	−0.058	−0.108	0.132
Ace	−0.104	**0.962**	0.093	−0.025	−0.120	0.028	0.037	0.052
Fl	0.088	**0.973**	−0.026	−0.063	−0.030	−0.060	−0.087	−0.065
PH	0.068	**0.987**	−0.018	−0.060	0.027	−0.027	−0.049	−0.057
Ant	0.360	**0.632**	−0.021	−0.112	0.462	0.123	−0.027	−0.105
Flu	**0.913**	0.367	0.003	−0.016	0.113	0.013	0.021	−0.055
Pyr	**0.975**	0.115	0.071	−0.007	−0.027	0.035	−0.047	−0.007
Ba[A]	**0.987**	0.008	0.026	0.064	0.067	0.002	0.057	0.037
Chr	**0.974**	0.058	0.086	−0.001	−0.058	0.052	0.032	−0.042
B[b]F	**0.973**	0.111	0.011	−0.009	0.116	0.020	0.032	0.069
B[k]F	**0.991**	0.061	0.034	0.036	0.041	0.016	0.027	0.018
B[a]P	**0.929**	−0.049	−0.047	0.020	0.133	0.018	0.123	0.128
Ind	**0.986**	0.050	0.004	0.011	0.077	−0.003	0.061	0.041
D[ah]A	**0.971**	0.033	−0.070	0.004	0.033	0.053	−0.006	−0.023
B[ghi]P	**0.940**	0.018	−0.078	−0.039	0.145	−0.052	0.191	0.029
Zn	0.261	−0.091	0.310	**0.613**	**0.563**	−0.159	0.050	0.216
Fe	0.152	−0.158	0.032	0.048	0.133	−0.080	**0.910**	0.015
Al	−0.078	−0.166	**0.933**	−0.056	0.202	0.032	−0.131	0.037
Hg	**0.883**	−0.016	0.013	−0.039	−0.069	0.383	0.044	−0.048
V	0.218	−0.065	**0.750**	0.460	0.108	0.015	0.260	0.099
Cr	0.276	0.188	0.491	0.008	0.133	**0.761**	−0.059	0.100
Mn	0.203	−0.050	0.188	0.314	−0.461	−0.273	0.460	−0.176
Ni	**0.535**	0.016	−0.042	0.203	0.468	−0.004	**0.513**	−0.002
Cu	0.146	0.019	0.153	−0.246	**0.859**	−0.123	0.187	−0.160
As	−0.017	−0.027	0.026	**0.963**	−0.172	−0.031	−0.014	−0.016
Se	0.203	0.298	−0.127	0.112	0.049	−0.299	−0.225	−**0.635**
Sr	−0.077	0.263	**0.897**	0.177	−0.138	−0.066	0.083	−0.120
Cd	0.123	−0.158	−0.195	−0.033	−0.145	**0.893**	−0.103	−0.039
Ba	−0.184	−0.276	0.357	**0.691**	−0.156	0.070	0.259	−0.174
Pb	0.198	0.154	−0.074	0.022	−0.044	−0.117	−0.130	**0.802**

Concerning PAHs, the first column includes Flut, Pyr, B [a] A, Chr, B [b] F, B [k] F, B [a] P, Ind, D [ah] A and B [ghi] P, while of the metals mainly includes Hg and less Ni. Thus, Factor 1 appears to be related to smoking indoors at home since the aforementioned PAHs belong to the COMPAHs category and are related to combustion processes, while both the Hg and Ni have also been suggested as derivatives of smoking [38]. This is also confirmed by Figure 2, where the concentrations of COMPAHS are considerably high in the houses of smokers compared to those of nonsmokers. Similar results were obtained for Hg (smokers' houses: 0.5, nonsmokers' houses: 0.2 µg/g) and to a lesser extent for Ni (smokers' houses: 24, nonsmokers' houses: 22 µg/g). On Factor 2, the contribution of Acy, Ace, Fl, and Phe is important and to a lesser extent the one of Ant, indicating that their common source is the combustion from vehicle emissions. Therefore, this factor is related to traffic and the aforementioned species originating from ambient air penetrate indoors. On factor 3, the main contribution of Al, followed by Sr and V, indicates that the above elements have mainly natural origin. On factor 4, the main contribution of As and to a lesser extent of Ba and Zn implies the existence of sources related to vehicle traffic, while the presence of Cu and Zn on factor 5 is associated with the road dust resuspension [39]. Regarding factor 6, the significant contribution of Cd and Cr may be attributed to abrasion of plated surfaces (fragments of the coating of metal surfaces or galvanized paints). Finally, on factor 7 the main contribution of Fe indicated that this factor is soil-related, while on factor 8 the great loading of Pb is mainly related to vehicle emissions. The latter is in agreement with the literature since it has been reported that Pb has anthropogenic contamination rather than soil-related origin [31].

3.5. Health Risk Assessment of PAHs and Heavy Metals in Settled House Dust

3.5.1. Cancer Risk Assessment of PAHs in Settled House Dust

A risk assessment using the concentration data of carcinogenic PAHs was conducted in order to assess the potentially detrimental effects associated with exposure to B2PAHs. For the moderate exposure scenario (0.05 g day^{-1}) the results indicated that the risk of carcinogenesis was 1.2×10^{-7}, while for a residence, in which the highest concentrations of B2 PAHs were observed, the risk was 5.4×10^{-7}. These values were found to be one order of magnitude smaller than those of Maertens et al. [18] (1.6×10^{-6}) and Roberts et al. [40] (7.8×10^{-6}). The results of the present study were comparable but slightly higher than those of Langer et al. [7]. For the high exposure scenario (0.1 g day-1), the risk of carcinogenesis was estimated as 2.5×10^{-7} and compared with the values reported in Maertens et al. [18] (3.2×10^{-6}) and Roberts et al. [40] (1.6×10^{-5}) was one and two orders of magnitude lower, respectively. However, in the case of the residence, where the highest concentrations of carcinogenic PAHs were observed, the risk was 1.1×10^{-6}, and it was comparable to the one of Maertens et al. [18] and an order of magnitude lower than that of Roberts et al. [40]. Although a strict threshold for risk of carcinogenicity that corresponds to safe exposure has not been established yet, one cancer case per million people (1×10^{-6}) is commonly used as the baseline level of acceptable risk. The values that were obtained by the present study are lower than the above limit for both exposure scenarios. As for the house with the highest levels of exposure, the risk of carcinogenesis for the high exposure scenario marginally exceeded the acceptable risk.

3.5.2. Heavy Metal Risk Exposure in Settled House Dust

Table 4 presents the doses of metals that children and their parents received through all the three routes of exposure, as well as the total dose. According to the results, the ingestion of house dust was found to be the main exposure pathway for metals to children, followed by dermal contact. This finding is in agreement with previously conducted studies [15,41]. The results also indicated that the doses received by the children were one or two orders of magnitude lower than those of their parents. Depending on the magnitude of exposure dose the heavy metals for both children and their parents were sorted as follows: Zn > Cu > Cr > Pb > Ni > Ni > As > Se > Cd > Hg.

Table 4. Exposure to heavy metals in SHD through the different exposure pathways.

Element	$D_{ingestion}$		$D_{inhalation}$		D_{Dermal}		D_{total}	
	Child	Adult	Child	Adult	Child	Adult	Child	Adult
Cr	1.00×10^{-3}	9.32×10^{-5}	2.11×10^{-8}	1.30×10^{-8}	1.15×10^{-6}	1.40×10^{-7}	1.00×10^{-3}	9.33×10^{-5}
Ni	4.60×10^{-4}	4.27×10^{-5}	9.65×10^{-9}	5.97×10^{-9}	5.29×10^{-7}	6.41×10^{-8}	4.60×10^{-4}	4.27×10^{-5}
Cu	5.22×10^{-3}	4.85×10^{-4}	1.10×10^{-7}	6.78×10^{-8}	6.00×10^{-6}	7.28×10^{-7}	5.22×10^{-3}	4.85×10^{-4}
Cd	8.23×10^{-6}	7.65×10^{-7}	1.73×10^{-10}	1.07×10^{-10}	9.47×10^{-9}	1.15×10^{-9}	8.24×10^{-6}	7.66×10^{-7}
Pb	7.09×10^{-4}	6.58×10^{-5}	1.49×10^{-8}	9.20×10^{-9}	8.15×10^{-7}	9.88×10^{-8}	7.09×10^{-4}	6.59×10^{-5}
Zn	6.16×10^{-3}	5.72×10^{-4}	1.29×10^{-7}	8.01×10^{-8}	7.09×10^{-6}	8.60×10^{-7}	6.17×10^{-3}	5.73×10^{-4}
Hg	5.79×10^{-6}	5.38×10^{-7}	1.22×10^{-10}	7.53×10^{-11}	6.66×10^{-9}	8.08×10^{-10}	5.80×10^{-6}	5.39×10^{-7}
As	6.08×10^{-5}	5.64×10^{-6}	1.28×10^{-9}	7.89×10^{-10}	6.99×10^{-8}	8.47×10^{-9}	6.08×10^{-5}	5.65×10^{-6}
Se	1.61×10^{-5}	1.50×10^{-6}	3.39×10^{-10}	2.09×10^{-10}	1.85×10^{-8}	2.25×10^{-9}	1.61×10^{-5}	1.50×10^{-6}

According to Table 5, which presents a comparison between the total doses for children and their parents with the reference doses, the total dose of Cr for children was the same order of magnitude as the reference dose. The doses for all the other elements were found 1–2 order of magnitude lower than the corresponding reference doses. As for the parents, the doses were significantly lower with respect to the children, and the differences between the reference doses were larger (2–4 order of magnitude).

Table 5. Comparison between the total doses for children and their parents with the reference doses.

Element	D_{total}		D_{Ref}
	Child	Adult	Child & Adult
Cr	1.00×10^{-3}	9.33×10^{-5}	3.0×10^{-3}
Ni	4.60×10^{-4}	4.27×10^{-5}	2.0×10^{-2}
Cu	5.22×10^{-3}	4.85×10^{-4}	2.0×10^{-2}
Cd	8.24×10^{-6}	7.66×10^{-7}	1.0×10^{-3}
Pb	7.09×10^{-4}	6.59×10^{-5}	3.0×10^{-3}
Zn	6.17×10^{-3}	5.73×10^{-4}	3.0×10^{-1}
Hg	5.80×10^{-6}	5.39×10^{-7}	3.0×10^{-4}
As	6.08×10^{-5}	5.65×10^{-6}	3.0×10^{-4}
Se	1.61×10^{-5}	1.50×10^{-6}	5.0×10^{-3}

All HQ values were found to be less than one, and therefore, the risk is limited or even negligible, while as expected all values were found to be higher for children than for adults (Table 6). The element of the house dust associated with the highest noncarcinogenic risk was Cr, for both children and their parents. This finding is in agreement with Shao et al. [27], who also found that Cr was linked with the highest noncarcinogenic risk. On the scales of dangerousness Cu and Pb followed, while Cd was the element that was linked to the lowest risk, probably because it was detected only in low concentrations.

Table 6. Noncarcinogenic exposure risk values for children and their parents.

Element	HQ_{child}	HQ_{adult}
Cr	0.335	0.031
Ni	0.023	0.002
Cu	0.261	0.024
Cd	0.008	0.001
Pb	0.236	0.022
Zn	0.021	0.002
Hg	0.019	0.002
As	0.203	0.019
Se	0.003	0.000

Table 7 presents the lifetime average exposure doses of carcinogenic heavy metals (Cr, Ni, Cd, As) through the inhalation pathway, indicating that the elements are sorted as follows: Cr > Ni > As > Cd. According to the results, the carcinogenic risk exposure dose Exposure Risk was found to be too low to constitute a significant hazard for children and their parents.

Table 7. Carcinogenic risk values of trace metal in inhalation exposure pathway.

Element	LD_{inh}	SF_{inh} (mg·kg^{-1}·d^{-1})	Exposure Risk
Cr	9.43×10^{-8}	42.00	3.96×10^{-6}
Ni	4.32×10^{-8}	0.84	3.63×10^{-8}
Cd	7.73×10^{-10}	6.30	4.87×10^{-9}
As	5.71×10^{-9}	1.50	8.56×10^{-9}

4. Conclusions

In summary, the dust samples collected from houses of young children in Athens, Greece, are indicative of relatively clean or moderately polluted indoor environments. This is why, the ΣPAH concentrations reported in the present study were in general lower with respect to those recorded in literature, while the metal concentrations were in some cases higher compared to those of previous studies. The calculated EFs of the heavy metals indicated that the elements Cu, Se, Zn, Hg, Cd, and Pb (EF > 40) were mainly of anthropogenic origin. PCA analysis on the chemical composition matrix of house dust showed that the sources that contributed the most to the concentrations of household dust were: smoking, combustion processes, resuspension of soil dust, traffic, and the abrasion of plated surfaces. The cancer risk of PAHs exposure was found lower than the threshold value, with the exception of one residence where high concentrations of carcinogenic PAHs were detected. Health risk assessment indicated that the ingestion of house dust is the most important route of exposure to heavy metals, followed by dermal absorption. As for the children, except of Cr, the doses of all elements were found 1–2 orders of magnitude lower than the corresponding reference doses, and the doses of parents were significantly lower with respect to the children. Regarding the noncarcinogenic risk, the heavy metals were sorted as Cr > Cu > Pb, while the risk of exposure of children and their parents was within the safety limits. Concerning the carcinogenic risk, the heavy metals were sorted as Cr > Ni > As > Cd, while the carcinogenic risk was found too low to be a risk to human health.

Author Contributions: Conceptualization, A.S. (A. Stamatelopoulou) and T.M.; Data curation, A.S. (A. Stamatelopoulou); Formal analysis, A.S. (A. Stamatelopoulou); Funding acquisition, T.M.; Investigation, A.S. (A. Stamatelopoulou), M.D., K.B., S.K., A.S. (A. Sakellari) and T.M.; Methodology, A.S. (A. Stamatelopoulou) and T.M.; Resources, T.M.; Supervision, T.M.; Validation, A.S. (A. Stamatelopoulou); Writing—review & editing, S.K. and T.M. All authors have read and agreed to the published version of the manuscript.

Funding: This research was funded by the European Union's Horizon 2020 research and innovation programme under grant agreement No 690105.

Institutional Review Board Statement: The study was conducted according to the guidelines of the Declaration of Helsinki, and approved by the Institutional Review Board of NCSR "DEMOKRITOS".

Informed Consent Statement: Informed consent was obtained from all subjects involved in the study.

Acknowledgments: This work was supported by ICARUS; this project has received funding from the European Union's Horizon 2020 research and innovation programme under grant agreement No 690105.

Conflicts of Interest: The authors declare no conflict of interest.

References

1. Matz, C.J.; Stieb, D.M.; Davis, K.; Egyed, M.; Rose, A.; Chou, B.; Brion, O. Effects of age, season, gender and urban-rural status on time-activity: CanadianHuman Activity Pattern Survey 2 (CHAPS 2). *Int. J. Environ. Res. Public Health* **2014**, *11*, 2108–2124. [CrossRef]
2. Qi, H.; Li, W.L.; Zhu, N.Z.; Ma, W.L.; Liu, L.Y.; Zhang, F.; Li, Y.F. Concentrations and sources of polycyclic aromatic hydrocarbons in indoor dust in China. *Sci. Total Environ.* **2014**, *491–492*, 100–107. [CrossRef] [PubMed]
3. Stamatelopoulou, A.; Asimakopoulos, D.N.; Maggos, T. Effects of PM, TVOCs and comfort parameters on indoor air quality of residences with young children. *Build. Environ.* **2019**, *150*, 233–244. [CrossRef]
4. Landrigan, P.J.; Kimmel, C.A.; Correa, A.; Eskenazi, B. Children's health and the environment: Public health issues and challenges for risk assessment. *Environ. Health Perspect.* **2004**, *112*, 257–265. [CrossRef] [PubMed]
5. Pickett, A.R.; Bell, M.L. Assessment of indoor air pollution in homes with infants. *Int. J. Environ. Res. Public Health* **2011**, *8*, 4502–4520. [CrossRef] [PubMed]
6. Stamatelopoulou, A.; Pyrri, I.; Asimakopoulos, D.N.; Maggos, T. Indoor air quality and dustborne biocontaminants in bedrooms of toddlers in Athens, Greece. *Build. Environ.* **2020**, *173*, 106756. [CrossRef]
7. Langer, S.; Weschler, C.J.; Fischer, A.; Bekö, G.; Toftum, J.; Clausen, G. Phthalate and PAH concentrations in dust collected from Danish homes and daycare centers. *Atmos. Environ.* **2010**, *44*, 2294–2301. [CrossRef]

8. Wang, W.; Wu, F.; Zheng, J.; Wong, M.H. Risk assessments of PAHs and Hg exposure via settled house dust and street dust, linking with their correlations in human hair. *J. Hazard. Mater.* **2013**, *263 Pt 2*, 627–637. [CrossRef]
9. Wang, Z.; Wang, S.; Nie, J.; Wang, Y.; Liu, Y. Assessment of polycyclic aromatic hydrocarbons in indoor dust from varying categories of rooms in Changchun city, northeast China. *Environ. Geochem. Health* **2017**, *39*, 15–27. [CrossRef]
10. Pateraki, S.; Bairachtari, K.; Stamatelopoulou, A.; Panagopoulos, P.; Markellou, C.; Vasilakos, C.; Mihalopoulos, N.; Maggos, T. Vertical characteristics of the pm10 and pm2.5 profile in an a real urban street canyon: Concentrations, chemical composition and associated health risks. *Fresenius Environ. Bull.* **2017**, *26*, 283–291.
11. Miller, R.L.; Garfinkel, R.; Horton, M.; Camann, D.; Perera, F.P.; Whyatt, R.M.; Kinney, P.L. Polycyclic aromatic hydrocarbons, environmental tobacco smoke, and respiratory symptoms in an inner-city birth cohort. *Chest* **2004**, *126*, 1071–1078. [CrossRef]
12. Hejami, A.A.; Davis, M.; Prete, D.; Lu, J.; Wang, S. Heavy metals in indoor settled dusts in Toronto, Canada. *Sci. Total Environ.* **2020**, *703*, 134895. [CrossRef]
13. Turner, A. Oral bioaccessibility of trace metals in household dust: A review. *Environ. Geochem. Health* **2011**, *33*, 331–341. [CrossRef]
14. Morawska, L.; Salthammer, T. *Indoor Environments: Airborne Particles and Settled Dust*; Wiley-VCH: Weinheim, Germany, 2003.
15. Lu, X.; Zhang, X.; Li, L.Y.; Chen, H. Assessment of metals pollution and health risk in dust from nursery schools in Xi'an, China. *Environ. Res.* **2014**, *128*, 27–34. [CrossRef]
16. Meza-Figueroa, D.; De la O-Villanueva, M.; De la Parra, M.L. Heavy metal distribution in dust from elementary schools in Hermosillo, Sonora, México. *Atmos. Environ.* **2007**, *41*, 276–288. [CrossRef]
17. Saeedi, M.; Li, L.Y.; Salmanzadeh, M. Heavy metals and polycyclic aromatic hydrocarbons: Pollution and ecological risk assessment in street dust of Tehran. *J. Hazard. Mater.* **2012**, *227–228*, 9–17. [CrossRef]
18. Maertens, R.M.; Yang, X.; Zhu, J.; Gagne, R.W.; Douglas, G.R.; White, P.A. Mutagenic and Carcinogenic Hazards of Settled House Dust I: Polycyclic Aromatic Hydrocarbon Content and Excess Lifetime Cancer Risk from Preschool Exposure. *Environ. Sci. Technol.* **2008**, *42*, 1747–1753. [CrossRef]
19. Mannino, M.R.; Orecchio, S. Polycyclic aromatic hydrocarbons (PAHs) in indoor dust matter of Palermo (Italy) area: Extraction, GC–MS analysis, distribution and sources. *Atmos. Environ.* **2008**, *42*, 1801–1817. [CrossRef]
20. Lucas, J.P.; Le Bot, B.; Glorennec, P.; Etchevers, A.; Bretin, P.; Douay, F.; Sebille, V.; Bellanger, L.; Mandin, C. Lead contamination in French children's homes and environment. *Environ. Res.* **2012**, *116*, 58–65. [CrossRef] [PubMed]
21. Zhu, S.; Cai, W.; Yoshino, H.; Yanagi, U.; Hasegawa, K.; Kagi, N.; Chen, M. Primary pollutants in schoolchildren's homes in Wuhan, China. *Build. Environ.* **2015**, *93*, 41–53. [CrossRef]
22. Saraga, D.; Maggos, T.; Sadoun, E.; Fthenou, E.; Hassan, H.; Tsiouri, V.; Karavoltsos, S.; Sakellari, A.; Vasilakos, C.; Kakosimos, K. Chemical Characterization of Indoor and Outdoor Particulate Matter (PM2.5, PM10) in Doha, Qatar. *Aerosol Air Qual. Res.* **2017**, *17*, 1156–1168. [CrossRef]
23. Yoshinaga, J.; Yamasaki, K.; Yonemura, A.; Ishibashi, Y.; Kaido, T.; Mizuno, K.; Takagi, M.; Tanaka, A. Lead and other elements in house dust of Japanese residences-source of lead and health risks due to metal exposure. *Environ. Pollut.* **2014**, *189*, 223–228. [CrossRef]
24. Al-Momani, I.F. Trace Elements in Street and Household Dusts in Amman, Jordan. *Soil Sediment Contam. Int. J.* **2007**, *16*, 485–496. [CrossRef]
25. U.S. EPA. *ethod 3050B: Acid Digestion of Sediments, Sludges and Soils (Revision 2)*; U.S. EPA: Washington, DC, USA, 1996.
26. Shao, T.; Pan, L.; Chen, Z.; Wang, R.; Li, W.; Qin, Q.; He, Y. Content of Heavy Metal in the Dust of Leisure Squares and Its Health Risk Assessment—A Case Study of Yanta District in Xi'an. *Int. J. Environ. Res. Public Health* **2018**, *15*, 394. [CrossRef] [PubMed]
27. U.S. EPA. *Supplemental Guidance for Developing Soil Screening Levels for Superfund Sites*; Office of Emergency and Remedial Response: Washington, DC, USA, 2002.
28. Mahler, B.; Vanmetre, P.; Wilson, J.; Musgrove, M. Coal-Tar-Based Parking Lot Sealcoat: An Unrecognized Source of PAH to Settled House Dust. *Environ. Sci. Technol.* **2010**, *44*, 894–900. [CrossRef] [PubMed]
29. Wilson, N.K.; Chuang, J.C.; Lyu, C.; Menton, R.; Morgan, M.K. Aggregate exposures of nine preschool children to persistent organic pollutants at day care and at home. *J. Expo. Anal. Environ. Epidemiol.* **2003**, *13*, 187–202. [CrossRef]
30. Morgan, M.K.; Wilson, N.K.; Chuang, J.C. Exposures of 129 preschool children to organochlorines, organophosphates, pyrethroids, and acid herbicides at their homes and daycares in North Carolina. *Int. J. Environ. Res. Public Health* **2014**, *11*, 3743–3764. [CrossRef] [PubMed]
31. Amato, F.; Schaap, M.; A.C. Denier van der Gon, H.; Pandolfi, M.; Alastuey, A.; Keuken, M.; Querol, X. Short-term variability of mineral dust, metals and carbon emission from road dust resuspension. *Atmos. Environ.* **2013**, *74*, 134–140. [CrossRef]
32. Shakya, P. Chemical Associations of Lead, Cadmium, Chromium, Nickel and Zinc in Household Dust of Kathmandu Metropolitan Area. *Pak. J. Anal. Environ. Chem.* **2013**, *14*, 26–32.
33. Fromme, H.; Lahrz, T.; Kraft, M.; Fembacher, L.; Mach, C.; Dietrich, S.; Burkardt, R.; Volkel, W.; Goen, T. Organophosphate flame retardants and plasticizers in the air and dust in German daycare centers and human biomonitoring in visiting children (LUPE 3). *Environ. Int.* **2014**, *71*, 158–163. [CrossRef]
34. Manousakas, M.; Papaefthymiou, H.; Diapouli, E.; Migliori, A.; Karydas, A.G.; Bogdanovic-Radovic, I.; Eleftheriadis, K. Assessment of PM2.5 sources and their corresponding level of uncertainty in a coastal urban area using EPA PMF 5.0 enhanced diagnostics. *Sci. Total Environ.* **2017**, *574*, 155–164. [CrossRef]

35. Amato, F.; Alastuey, A.; Karanasiou, A.; Lucarelli, F.; Nava, S.; Calzolai, G.; Severi, M.; Becagli, S.; Gianelle, V.L.; Colombi, C.; et al. AIRUSE-LIFE+: A harmonized PM speciation and source apportionment in five southern European cities. *Atmos. Chem. Phys.* **2016**, *16*, 3289–3309. [CrossRef]
36. Ridley, I.; Fox, J.; Oreszczyn, T.; Hong, S.H. The Impact of Replacement Windows on Air Infiltration and Indoor Air Quality in Dwellings. *Int. J. Vent.* **2016**, *1*, 209–218. [CrossRef]
37. Maertens, R.M.; Bailey, J.; White, P.A. The mutagenic hazards of settled house dust: A review. *Mutat. Res. Rev. Mutat. Res.* **2004**, *567*, 401e425. [CrossRef]
38. Whitehead, T.P.; Metayer, C.; Petreas, M.; Does, M.; Buffler, P.A.; Rappaport, S.M. Polycyclic aromatic hydrocarbons in residential dust: Sources of variability. *Environ. Health Perspect.* **2013**, *121*, 543–550. [CrossRef]
39. Vasilakos, C.; Pateraki, S.; Veros, D.; Maggos, T.; Michopoulos, J.; Saraga, D.; Chelmis, C. Temporal determination of heavy metals in PM2.5 aerosols in a suburban site of Athens, Greece. *J. Atmos. Chem.* **2007**, *57*, 1–17. [CrossRef]
40. Psanis, C.; Triantafyllou, E.; Giamarelou, M.; Manousakas, M.; Eleftheriadis, K.; Biskos, G. Particulate matter pollution from aviation-related activity at a small airport of the Aegean Sea Insular Region. *Sci. Total Environ.* **2017**, *596–597*, 187–193. [CrossRef]
41. Roberts, J.W.; Budd, W.T.; Ruby, M.G.; Camann, D.; Fortmann, R.C.; Lewis, R.G.; Wallace, L.A.; Spittler, T.M. Humanexposure to pollutants in the floor dust of homes and offices. *J. Expo. Anal. Environ. Epidemiol.* **1992**, *2*, 127–146.

Article

Atmospheric Concentrations and Health Implications of PAHs, PCBs and PCDD/Fs in the Vicinity of a Heavily Industrialized Site in Greece

Konstantinos G. Koukoulakis [1], Panagiotis George Kanellopoulos [1], Eirini Chrysochou [1], Danae Costopoulou [2], Irene Vassiliadou [2], Leondios Leondiadis [2] and Evangelos Bakeas [1,*]

[1] Laboratory of Analytical Chemistry, Department of Chemistry, National and Kapodistrian University of Athens, Zografos, Panepistimiopolis, 15784 Athens, Greece; kkoukoulakis@chem.uoa.gr (K.G.K.); gpkan@chem.uoa.gr (P.G.K.); eirinichr@chem.uoa.gr (E.C.)

[2] Mass Spectrometry and Dioxin Analysis Laboratory, INRASTES, NCSR "Demokritos", Neapoleos 27, 15341 Athens, Greece; costodan@rrp.demokritos.gr (D.C.); vasilirn@rrp.demokritos.gr (I.V.); leondi@rrp.demokritos.gr (L.L.)

* Correspondence: bakeas@chem.uoa.gr

Received: 31 October 2020; Accepted: 15 December 2020; Published: 17 December 2020

Abstract: Background: Thriassion Plain is considered the most industrialized area in Greece and thus a place where emissions of pollutants are expected to be elevated, leading to the degradation of air quality. Methods: Simultaneous determination of polycyclic aromatic hydrocarbons (PAHs), polychlorinated dibenzo-p-dioxins/dibenzofurans (PCDD/Fs), and polychlorinated biphenyls (PCBs) was performed in PM10 samples. SPSS statistical package was employed for statistical analysis and source apportionment purposes. Cancer risk was estimated from total persistent organic pollutants' (POPs) dataset according to the available literature. Results: POPs concentrations in particulate matter were measured in similar levels compared to other studies in Greece and worldwide, with mean concentrations of ΣPAHs, ΣPCDD/Fs, dioxin like PCBs, and indicator PCBs being 7.07 ng m^{-3}, 479 fg m^{-3}, 1634 fg m^{-3}, and 18.1 pg m^{-3}, respectively. Seasonal variations were observed only for PAHS with higher concentrations during cold period. MDRs, D/F ratios, and principal component analysis (PCA) highlighted combustions as the main source of POPs' emissions. Estimation of particles' carcinogenic and mutagenic potential indicates the increased toxicity of PM10 during cold periods, and cancer risk assessment concludes that 3 to 4 people out of 100,000 may suffer from cancer due to POPs' inhalation. Conclusions: Increased cancer risk for citizens leads to the necessity of chronic POPs' monitoring in Thriassion Plain, and such strategies have to be a priority for Greek environmental authorities.

Keywords: POPs; PAHs; PCDD/Fs; PCBs; industrial site; Greece; air quality

1. Introduction

Polycyclic aromatic hydrocarbons (PAHs), polychlorinated dibenzo-p-dioxins and dibenzofurans (PCDDs and PCDFs or PCDD/Fs), along with, polychlorinated biphenyls (PCBs), both dioxin like (dlPCBs) and non-dioxin like (ndlPCBs), are ubiquitous semi volatile persistent organic pollutants (POPs) that can be found far from their emission sources through long range transport of air masses either as gas molecules or bounded to particulate matter [1–3]. POPs are also characterized as low soluble compounds that tend to bioaccumulate and biomagnify in biota [4], and due to their potential health implications and especially their mutagenic, teratogenic, and carcinogenic effects, they have attracted global research attention [5–8].

PAHs constitute byproducts of incomplete fossil fuel combustion or biomass burning [9], and they could also be emitted by solid waste incineration [10] and aluminum production, and they can be found in crude oil, asphalt coal, and tar [11]. The most dominant routes of human exposure to PAHs are via ingestion and inhalation [12–14]. On the other hand, PCDD/Fs are not produced intentionally, besides using them for research scope, but they are unintentionally formed as by-products of chlorinated compounds in combustion and industrial thermal processes like waste incineration, ferrous and secondary nonferrous smelting, cement kilning, and also from fuel combustion, pulp production, chlorinated substances production, and chemical and petrochemical industries [4,15–20]. PCBs have similar emission sources as PCDD/Fs, and they were used in transformer, paint, and capacitor production. Although PCBs have been phased out from production processes in most countries in the last decades, many studies investigate PCB contaminants in air, especially in industrialized sites [2,16,20,21]. PCBs are divided into dlPCBs, due to their similar metabolism in humans with dioxins [22] and to non-dioxin like PCBs (ndlPCBs), of which six congeners have often been chosen as indicators PCBs (indPCBs) for the evaluation of ndlPCBs' contamination in the atmosphere [23]. Due to their presence in many commercial PCB mixtures [24], their predominance in air samples from industrial sites is an indication of their impact on the atmospheric environment.

There is a general acceptance that either long- or short-term exposure to POPs may lead to adverse health effects [25–27]. Therefore, many monitoring programs have been undertaken by developed countries' authorities under the prism of the Stockholm Convention for the cooperation among nations to eliminate unintentionally emitted POPs including PCDDs/Fs and dlPCBs. Intense POPs research provides evidence that some compounds (e.g., PAHs, PCDDs/Fs, and dlPCBs) exert intense carcinogenic, mutagenic, or teratogenic effects on humans, and therefore IARC [28,29] and the United States Environmental Protection Agency [12] have classified them as probable human carcinogens. PCBs have also been included in IARC's latest report as substances with carcinogenic impacts on humans due to their relation with melanoma cancer [30].

In this study, the most heavily industrialized area of Greece has been selected for the simultaneous monitoring of PAHs, PCDD/Fs, dlPCBs, and indPCBs bonded to particulate matter (PM_{10}) and for the estimation of their health risks for nearby citizens. Although some toxic PCDDF/s and PCBs were found to be particularly in gas phase [31,32], according to several other works, PCDD/Fs participated majorly in particulate form, especially the congeners with increased chlorine atoms (penta, hexa, hepta, and octa compounds) [33,34]. According to the study by Lee and Jones (1999) [33] about the partitioning behavior of PCDD/Fs in gas and particulate phase, PCDDs tended to be more associated with atmospheric particulates than the equivalent PCDF homologue groups, probably reflecting the slightly lower vapor pressures of PCDDs. Our choice to study these POPs simultaneously in PM10 was made to find out the levels of POPs in samples already legislated and monitored for other pollutants. To our knowledge, this is the first study in this direction in Greece, and one of the few globally, and thus it may be helpful for the development and implementation of strategies for the regulation of emissions in this site, only a few kilometers east, southeast from the Greek capital, Athens.

2. Materials and Methods

2.1. Site Description and Sampling Procedure

The sampling campaign was performed in the industrial city of Eleusis located at central Greece, with a population reaching approximately 25,000 people according to a 2011 census. Eleusis is in the heart of Thriassion Plain, the largest industrial area in Greece, approximately 18 km northwest from the center of Athens (the capital of Greece) (Figure 1). It is the place where the majority of crude oil in Greece is imported and refined. Within this area, the largest crude oil refineries are located, and over 300 industrial plants, referring to metallurgical processes, cement, chemical and food production plants, shipyards, etc., are situated. A recently imposed environmental pressure on the surrounding area is the illegal uncontrolled combustions that take place in the neighboring industrial site of Aspropirgos for the recovery of raw materials by burning tires, electronics, plastics, etc. [35,36].

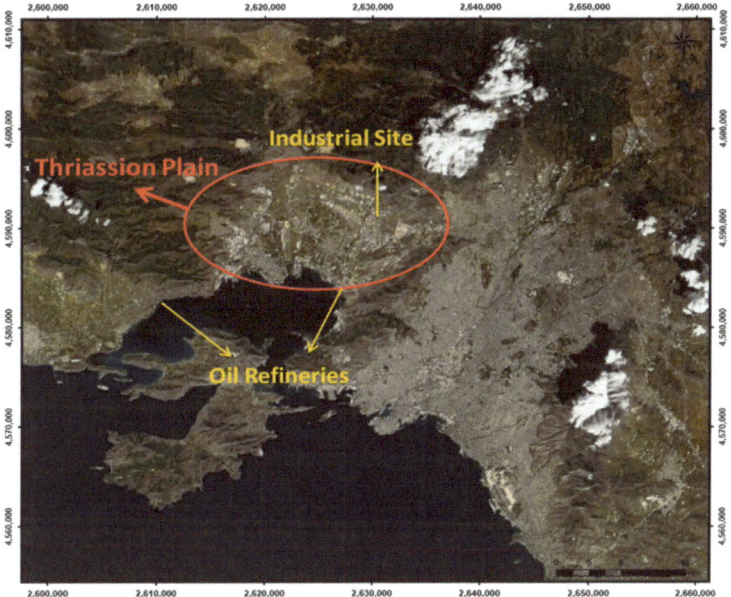

Figure 1. The sampling point in the Thriassion Plain and the nearby capital Athens.

The sampling point was located in the center of Eleusis at a height of 4 m from the ground. A medium volume sampler (MVS) equipped with a PM10 cutoff inlet was used to collect particulate phase on Quartz fiber filters with a diameter of 47 mm, at a flow rate of 2.3 m^3 h^{-1} for 24 h. The sampling procedure was performed according to EN12341. Filter blanks were included in each sampling campaign, and if necessary, appropriate corrections of the results were performed. The sampling duration was from December 2018 to July 2019, and 30 samples were collected in total. The sampling procedure was apportioned in order to study possible seasonal variations, and fifteen samples were collected during a cold period (December 2018 to February 2019) and another fifteen during a warm period (May 2019 to July 2019).

2.2. Materials

A PAH-determination procedure was validated using the Polynuclear Aromatic Hydrocarbons Mix, a standard solution of the compounds studied including naphthalene (NAP), acenaphthylene (ACY), acenaphthene (ACE), fluorene (FL), phenanthrene (PHE), anthracene (ANT), fluoranthene (FLT), pyrene (PYR), chrysene (CHR), benzo[a]anthracene (BaA), benzo[b,k]fluoranthenes (BbkF), benzo[a]pyrene (BaP), indeno[1,2,3 cd]pyrene (IPY), dibenzo[a,h]anthracene (DBaA), and benzo[ghi]perylene (BPE), and purchased by Sigma Aldrich. A mix of phenanthrene D10 and perylene D12 (Supelco) was also prepared for quantitation and quantification of PAHs, and all solvents were appropriate for residue analysis purchased from Carlo Erba and Macron Fine Chemicals. For the PCDD/Fs and PCBs analysis, all solvents used were residue analysis picograde, and were purchased from Promochem. Activated carbon FU 4652 was purchased from Schunk Kohlenstofftechnik GmbH. Basic Alumina for dioxin analysis was purchased from MP Biochemicals GmbH and silica gel 60–200 mesh from Merck. The internal standards used were 13 C-labelled solutions of PCDD/Fs and PCBs in toluene, and were added to each sample prior to extraction. They contained a mixture of [13]

C_{12} isomers of all the 17 PCDD/F congeners except OCDF, the four $^{13}C_{12}$ non-ortho PCBs, the eight $^{13}C_{12}$ mono ortho PCBs, and the six $^{13}C_{12}$ indicator PCBs. The isomers for the preparation of the $^{13}C^{12}$ internal standard solutions, the injection standards $^{13}C_{12}$ 1,2,3,4-TCDD, and $^{13}C_{12}$ PCB-80 were purchased from Wellington Laboratories.

2.3. Sample Extraction and Clean Up

After sampling, the filter was cut in two symmetrical halves. The first half was used for PAHs analysis, while the second half was used for the determination of PCDD/Fs. The procedure, carried out for PAHs analysis, was fully described in our previous study [36]. In general, the filters were spiked with the mix of internal standards (phenanthrene D10 and perylene D12), and then were extracted with dichloromethane in an ultrasonic bath (Ultrasonic LC 130H, Elma, Germany). The extraction procedure was performed in triplicate. The obtained extracts were combined and concentrated in volume using a rotary evaporator (Rotavapor R-210, Buchi, Switzerland) at 28 °C nearly to 2–3 mL. Then, a solvent change step was performed with the addition of hexane. Subsequently a purification step was followed. A 30 cm × 1 cm i.d. glass column chromatography was used. The column was packed with anhydrous sodium sulphate and silica gel, and was activated with hexane before the sample was placed on the top of the column. The clean-up was performed using initially hexane and then a mixture of CH_2Cl_2: n-hexane 3:2. This eluted fraction was collected and finally evaporated under a flow of nitrogen.

The analysis of PCDD/Fs and PCBs was performed according to EPA T0-9A protocol. Quartz filters were extracted overnight with toluene in a soxhlet apparatus. Prior to extraction, samples were spiked with 0.1 ng $^{13}C_{12}$ labeled PCDD/Fs and non-ortho PCBs (n.o.-PCBs), and 1.0 ng mono-ortho PCBs (m.o.-PCBs) and indPCBs as quantitative standards. The extract was subjected to clean-up. Active carbon Carbosphere or FU4652 was used for the separation of PCDD/Fs and n.o.-PCBs, in two different fractions, with toluene as elution solvent. Further clean-up of fractions was performed by column chromatography with basic alumina and 44% H_2SO_4-silicagel eluted with different mixtures of hexane:diclomethane. After evaporation, the eluate containing the PCDD/Fs was re-dissolved in n-nonane containing 2 ng mL^{-1} of injection standard $^{13}C_{12}$ 1,2,3,4-TCDD, while that containing the n.o. PCBs was re-dissolved in n-nonane containing 2 ng mL^{-1} $^{13}C_{12}$ PCB80. For mono-ortho and indicator PCBs, 10% of the fraction obtained from Soxhlet extraction was dissolved in hexane and brought onto a column chromatography filled with 10 g of 44% H_2SO_4-silica. Further clean-up was performed by column chromatography with basic alumina and 44% H_2SO_4-silica and elution with hexane:diclomethane. The eluate was evaporated to dryness and re-dissolved in n-nonane containing 20 ng mL^{-1} of injection standard ($^{13}C_{12}$ PCB 80). A detailed description of the whole clean-up procedure has been given elsewhere [7].

2.4. Instrumental Analysis and Quality Assurance

A GC/MS (6890N/5975B, Agilent Technologies, Santa Clara, CA 95051, USA) was employed for PAHs determination. The GC instrument was equipped with a split/splitless injector and a HP-5ms [5%-(phenyl)-methylpolysiloxane] (Agilent J&W GC Columns, Agilent Technologies, Santa Clara, CA, USA) capillary column. High purity Helium was the carrier gas with a velocity of 1.5 mL min^{-1}. A pulsed splitless mode was used for the injection, and the injector's temperature was set at 280 °C. The GC oven temperature program was: 65 °C (hold for 1 min) to 320 °C at 15 °C min with a final isothermal hold for 3 min. Inlet and MS source temperatures were 280 °C and 230 °C, respectively. Selected ion monitoring (SIM) mode was used for the quantification of the analytes. Detection limits (DLs) of the studied PAHs ranged from 0.0002 (ANT and DBA) to 0.002 (FLT and PYR) ng m^{-3}. Recoveries varied from 82% (FL) to 117% (CHR) calculated from spiked filters determination.

The quantification of PCDD/Fs and PCBs was performed by High Resolution Gas Chromatography–High Resolution Mass Spectrometry (Electron-Impact) (HRGC-HRMS, EI), on Multiple Ion Detection (MID) mode, on a Trace 1310 gas chromatograph (ThermoScientific,

Waltham, MA 02451, USA) equipped with an Agilent DB-5MS GC column, 60 m length, 0.25 mm I.D., 0.10 µm film, a TriPlus RSH autosampler, coupled to a DFS mass spectrometer (ThermoScientific, Waltham, MA, USA) performing at 10,000 resolving power (10% valley definition). Instrumental conditions and quality control criteria are according to US EPA Method 1613 and European Standard EN 1948. The quantification was carried out by the isotopic dilution method. According to the European guidance, in the field of PCDD/Fs and PCBs analysis, results are calculated as sum-parameters based on concentrations and on limits of quantitation (LOQs) only, while limits of detection (LODs) do not carry any relevant information, and due to high precision in measurement, are considered equal to LOQs. The limit of quantitation (LOQ) for each congener was determined as the concentration in the extract which produced an instrumental response at two different ions to be monitored with a signal to noise ratio of 3:1 for the less sensitive signal. LOQ values were 0.1 (PCDD/Fs and n.o.-PCBs) and 2 (m.o.-PCBs) pg/sample. LOQs were evaluated during accreditation of the method using different reference materials and the respective higher LOQ value is used for each group of congeners for all samples. These values are low enough to ensure that the difference between the upper-bound level and lower-bound level does not exceed 20%. Recovery rates of 70–120% were also calculated by spiked filters analysis.

2.5. Statistical Analysis

Statistical analysis was performed using SPSS software package (IBM SPSS statistics version 24). This statistical software package was suitable for multivariate analysis of the environmental data [16,35–39]. Shapiro-Wilk and Kolmogorov–Smirnov tests used to study whether the data followed normal distribution with a value of $p > 0.05$ indicated normal distribution. As no variable of the dataset was normally distributed, the Mann–Whitney test for 2 independents was employed to carry out if there was a statistically significant difference. A value of $p < 0.05$ (95% confidence level) was considered to indicate a significant difference in the statistical analysis of the data. Principal Component Analysis (PCA) was used for the investigation of any possible associations and source apportionment among PAHs, PCDD/Fs, and PCBs. PCA consists of eigenvalue decomposition of the covariance matrix of Gaussian distributed random variables. However, in environmental studies, PCA is used as a tool for data compression, dimension reduction, or even filtering method for non-Gaussian (non-normal) distributed and/or nonlinear data. Application of PCA for source apportionment purposes has been performed in many POPs' studies [16,23,39].

2.6. Health Risk Estimation

BaPE is the first parameter to estimate carcinogenicity of total PAHs. BaPE values above 1.0 ng m^{-3} represent an increased cancer risk. BaPE is calculated according to Equation (1):

$$\text{BaPE} = ([\text{BaA}] * 0.06) + ([\text{BbF}] * 0.07) + ([\text{BkF}] * 0.07) + ([\text{BaP}] * 1) + ([\text{DBA}] * 0.6) \quad (1)$$

Total carcinogenic and mutagenic potential of particulate bound PAHs, ΣBaP_{TEQ} and ΣBaP_{MEQ}, were calculated as described elsewhere [40–42], using Equations (2) and (3):

$$\Sigma\text{BaPTEQ} = \Sigma\{([\text{BaA}] * 0.1) + ([\text{CHR}] * 0.01) + ([\text{BbF}] * 0.1) + ([\text{BkF}] * 0.1) \\ + ([\text{BaP}] * 1) + ([\text{IPY}] * 0.1) + ([\text{DBA}] * 5) + ([\text{BPE}] * 0.01)\} \quad (2)$$

$$\Sigma\text{BaPMEQ} = \Sigma\{([\text{BaA}] * 0.082) + ([\text{CHR}] * 0.017) + ([\text{BbF}] * 0.25) + ([\text{BkF}] * 0.11) \\ + ([\text{BaP}] * 1) + ([\text{IPY}] * 0.31) + ([\text{DBA}] * 0.29) + ([\text{BPE}] * 0.19)\} \quad (3)$$

The relation between exposure and cancer risk is considered linear in low doses, and thus inhalation cancer risk associated to PAHs could be calculated from ΣBAP_{TEQ} using the inhalation unit risk ($IUR_{BaP} = 1.1 \times 10^{-3}$ (µg m^{-3})$^{-1}$ [43] for BaP and according to the following equation (Equation (4)):

$$ICR = \Sigma BaPTEQ * IURBaP \quad (4)$$

In general, exposure to toxic substances in the ambient air depends on the chronic daily intake (CDI) of each pollutant emitted by the source. CDI (mg/kg/day) could be calculated as Life Averaged Daily Dose (LADD) using Equation (5) [43–45]:

$$CDI = C_{air} * IF \quad (5)$$

where C_{air} is concentration of pollutant (mg m^{-3}), and *IF* is Intake Factor (mg^3/kg/day) derived from Equation (6):

$$IF = \frac{IR * EF * ED * ET}{BW * AT} \quad (6)$$

where according to EPA (1998) Inhalation Rate (*IR*) = 20 m^3/day; Exposure Frequency (*EF*) = 365 days; Exposure Duration (*ED*) = 70 years; Exposure Time (*ET*) = 24 h/day; Body Weight (*BW*) = 70 kg; and Average Time (*AT*) = 35,500 days for exposure to carcinogenic pollutants. Cancer risk of the specific substances like POPs is calculated using Equation (7):

$$\text{Cancer Risk} = LADD * SF \quad (7)$$

where Slope Factor (*SF*) (mg/kg·day)$^{-1}$ values were calculated by Equation (8):

$$SF = \frac{IUR\ m^3 \mu g - 1 \times 70\ kg \times 10^3\ \mu g\ mg - 1}{20\ m^3\ day - 1} \quad (8)$$

Inhalation Unit Risk values used for the estimation of Cancer Risk by each PCDD/Fs, PCBs, and PAHs were obtained from OEHHA [46]. Cancer risk values in our study were compared to upper-bound cancer risk of 1×10^{-6} (one person per million could develop cancer from the inhalation of this pollutant). Cancer risk values over this benchmark level are considered significant, and risks over 1×10^{-4} are unacceptable by EPA (2012). Finally, a total risk related to the sampling site was calculated, summarizing the risk from each pollutant.

3. Results and Discussion

3.1. Results

3.1.1. PAHs and Indpcbs

PAHs concentrations are presented in Table 1. ΣPAHs ranged from 1.27 to 16.5 ng m^{-3} with a median value of 6.12 ng m^{-3}. PAHs levels were found in the same levels with a previous research in the same site [47] (ΣPAHs = 7.9 ng m^{-3}) and slightly lower than our previous study in the nearby area of Aspropirgos (mean value 9.8 ng m^{-3}) [36]. PAHs in other Greek cities were measured in lower concentrations, with mean values 3.34 and 6.46 ng m^{-3} in the harbor of Volos [48], 3.08 ng m^{-3} in Spata, and 3.21 ng m^{-3} in Koropi [49], but significantly higher mean concentrations were reported in industrial sites of Istanbul (60.5 ng m^{-3}) [39] and Aliaga (218 ng m^{-3}) [21], Turkey.

Table 1. Average, median, and ranged concentrations of polycyclic aromatic hydrocarbons (PAHs) (ng m^{-3}) for the overall and seasonally divided sampling period. NAP: Naphthalene; ACY: Acenaphthylene; ACE: Acenaphthene; FL: Fluorene; PHE: Phenanthrene; ANT: Anthracene; FLT: Fluoranthene; PYR: Pyrene; CHR: Chrysene; BaA: Benzo[a]anthracene; BbkF: Benzo[b,k]fluoranthenes; BaP: Benzo[a]pyrene; IPY: Indeno[1,2,3-cd]pyrene; DBaA: Dibenzo[a,h]anthracene; BPE: Benzo[ghi]perylene.

	NAP	ACY	ACE	FL	PHE	ANT	FLT	PYR	CHR	BaA	BbkF	BaP	IPY	DBaA	BPE	ΣPAHs
							Overall Sampling Campaign (N = 30)									
Average	0.22	0.03	0.03	0.06	0.35	0.29	0.23	0.34	0.60	0.50	1.82	0.93	0.80	0.20	0.65	7.07
Median	0.17	0.007	0.0001	0.06	0.30	0.30	0.18	0.34	0.49	0.39	1.54	0.66	0.69	0.19	0.59	6.12
Range	0.05–0.67	0.0002–0.21	0.0001–0.20	0.02–0.12	0.19–0.79	0.06–0.51	0.05–0.69	0.05–0.73	0.00009–2.94	0.0002–1.63	0.0001–4.38	0.04–3.07	0.00005–1.86	0.00002–0.50	0.00003–1.59	1.27–16.5
							Cold Period (N = 15)									
Average	0.24	0.03	0.04	0.06	0.37	0.30	0.26	0.34	0.74	0.63	2.32	1.03	1.00	0.25	0.82	8.44
Median	0.19	0.002	0.0001	0.06	0.31	0.32	0.21	0.30	0.60	0.73	2.39	0.90	0.99	0.24	0.76	8.88
Range	0.07–0.67	0.002–0.21	0.0001–0.20	0.02–0.12	0.19–0.79	0.06–0.51	0.05–0.69	0.05–0.73	0.00009–2.94	0.09–1.63	0.36–4.38	0.23–2.34	0.32–1.86	0.10–0.50	0.24–1.59	2.88–16.5
							Warm Period (N = 15)									
Average	0.15	0.04	0.03	0.04	0.27	0.28	0.11	0.34	0.05	0.03	0.0001	0.57	0.05	0.00002	0.04	2.01
Median	0.13	0.03	0.0001	0.04	0.28	0.28	0.11	0.34	0.00009	0.04	0.0001	0.07	0.00005	0.00002	0.04	1.60
Range	0.05–0.37	0.01–0.07	0.0001–0.18	0.02–0.06	0.21–0.31	0.22–0.33	0.09–0.15	0.32–0.36	0.00009–0.17	0.0002–0.06	0.0001–0.0001	0.04–3.07	0.0005–0.14	<0.0002	0.00003–0.09	1.27–4.46

BaP is considered the most toxic PAHs compound, and its values ranged from below LOD to 3.04 ng m^{-3} with an average value of 0.93 ng m^{-3}, which is close to the target value of 1 ng m^{-3} from the 4th Daughter Directive [50,51]. BaP highest average concentration in Aspropirgos was 3.6 ng m^{-3}, while in Eleusis, the mean concentration was 0.71 ng m^{-3} [47]. Similar levels of BaP were also recorded in Thessaloniki, with the average values for the cold and warm periods being 0.12 and 0.86 ng m^{-3}, respectively [42], while BaP concentration in Volos measured 0.60 ng m^{-3} [49]. Lower BaP concentrations were found in industrial sites of Istanbul, Turkey (0.39 ng m^{-3}) [39], Dunkirk, France (0.29 ng m^{-3}) [31], and Jiangsu Province, China (up to 0.35 ng m^{-3}) [16], whereas much higher concentrations were presented in the industrial site of Shanghai, with a mean value of 5.95 ng m^{-3} [52].

In Table 2, concentrations of indPCBs are presented with values ranging from 11.4–26.6 pg m^{-3}, with a median value of 18.1 pg m^{-3}. The most abundant compound was PCB28 (7.96 pg m^{-3}), followed by PCB101 (2.86 pg m^{-3}). Levels of PM$_{10}$-bound indPCBs were similar to that reported in Thessaloniki (0.5–29.2 pg m^{-3}) [53], but relatively lower than in Rome (163 pg m^{-3}) [54] and Brescia (474 pg m^{-3}) [23], Italy.

Table 2. Average, median, and ranged concentrations of indicator polychlorinated biphenyls (indPCBs) (pg m^{-3}) for the overall and seasonally divided sampling period.

	PCB-28	PCB-52	PCB-101	PCB-138	PCB-153	PCB-180	ΣPCBs
	Overall Sampling Campaign (N = 30)						
Average	7.96	1.82	2.86	2.30	2.01	1.18	18.1
Median	8.25	1.71	2.56	2.26	1.83	1.11	17.4
Range	4.53–11.7	1.09–2.94	1.34–5.78	1.68–3.80	1.18–3.30	0.66–1.87	11.4–26.6
	Cold Period (N = 15)						
Average	7.36	1.93	2.47	2.26	2.34	1.29	17.6
Median	7.49	1.86	2.53	2.35	2.30	1.21	17.1
Range	4.53–9.37	1.09–2.94	1.63–3.37	1.72–3.07	1.56–3.30	0.80–1.87	11.6–22.4
	Warm Period (N = 15)						
Average	8.56	1.72	3.24	2.34	1.67	1.07	18.6
Median	8.46	1.62	2.80	2.13	1.58	0.94	17.6
Range	4.60–11.7	1.10–2.32	1.34–5.78	1.68–3.80	1.18–2.33	0.66–1.73	11.4–26.6

3.1.2. PCDD/Fs and dlPCBs

The PCDD/F and dlPCB results obtained in this study are summarized in Tables 3 and 4, respectively. Toxic equivalency factors that were reconsidered by World Health Organization in 2005 were applied to calculate PCDD/F and dlPCB toxic equivalent (TEQ) concentrations [22]. Values below LOQ were assumed to be equal to LOQ (upper-bound concentrations) to evaluate the worst case scenario of exposure. PCDD/F concentrations in PM$_{10}$ collected from the industrial area of Eleusis were 145–3472 fg m^{-3} (mean 696 fg m^{-3}) and the PCDD/F TEQs ranged from 287 to 2560 fg WHOTEQ$_{2005}$ m^{-3} (mean 656 fg WHOTEQ$_{2005}$ m^{-3}). Lower atmospheric (both particulate and gas phase) concentrations have been reported in a few studies around industrial sites like Shanghai, China where TEQs were calculated between 9.28–423 fg WHO-TEQ$_{2005}$ m^{-3} (mean 88.9 fg WHOTEQ$_{2005}$ m^{-3}) [55]. Additionally, in an urban site of Brno, Czech Republic, particulate-bound concentrations of PCDD/Fs ranged from 4.63 to 661 fg m^{-3} [56] while in gas phase of background areas in Spain (2.18–19.1, mean 656 fg WHOTEQ$_{2005}$ m^{-3}) [57]. Similar levels of PCDDFs were measured in the atmosphere (particulate and gas phase) of Umm-Al-Aish oil field, with values of 31.2–516 fg WHOTEQ$_{2005}$ m^{-3} [20], and in PM10 from Rome with 5.43–734 fg WHOTEQ$_{98}$ m^{-3} [54]. Higher concentrations were also presented in many industrial sites worldwide (Table 5), and our results were classified among the lowest in global literature. The comparison emphasizes in studies performed in industrial sites even the samples derived from particulate, gas, or both atmospheric phases. To our knowledge, in Greece, only two studies referred to ambient concentrations of PCDD/Fs, with our results being in the same levels with those found in particulate phase from the center of Athens (73 fg m^{-3} in the background site and 462 fg m^{-3} in the urban site) [32] and lower than the particle-bound concentrations measured in Thessaloniki (150–12,890 fg m^{-3}) [53]. There is no international standard about reference or target values of PCDD/Fs in the air, however, a review by Lohmann and Jones (1998) [58] reported some typical values according to the sampling point (background—remote sites: <10 fg I-TEQ m^{-3}, rural sites: 20–50 fg I-TEQ m^{-3}), urban or industrial sites: 100–400 fg I-TEQ m^{-3}). Comparing our results with these values, it is obvious that the industrial area of Thriassion Plain has a significant impact on the ambient air concentration of PCDD/Fs. The mean values in our study (mean 656 fg WHO-TEQ$_{2005}$/m^{-3}) are also below the recommendations by the Environment Minister in Ontario (Canada) (5000 fg I-TEQ m^{-3}) [3] and slightly over the atmospheric standard of Japan (600 fg WHOTEQ$_{2005}$ m^{-3}), but far higher than the standard of Germany (150 fg WHOTEQ$_{2005}$ m^{-3}) [59,60]. It is noteworthy that the results in this study referred only to the particulate phase, and the calculated WHO-TEQ values may be underestimated.

For the dlPCBs, the total particulate concentrations (expressed as the sum of m.o. and n.o. congeners) were 975–4083 fg m^{-3} (median 1438 fg m^{-3}), and the WHOTEQ$_{2005}$ concentrations were 53–512 fg m^{-3}. As it is clear from Table 5, the concentrations and WHOTEQ$_{2005}$ of dlPCBs are in comparable values with studies in industrial sites of Dunkirk, France (743–1747 fg m^{-3}, particulate phase) [31], and Shanghai, China (340–7607 fg m^{-3}, particulate plus gas phase) [55], but significantly higher concentrations have been reported in gas phase under the vicinity of a steel complex in Korea (6100–61,800 fg m^{-3}) [61] or in the urban atmospheres (particulates and gas phase) in Hochiminh, Vietnam (1570–8300 fg m^{-3}) [62] and Brescia, Italy (11,600–708,400 fg m^{-3}) [23].

Table 3. Average, median, and ranged concentrations of polychlorinated dibenzo-p-dioxins/dibenzofurans (PCDD/Fs) (fg m^{-3}) for the overall and seasonally divided sampling period.

	2,3,7,8-TCDD	1,2,3,7,8-PeCDD	1,2,3,4,7,8-HxCDD	1,2,3,6,7,8-HxCDD	1,2,3,7,8,9-HxCDD	1,2,3,4,6,7,8-HpCDD	OCDD	ΣPCDDs	2,3,7,8-TCDF	1,2,3,7,8-PeCDF	2,3,4,7,8-PeCDF	1,2,3,4,7,8-HxCDF	1,2,3,6,7,8-HxCDF	2,3,4,6,7,8-HxCDF	1,2,3,7,8,9-HxCDF	1,2,3,4,6,7,8-HpCDF	1,2,3,4,7,8,9-HpCDF	OCDF	ΣPCDFs	WHO-TEQ PCDD/Fs
								Overall Sampling Campaign (N = 30)												
Average	1.81	2.35	2.58	2.69	3.19	33.2	172	217	1.83	2.64	6.93	13.8	10.4	21.6	4.25	129	9.93	279	479	656
Median	1.81	1.81	1.81	1.81	1.81	10.8	111	167	1.81	1.81	1.81	3.93	1.92	2.88	2.14	55.1	3.89	94.0	325	444
Range	<1.81	1.81–4.75	1.81–7.70	1.81–7.53	1.81–7.53	1.81–167	10.4–510	29.8–528	1.81–2.07	1.81–14.0	1.81–74.5	1.81–99.3	1.81–77.2	1.81–172	1.81–18.3	1.81–569	1.81–39.6	2.75–2912	52.8–2945	287–2560
								Cold Period (N = 15)												
Average	1.81	2.53	2.41	2.96	2.45	50.2	151	427	1.83	3.43	11.2	24.8	18.0	39.6	5.84	193	12.8	128	439	914
Median	1.81	1.81	1.81	2.07	2.07	34.0	114	327	1.81	1.81	2.07	8.50	10.5	8.59	3.28	141	6.84	86.4	311	692
Range	<1.81	1.81–4.67	1.81–5.01	1.81–7.35	1.81–5.36	5.43–167	10.4–306	157–861	1.81–2.07	1.81–74.5	1.81–74.5	1.81–99.3	1.81–77.2	1.81–172	1.81–18.3	5.06–569	1.81–32.7	20.8–408	52.9–1176	332–2560
								Warm Period (N = 15)												
Average	1.81	2.17	2.75	2.42	3.93	16.3	192	221	1.83	1.85	2.66	2.79	2.78	3.57	2.66	64.1	7.06	429	519	398
Median	1.81	1.81	1.81	1.81	1.81	3.51	96.6	170	1.81	1.81	1.81	1.81	1.81	2.02	1.81	26.3	1.81	181	339	366
Range	<1.81	1.81–4.75	1.81–7.70	1.81–7.53	1.81–20.6	1.81–97.6	17.2–510	29.8–527	1.81–2.03	1.81–2.21	1.81–8.85	1.81–5.36	1.81–7.15	1.81–11.8	1.81–8.66	1.81–279	1.81–39.6	2.75–2912	63–2945	287–544

Table 4. Average, median, and ranged concentrations of dioxin like PCBs (dlPCBs) (fg m^{-3}) for the overall and seasonally divided sampling period.

	PCB-77	PCB-81	PCB-126	PCB-169	PCB-105	PCB-114	PCB-118	PCB-123	PCB-156	PCB-157	PCB-167	PCB-189	ΣdlPCBs	WHO-TEQ dlPCBs
		Non-Ortho PCBs						Mono-Ortho PCBs						
					Overall Sampling Campaign (N = 30)									
Average	85.8	6.28	22.4	6.37	238	36.3	752	62.4	101	75.7	208	40.5	1634	137
Median	72.9	5.45	17.1	5.25	186	36.3	594	60.8	92.9	74.5	156	36.3	1438	105
Range	41.2–268	1.81–19.0	8.26–86.6	1.81–22.0	113–634	<36.3	453–2029	36.3–143	36.0–259	36.3–188	36.3–969	36.3–116	975–4083	53.1–512
					Cold Period (N = 15)									
Average	75.4	5.72	21.4	7.51	190	36.3	637	61.5	104	93.3	112	37.4	1382	136
Median	74.0	5.61	17.2	6.40	184	36.3	586	63.7	93.5	87.1	74.3	36.3	1290	105
Range	41.2–113	1.99–9.33	8.47–35.6	1.81–14.7	146–287	<36.3	505–871	41.0–94.0	36.0–195	36.3–188	36.3–324	36.3–43.9	1092–1893	62.5–217
					Warm Period (N = 15)									
Average	96.3	6.80	23.3	5.24	287	36.3	867	63.3	97.2	58.0	303	43.5	1887	139
Median	71.6	5.16	16.0	2.76	237	36.3	815	42.5	85.1	51.6	238	36.3	1592	93.9
Range	50.8–268	1.81–19.0	8.26–86.6	1.81–22.0	113–634	<36.3	453–2029	36.3–143	37.6–259	36.3–81.2	36.3–969	36.3–116	975–4083	53.1–512

Table 5. Worldwide concentrations of persistent organic pollutants (POPs) according to literature in comparison with this study. (*p* refers to particulate phase and G to gas phase samples).

Area	Area Description	Sampling (Year/Phase)	ΣPCDD/Fs (fg m^{-3})	WHO (TEQ fg m^{-3})	ΣdlPCBs	WHO (TEQ fg m^{-3})	ΣindPCBs (pg m^{-3})	ΣPAHs (ng m^{-3})	Reference
Eleusis, Greece	Industrial	2018–2019/P	145–3472	287–2560	975–4083	53.1–512	22.9–53.2	1.3–16.5	This study
Athens, Greece	Background/Urban	2000/P	73–462 (P)	73 (P + G)				1.79–4.94	[32]
Thessaloniki, Greece	Semirural/Urban	1999/P	150–12,890	4–119			0.5–29.2		[53]
Porto, Portugal	Urban/Industrial	1999–2004/P + G	200–15,000	9.8–817					[63]
Rome, Italy	Urban	2000–2001/P		5.43–734		0.66–7.28	88.9–372		[54]
Steel Complex, Korea	Industrial	2006/G			6100–61,800				[61]
Anshan, China	Industrial	2008/G	20–9790		4560	0.3–23			[38]
Satellite cities of Seoul, Korea	Industrial	2003–2009/P + G	360–55,755	310–3143	678–40,968	200–1712			[64]
Brescia, Italy	Urban/Industrial	/P	380–11,390	10–190	11,600–708,400	4–130	92.7–8566		[23]
Brno, Czech Republic	Urban	2009–2010/P	4.3–661-		14.2–614			0.04–14.4	[56]
Aliaga, Turkey	Industrial	2009–2010/G						1.6–838	[21]
Shanghai, China	Industrial	2013/P + G	258–4928	9.28–423	340–7607	1.08–55.9			[55]
Dunkirk, France	Industrial	2008–2009/P	718–1070	12,990–25,430	744–1747			0.29–1.95	[31]
Istanbul, Turkey	Industrial	2011–2012/P + G		0.7–27.9				0.04–445	[39]
Umm Al-Aish, Kuwait	Industrial	2014–2015/P + G	410–52,700	31.2–516	1570–8300	3.9–36.8			[20]
Hochiminh, Vietnam	Urban	2016–2017/P + G	1180–2180	22.2–1530	490–900	1.77–44.1			[62]
Tibet-Qinghai Plateau, China	Industrial	2015/G–	11,900–57,700	113–242	580–2710	3.18–7.12			[4]
Jiangsu, China	Industrial	2018/P + G		136–597		0.64–11.3		11–18	[16]

3.2. Seasonal Variations and Congeners' Contributions

In Tables 1–4, the results were divided by their seasonality in order to reveal any possible variation between measurements during cold and warm periods. The Mann–Whitney test was applied in each dataset for this purpose.

Regarding PAHs, seasonal variations were observed for PAHs with 4 or more rings in their molecules. These high molecular weight PAHs presented significantly higher concentrations during cold months ($p < 0.05$). PAHs' seasonal variations are described also in Figure 2 and it is worth mentioning that ΣPAHs was 4 times higher during cold months (8.44 ng m^{-3} compared to 2.01 ng m^{-3}), and the most carcinogenic among the compounds, BaP, exceeded the target limit of the European Commission with an average value of 1.03 ng m^{-3}, while during warm period, it was 0.57 ng m^{-3}. Such differences can be explicated by the increased direct emissions sources (fossil fuel combustion, biomass burning, etc.) during cold months, and by the enhanced condensation of PAHs in the particulate matter due to the lower temperatures [41], whereas increased solar radiation and atmospheric oxidants during warm months lead to PAHs' degradation [65]. Similar seasonal trends, especially for high molecular weight PAHs, have also been reported in other studies in Greece [36,47,66] and globally [21,39,41].

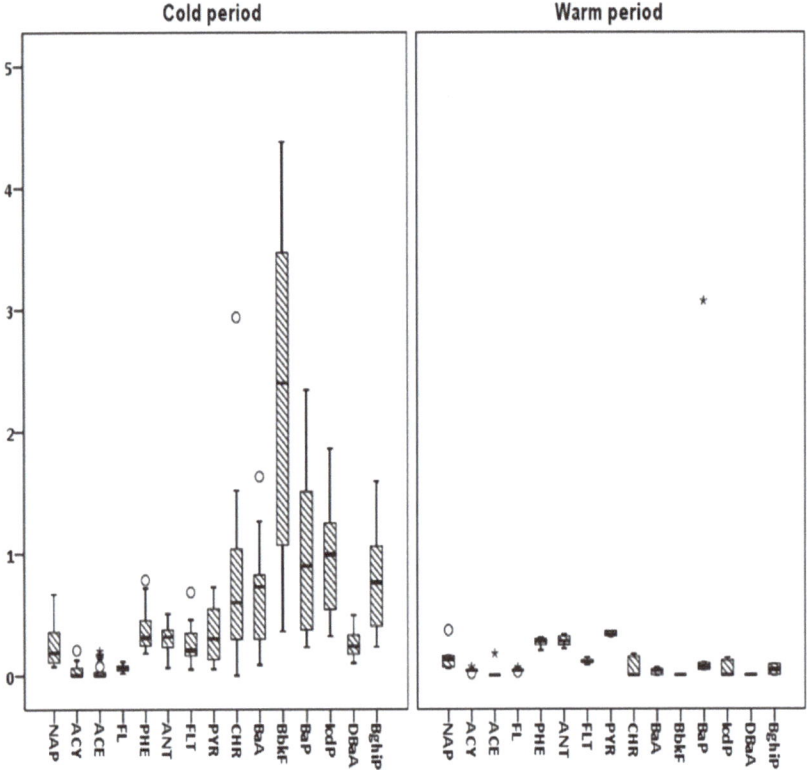

Figure 2. PAH seasonal variations in ng m^{-3} (* outliers and ° values above 3rd quartile).

In Figure 3, the relative contribution of each compound is presented for both sampling periods. Obviously, 5–6 rings PAHs were the most abundant species during cold period, accounting for about 80% of the ΣPAHs, while the percentage is about 30.4% in the warm period dominating by the low molecular weight PAHs. This inversion may be a result of the decreased emissions from combustions related with central heating during the warm period. The dominance of petrogenic low molecular weight PAHs, especially with 3–4 rings, in the warm period may be affected also by emissions from petroleum depositories from oil refineries and cargo ships [49,67,68] that both constitute a constant factor in the area working all year long. The most abundant PAH during the cold period was BbkF 27.5% followed by BaP 12.2%, while in summer BaP remained the dominant PAH with 28.2% relative abundance, followed this time by petrogenic ANT and PHE with 13.9 and 13.5%, respectively (Figure 3). BbkF and BaP were also the most abundant compounds in other studies in Aspropirgos [36], Athens [66], Volos [49], Greece, in Dunkirk, France [31], and Sao Paolo, Brazil [41].

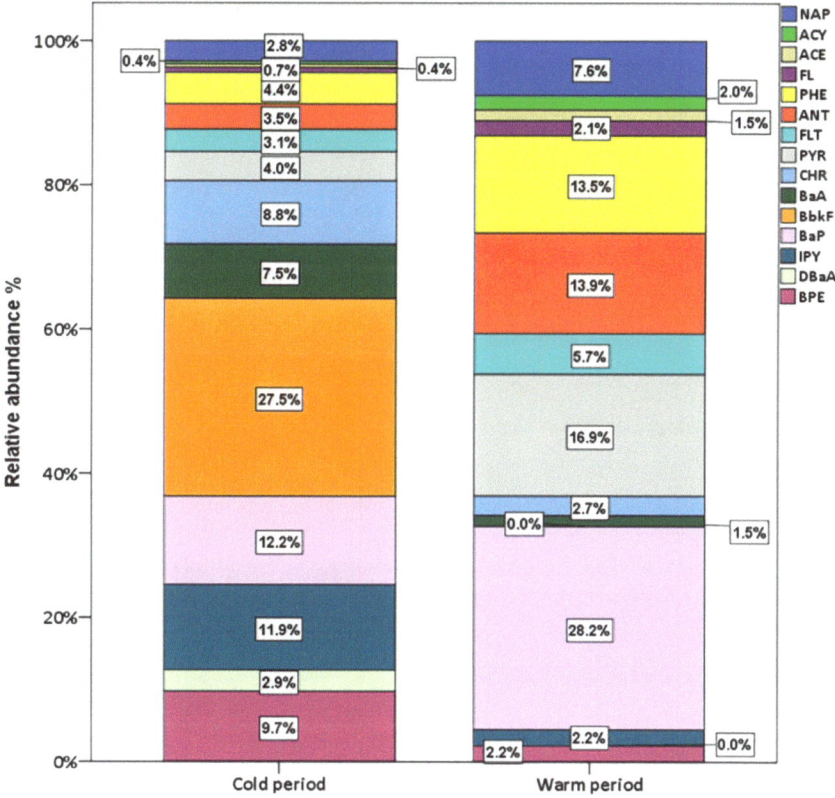

Figure 3. Relative abundance among PAH compounds for the cold and warm periods.

As for the indPCB's seasonal variations, none of the six congeners presented statistically significant differences (p values were >0.05) between the cold and warm periods, with the concentrations for both seasons being included in Table 2. The indPCBs also did not present any alteration regarding their relative abundance, with PCB28 being the most abundant congener in both sampling periods (41.7 and 46.0%, respectively) (Figure 4). The indPCBs profile also indicates a marked decrease in concentration with increasing chlorination for different congeners. The same outcome was highlighted also by Colombo et al. (2013) in the highly industrialized city of Brescia, in northern Italy [23].

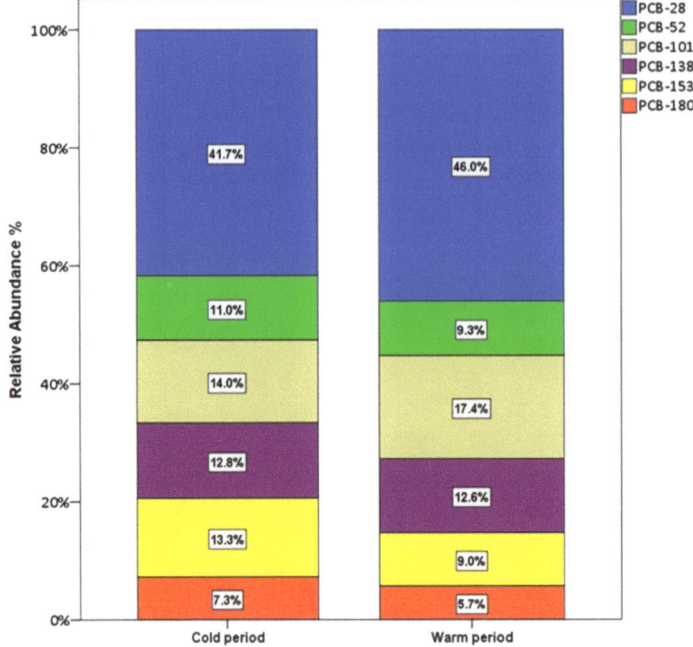

Figure 4. Relative abundance among indPCB congeners for the cold and warm periods.

Seasonal variations of dioxins are presented in Figure 5 and according to the Mann–Whitney test only 1,2,3,4,6,7,8-HpCDD, 1,2,3,4,7,8-HxCDF were found significantly higher during the cold period. Table 2 and Figure 5 show that there was not a clear seasonal pattern, with some compounds like OCDD (warm: 192 fg m^{-3}, cold: 151 fg m^{-3}) and OCDF (warm: 429 fg m^{-3}, cold: 128 fg m^{-3}) being relatively higher during the warm period. However, the WHOTEQ$_{2005}$ value was significantly higher in cold months, revealing that particulate PCDD/Fs pose a higher risk to human health during cold periods due to their higher concentrations (Figure 6). This outcome could be combined with our previous studies in this area [35,36], where uncontrolled combustions in Thriassion Plain play a potential role in atmospheric degradation and the increased toxicity of the particles in the whole area, including Eleusis and Aspropirgos town (Figure 1). In particular, during cold months the wind direction is usually north, northeast (N, NE) in the sampling area and as a result the particulate matter is enriched by the site where uncontrolled combustions take part [35]. Comparing with other studies, in some of them, seasonal variations were observed with higher concentrations during cold months [38,54,64], assuming either elevated combustions for heating or relatively low boundary layer heights in cold weather [58] being the dominant reason for that outcome. Nevertheless, other studies did not describe any seasonal trend for PCDD/Fs [69].

Discussion on the congener profile of PCDD/Fs during the warm and cold periods will be based on Figure 7. The most toxic congener, 2,3,7,8-TCDD, was not detected in any sample. The congener profiles demonstrated the prevalence of OCDF, OCDD, 1,2,3,4,6,7,8,-HpCDF, and 1,2,3,4,6,7,8-HpCDD for the entire sampling period. The relative abundances of these four compounds were 19.7–58%, 23.1–26.0%, 8.7–29.6%, and 2.2–7.7%, respectively. The results are in agreement with those from other studies in industrial sites in Shanghai [55] and Anshan [38], China, in Seoul, Korea [64], and in Umm-Al-Aish oil field, Kuwait [20].

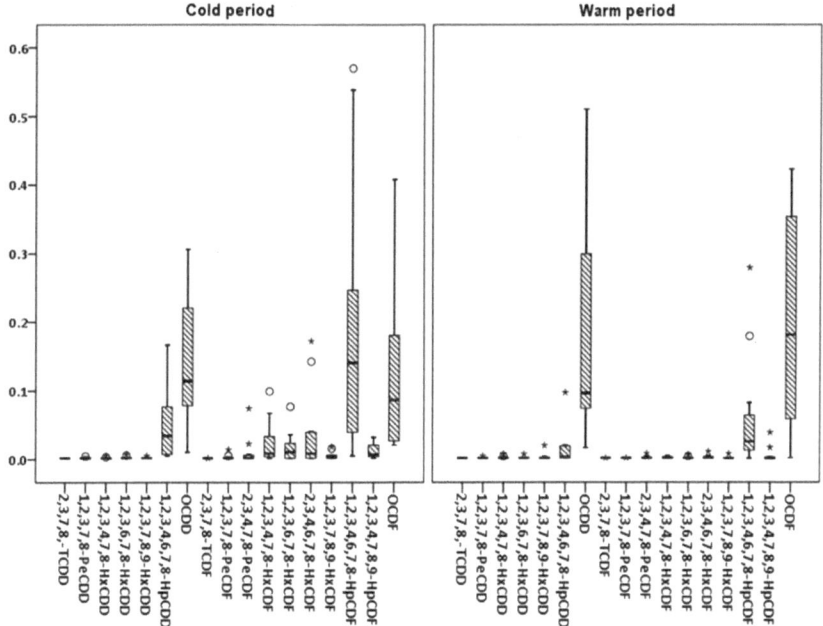

Figure 5. PCDD/F seasonal variations in pg m^{-3} (* outliers and ° values above 3rd quartile).

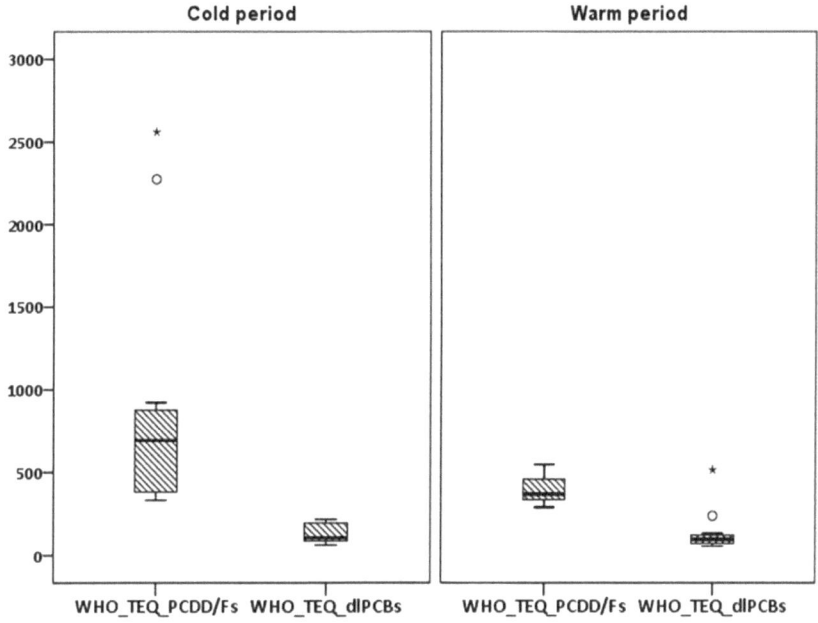

Figure 6. WHO-TEQ$_{2005}$ seasonal variations for PCDD/Fs and dlPCBs (* outliers and ° values above 3rd quartile).

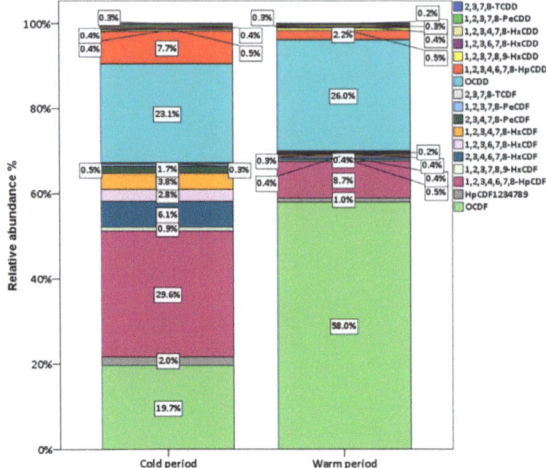

Figure 7. Relative abundance among PCDD/F congeners for the cold and warm periods.

In regard to particle-bound dlPCBs, no seasonal pattern was observed during the sampling campaign (Table 3), with concentrations and WHOTEQ$_{2005}$ values for each congener (Figure 6) being quite similar between the cold and warm periods. This result is in agreement with other studies [20], although there are papers reporting elevated concentrations during warm periods for PCBs in gas phase [38,64], assuming that PCBs evaporated more easily from soil to gaseous phase and could be transported long-range under high temperatures [38,70,71].

Although the sampling was performed only in the particulate phase, the dlPCBs congener fingerprint detected in our study complies with that reported in the literature. Thus, according to Figure 8, the most abundant congener was PCB118, accounting for 45.9–46.1% of the relative abundance followed by PCB105 (13.7–15.2%), PCB167 (8.1–16.7%), and PCB77 (5.1–5.7%). The outcome from our study matches the profile from other studies in both gas and particulate phase [6,20,23,72], except for PCB167, reinforcing the conclusion that the dlPCBs' profile does not depend on the sampling location.

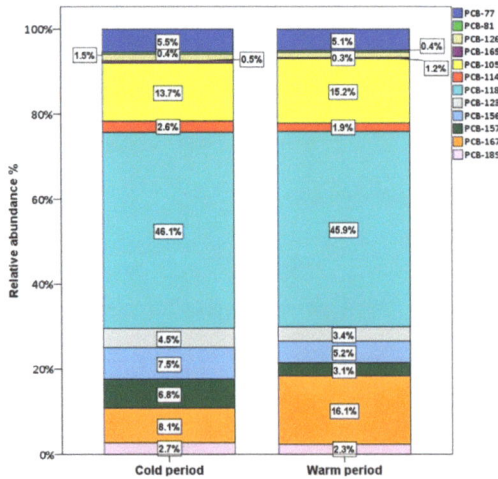

Figure 8. Relative abundance among dlPCB congeners for the cold and warm periods.

3.3. Source Apportionment

Calculation of different molecular diagnostic ratios (MDRs) has been applied in many studies for source apportionment of PAHs in the atmosphere [65,73,74]. The approach relies on the fact that emission of certain PAHs from characteristic sources tends to be constant, and thus the ratio receptor/source remains stable [67,68]. However, MDRs should be used with caution, as PAH components may differ in reactivity and atmospheric residence times [67]. In Table 6, the selected MDRs are presented for the whole period and for each season separately. All four MDRs conclude to the same outcome, suggesting strong pyrogenic source contributions. Moreover, there was no observed difference between the cold and warm periods, assuming that the sources are constant and strongly related with combustions in the area. The MDRs calculated here are parallel with those reported previously for Aspropirgos and Eleusis [36,47], which also suggest effects from pyrogenic sources.

Table 6. MDRs (molecular diagnostic ratios) for the whole sampling campaign and their interpretation from literature [67].

	Petrogenic Origin	Pyrogenic Origin	This Study		
			Mean	Cold	Warm
ANT/(ANT + PHE)	<0.1	>0.1	0.46	0.45	0.51
BaA/(BaA + CHR)	<0.2	>0.35	0.56	0.51	0.62
FLT/(FLT + PYR)	<0.4	>0.4	0.46	0.47	0.45
IPY/(IPY + BPE)	<0.2	>0.2	0.55	0.56	0.52

One of the tools to estimate possible common sources among PCDD/Fs is the calculation of PCDD to PCDF (D/F) ratio. In general, these ratios in ambient air indicate the degree of contamination from combustion sources. Values above 1.0 suggest less contribution, while values <0.5 show increased contribution of combustion sources [75]. In our study, the D/F ratio was found in an average value of 0.45, a value ranging from 0.49 during the cold period to 0.42 during the warm period. The values suggest that in the whole sampling period, the PCDD/Fs emissions were influenced dominantly by thermal processes. The same results were obtained from other studies in industrial sites in satellite cities of Seoul, Korea [64]. Steel and iron plants' emissions can also be considered as a continuous source of PCDD/Fs in Thriassion Plain, as reported also by Li et al. (2011) in northeast China [38].

In order to evaluate any potential associations and gain an overview of the relationships among the pollutants investigated, principal component analysis (PCA) was performed in the entire dataset (Table 7). PCA was also applied for PAHs, PCDD/Fs, and PCBs separately, but the results were overlapping so the combined PCA was preferred, including all the studied compounds. Three factors explained the 83.4% of the total variance. PC1 was heavily loaded with high molecular weight PAHs like BaP (0.805), IPY (0.821), and BbkF (0.759) together with most of the PCDFs and some of PCDDs. This factor agrees that PAHs and PCDD/Fs have common sources related with combustions, both controlled and illegal. Indeed, open burning for waste incineration in the studied area is a common practice, and these emissions combined with industrial ones could compose a constant source of both pollutants. Another possible common source for these POPs could be cement kiln factory where the conditions of the furnace; very high temperature, mixing and excess of oxygen; make it an ideal place for a 'perfect' combustion [39,76]. The PC2 (24.3% of variance) was tightly clustered with PCBs loadings (around 0.7) and is a factor indicating the common source of PCBs, both dlPCBs and indPCBs. PCBs were clearly distinct from PCDD/Fs and PAHs as also reported in other studies [23]. PCBs emissions may be affected of the industries and chemical plants of Thriassion Plain while uncontrolled combustions of plastic may also enrich the aerosols with PCBs. The third factor (22.9% of variance) was associated with petrogenic, low molecular weight PAHs and DBaA, suggesting a common petrogenic origin [16]. The oil refineries in this site are a dominant source of emissions, and volatile compounds could release into the atmosphere from their petroleum depositories.

Table 7. Principal component analysis in the whole sampling campaign.

	PC1	PC2	PC3
Explained Variance	36.2%	24.3%	22.9%
NAP	−0.399	0.329	−0.213
ACY	−0.223	−0.258	0.114
ACE	0.330	0.307	−0.205
FL	−0.198	0.210	0.763
PHE	−0.027	−0.048	0.805
ANT	0.439	−0.077	−0.260
FLT	0.116	0.204	0.893
PYR	0.132	0.370	0.848
CHR	0.755	0.420	0.525
BaA	0.729	0.398	0.501
BbkF	0.759	0.182	0.491
BaP	0.805	0.026	−0.005
IPY	0.821	0.124	0.448
DBaA	0.534	0.340	0.748
BPE	0.808	0.112	0.433
1,2,3,7,8- PeCDD	0.314	0.287	0.056
1,2,3,4,7,8- HxCDD	0.785	−0.119	0.444
1,2,3,6,7,8- HxCDD	−0.362	−0.029	0.338
1,2,3,7,8,9- HxCDD	−0.173	−0.388	0.345
1,2,3,4,6,7,8- HpCDD	0.768	0.268	0.015
OCDD	0.131	0.064	−0.580
2,3,7,8- TCDF	0.664	−0.362	−0.080
1,2,3,7,8- PeCDF	0.788	−0.227	−0.023
2,3,4,7,8- PeCDF	0.772	0.499	−0.174
1,2,3,4,7,8- HxCDF	0.815	−0.070	−0.161
1,2,3,6,7,8- HxCDF	0.743	0.554	−0.018
2,3,4,6,7,8- HxCDF	0.879	0.234	−0.135
1,2,3,7,8,9- HxCDF	0.622	0.531	−0.044
1,2,3,4,6,7,8- HpCDF	0.889	0.180	−0.104
1,2,3,4,7,8,9- HpCDF	0.497	−0.106	0.182
OCDF	−0.087	0.040	−0.452
PCB-77	0.270	0.113	0.082
PCB-81	−0.069	−0.251	0.010
PCB-126	0.079	0.779	0.223
PCB-169	0.132	0.600	0.570
PCB-105	−0.049	0.313	−0.360
PCB-118	−0.062	0.583	−0.388
PCB-123	−0.201	0.686	−0.184
PCB-156	0.169	0.720	0.128
PCB-157	0.286	0.706	0.234
PCB-167	−0.067	−0.489	−0.055
PCB-189	−0.082	0.455	0.294
PCB-28	0.108	0.384	−0.630
PCB-52	0.238	0.717	−0.269
PCB-101	−0.266	0.473	−0.472
PCB-138	−0.010	0.838	−0.236
PCB-153	0.163	0.832	0.216
PCB-180	−0.238	0.651	0.299

3.4. Health Risk Assessment

Starting with BaPE values, the average of 1.27 ng m^{-3} indicates increased toxicity of the particulate matter due to BaP related PAHs. The values were lower during the warm period (0.57 ng m^{-3}) and are in agreement with our previous study in the area with values ranging from 0.14–4.6 ng m^{-3}. Average ΣBaP$_{TEQ}$ in our study was found to be 2.29 ng m^{-3} (cold: 2.76 ng m^{-3}, warm: 0.58 ng m^{-3}), lower than in Aspropirgos (7.0 ng m^{-3} in autumn and 3.8 ng m^{-3} in winter months) [32], but higher than in Thessaloniki (1.5 ng m^{-3}) [42]. Similar values were presented in Mestre, Venice with a BaP$_{TEQ}$ value of 3.6 ng m^{-3} during winter and 1.7 ng m^{-3} in autumn months [77]. Parallel seasonal patterns were calculated for ΣBaP$_{MEQ}$ with an average value of 1.61 ng m^{-3} (cold: 1.89 ng m^{-3}, warm: 0.59 ng m^{-3}). As it is clear from Figure 9, the cold period was strongly associated with increased toxicity of aerosol due to PAHs levels.

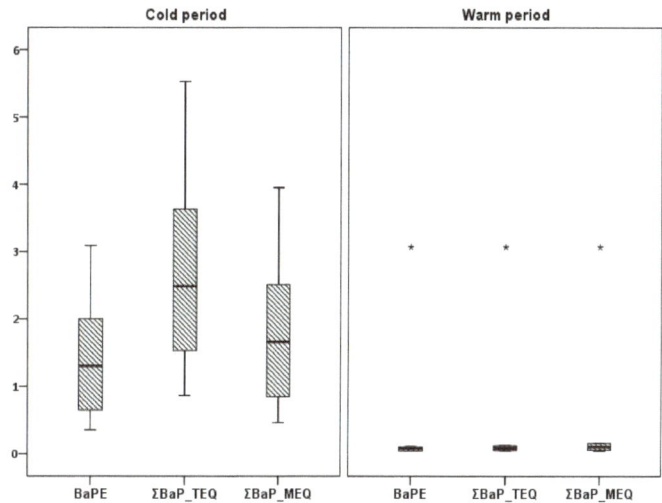

Figure 9. Seasonal variation of BaPE, ΣBaP$_{TEQ}$, and ΣBaP$_{MEQ}$ (* outliers).

ICR, due to the toxicity of PAHs bound to PM$_{10}$, was calculated with a mean value of 2.6 × 10^{-6} and 3.0 × 10^{-6} during the cold period and 1.1 × 10^{-6} during the warm period. ICR in Eleusis was higher than that calculated in Thessaloniki (1.6–1.7 × 10^{-6} during winter) [42], but in the same level with previous work in Thessaloniki with a value of 2.8 × 10^{-6} [78].

According to USEPA health risk evaluation procedure, the cancer risk from the investigated substances was calculated, and the total cancer risk from POPs in Eleusis was 3.6 × 10^{-5} (Table 6). This outcome could be translated as 3 to 4 cancer occurrences over 100,000 people living in Eleusis inside Thriassion Plain. The exposure in this area was considered significant as it exceeded the EPA limit of 1 × 10^{-6} and, as it is obvious in Table 8, cancer risks derived from PAHs, PCDD/Fs, and dlPCBs were individually above this threshold, with highest risk derived from PAHs (2.4 × 10^{-5}). Cancer risks calculated in this study were higher than in the literature [2,45,56], but lower than the cancer risk (over 6.2 × 10^{-5}) in Jiangsu Province, China, a place around chemical plants [16]. It has to be clarified here that risk evaluation for PCDD/Fs and dlPCBs may has been underestimated, as some particularly toxic congeners of PCDD/Fs (e.g., 2,3,7,8 TCDD, 1,2,3,7,8 PeCDD) and all dlPCBs are mostly presented in gas phase. Thus, the total cancer risk may be even higher than the calculated one.

Table 8. Cancer risk for PAHs, PCDD/Fs, and dlPCBs and total risk for citizens of Eleusis.

	Cancer Risk
ΣRisk$_{PAHs}$	2.4 × 10^{-5}
ΣRiskPCDD/Fs	9.9 × 10^{-6}
ΣRiskdlPCBs	1.6 × 10^{-6}
Total Cancer Risk	3.6 × 10^{-5}

4. Conclusions

An adequate descriptive profile of POPs contamination in particulate matter from the most affected atmosphere in Greece has been reported for the first time. Combustions were found to be the most important sources of atmospheric degradation by source apportionment tools. Although POPs concentrations were found at normal levels compared with studies from the available literature, the estimated cancer risk for the citizens was significantly elevated. As a result, the entire Thriassion Plain is a place that has to be fully monitored regarding POPs emissions, and strategies to this scope have to be scheduled and fulfilled.

Author Contributions: Conceptualization, E.B.; methodology, K.G.K., P.G.K.; software, K.G.K., P.G.K., E.C.; validation, K.G.K., P.G.K., D.C., I.V., L.L.; investigation, K.G.K., P.G.K., E.C.; data curation, K.G.K., P.G.K., E.B.; writing–original draft preparation, K.G.K., E.B.; writing–review & editing, K.G.K., P.G.K., E.C., D.C., I.V., L.L., E.B.; visualization, K.G.K., P.G.K.; supervision, E.B.; project administration, E.B. All authors have read and agreed to the published version of the manuscript.

Funding: This research received no external funding.

Conflicts of Interest: The authors declare no conflict of interest.

References

1. El-Shahawi, M.; Hamza, A.; Bashammakh, A.; Al-Saggaf, W. An overview on the accumulation, distribution, transformations, toxicity and analytical methods for the monitoring of persistent organic pollutants. *Talanta* **2010**, *80*, 1587–1597. [CrossRef] [PubMed]
2. Petrovic, M.; Sremacki, M.M.; Radonic, J.; Mihajlovic, I.; Obrovski, B.; Miloradov, M.V. Health risk assessment of PAHs, PCBs and OCPs in atmospheric air of municipal solid waste landfill in Novi Sad, Serbia. *Sci. Total Environ.* **2018**, *644*, 1201–1206. [CrossRef]
3. Weber, R.; Watson, A.; Forter, M.; Oliaei, F. Review Article: Persistent organic pollutants and landfills—A review of past experiences and future challenges. *Waste Manag. Res.* **2011**, *29*, 107–121. [CrossRef] [PubMed]
4. Hu, J.; Wu, J.; Xu, C.; Zha, X.; Hua, Y.; Yang, L.; Jin, J. Preliminary investigation of polychlorinated dibenzo-p-dioxin and dibenzofuran, polychlorinated naphthalene, and dioxin-like polychlorinated biphenyl concentrations in ambient air in an industrial park at the northeastern edge of the Tibet–Qinghai Plateau, China. *Sci. Total. Environ.* **2019**, *648*, 935–942. [PubMed]
5. [UNEP] United Nations Environment Program. Persistant Organic Pollutants (POPs). 2017. Available online: www.unenvironment.org/explore-topics/chemicals-waste/what-wedo/persistent-organic-pollutants/why-do-persistent-organic (accessed on 28 October 2020).
6. Zheng, G.J.; Leung, A.O.; Jiao, L.P.; Wong, M.H. Polychlorinated dibenzo-p-dioxins and dibenzofurans pollution in China: Sources, environmental levels and potential human health impacts. *Environ. Int.* **2008**, *34*, 1050–1061. [CrossRef] [PubMed]
7. Kedikoglou, K.; Costopoulou, D.; Vassiliadou, I.; Bakeas, E.; Leondiadis, L. An effective and low cost carbon based clean-up method for PCDD/Fs and PCBs analysis in food. *Chemosphere* **2018**, *206*, 531–538. [CrossRef]
8. Someya, M.; Ohtake, M.; Kunisue, T.; Subramanian, A.; Takahashi, S.; Chakraborty, P.; Ramachandran, R.; Tanabe, S. Persistent organic pollutants in breast milk of mothers residing around an open dumping site in Kolkata, India: Specific dioxin-like PCB levels and fish as a potential source. *Environ. Int.* **2010**, *36*, 27–35. [CrossRef]
9. Abdel-Shafy, H.I.; Mansour, M.S. A review on polycyclic aromatic hydrocarbons: Source, environmental impact, effect on human health and remediation. *Egypt. J. Pet.* **2016**, *25*, 107–123. [CrossRef]
10. Haritash, A.; Kaushik, C. Biodegradation aspects of Polycyclic Aromatic Hydrocarbons (PAHs): A review. *J. Hazard. Mater.* **2009**, *169*, 1–15. [CrossRef]
11. De Craemer, S.; Croes, K.; Van Larebeke, N.; Sioen, I.; Schoeters, G.; Loots, I.; Nawrot, T.; Nelen, V.; Campo, L.; Fustinoni, S.; et al. Investigating unmetabolized polycyclic aromatic hydrocarbons in adolescents' urine as biomarkers of environmental exposure. *Chemosphere* **2016**, *155*, 48–56. [CrossRef]
12. ATSDR Agency for Toxic Substances and Disease Registry. *Toxicological Profile for Polycyclic Aromatic Hydrocarbons*; US Department of Health and Human Services: Atlanta, GA, USA, 1995.

13. Boström, C.-E.; Gerde, P.; Hanberg, A.; Jernström, B.; Johansson, C.; Kyrklund, T.; Rannug, A.; Törnqvist, M.; Victorin, K.; Westerholm, R. Cancer risk assessment, indicators, and guidelines for polycyclic aromatic hydrocarbons in the ambient air. *Environ. Health Perspect.* **2002**, *110* (Suppl. 3), 451–488. [CrossRef]
14. *IARC Some Non-Heterocyclic Polycyclic Aromatic Hydrocarbons and Some Related Exposures*; IARC Monographs on the Evaluation of Carcinogenic Risks to Humans: Lyon, France, 2010.
15. Vassiliadou, I.; Papadopoulos, A.; Costopoulou, D.; Vasiliadou, S.; Christoforou, S.; Leondiadis, L. Dioxin contamination after an accidental fire in the municipal landfill of Tagarades, Thessaloniki, Greece. *Chemosphere* **2009**, *74*, 879–884. [CrossRef] [PubMed]
16. Liu, W.; Zhao, J.; Xu, S.; Liu, G.-B.; Tu, Y.; Shi, P.; Li, A. Concentrations, Sources, and Potential Human Health Risks of PCDD/Fs, dl-PCBs, and PAHs in Rural Atmosphere Around Chemical Plants in Jiangsu Province, China. *Bull. Environ. Contam. Toxicol.* **2020**, *104*, 846–851. [CrossRef] [PubMed]
17. Dat, N.-D.; Tsai, C.-L.; Hsu, Y.-C.; Chen, Y.-W.; Weng, Y.-M.; Chang, M.B. PCDD/Fs and dl-PCBs concentrations in water samples of Taiwan. *Chemosphere* **2017**, *173*, 603–611. [CrossRef]
18. Wu, J.; Hu, J.; Wang, S.; Jin, J.; Wang, R.; Wang, Y.; Jin, J. Levels, sources, and potential human health risks of PCNs, PCDD/Fs, and PCBs in an industrial area of Shandong Province, China. *Chemosphere* **2018**, *199*, 382–389. [CrossRef]
19. Zhao, Y.; Zhan, J.; Liu, G.; Ren, Z.; Zheng, M.; Jin, R.; Yang, L.; Wang, M.; Jiang, X.; Zhang, X. Field study and theoretical evidence for the profiles and underlying mechanisms of PCDD/F formation in cement kilns co-incinerating municipal solid waste and sewage sludge. *Waste Manag.* **2017**, *61*, 337–344. [CrossRef]
20. Martínez-Guijarro, K.; Ramadan, A.; Gevao, B. Atmospheric concentration of polychlorinated dibenzo-p-dioxins, polychlorinated dibenzofurans (PCDD/Fs) and dioxin-like polychlorinated biphenyls (dl-PCBs) at Umm-Al-Aish oil field-Kuwait. *Chemosphere* **2017**, *168*, 147–154. [CrossRef]
21. Aydin, Y.M.; Kara, M.; Dumanoglu, Y.; Odabasi, M.; Elbir, T. Source apportionment of polycyclic aromatic hydrocarbons (PAHs) and polychlorinated biphenyls (PCBs) in ambient air of an industrial region in Turkey. *Atmos. Environ.* **2014**, *97*, 271–285. [CrossRef]
22. Van den Berg, M.; Birnbaum, L.S.; Denison, M.; De Vito, M.; Farland, W.; Feeley, M.; Fiedler, H.; Hakansson, H.; Hanberg, A.; Haws, L.; et al. The 2005 World Health Organization Reevaluation of Human and Mammalian Toxic Equivalency Factors for Dioxins and Dioxin-Like Compounds. *Toxicol. Sci.* **2006**, *93*, 223–241. [CrossRef]
23. Colombo, A.; Benfenati, E.; Bugatti, S.G.; Lodi, M.; Mariani, A.; Musmeci, L.; Rotella, G.; Senese, V.; Ziemacki, G.; Fanelli, R. PCDD/Fs and PCBs in ambient air in a highly industrialized city in Northern Italy. *Chemosphere* **2013**, *90*, 2352–2357. [CrossRef]
24. Frignani, M.; Bellucci, L.G.; Carraro, C.; Raccanelli, S. Polychlorinated biphenyls in sediments of the Venice Lagoon. *Chemosphere* **2001**, *43*, 567–575. [CrossRef]
25. World Health Organization. Health Risks of Persistent Organic Pollutants from Long-Range Transboundary Air Pollution. 2003. Available online: http://www.euro.who.int (accessed on 28 October 2020).
26. World Health Organization. *Persistent Organic Pollutants: Impact on Child Health*; WHO Press: Geneva, Switzerland, 2010; ISBN 978 92 4 150110 1.
27. Koukoulakis, K.; Kanellopoulos, P.; Chrysochou, E.; Koukoulas, V.; Minaidis, M.; Maropoulos, G.; Nikoleli, G.-P.; Bakeas, V.B. Leukemia and PAHs levels in human blood serum: Preliminary results from an adult cohort in Greece. *Atmos. Pollut. Res.* **2020**, *11*, 1552–1565. [CrossRef]
28. IARC. IARC Monographs on the Evaluation of Carcinogenic Risks to Humans: Some Traditional Herbal Medicines, Some Mycotoxins, Naphthalene and Styrene. *IARC Monogr. Eval. Carcinog. Risks Chem. Hum.* **2002**, *82*, i-vi+1-551.
29. IARC. Benzo[a]pyrene. IARC Monogr. 100F. 2012. Available online: https://monographs.iarc.fr/wp-content/uploads/2018/06/mono100F-14.pdf (accessed on 28 October 2020).
30. IARC. IARC monographs on the evaluation of the carcinogenic risk of chemicals to humans. Polychlorinated biphenyls and polybrominated biphenyls. *IARC Monogr. Eval. Carcinog. Risks Chem. Hum.* **2016**, *107*, 1–501.
31. Cazier, F.; Genevray, P.; Dewaele, D.; Nouali, H.; Verdin, A.; LeDoux, F.; Hachimi, A.; Courcot, L.; Billet, S.; Bouhsina, S.; et al. Characterisation and seasonal variations of particles in the atmosphere of rural, urban and industrial areas: Organic compounds. *J. Environ. Sci.* **2016**, *44*, 45–56. [CrossRef]
32. Mandalakis, M.; Tsapakis, M.; Tsoga, A.; Stephanou, E.G. Gas–particle concentrations and distribution of aliphatic hydrocarbons, PAHs, PCBs and PCDD/Fs in the atmosphere of Athens (Greece). *Atmos. Environ.* **2002**, *36*, 4023–4035. [CrossRef]

33. Lee, R.G.M.; Jones, K.C. Gas–particle partitioning of atmospheric PCDD/Fs: Measurements and observations on modelling. *Environ. Sci. Technol.* **1999**, *33*, 3596–3604. [CrossRef]
34. Park, J.S.; Kim, J.G. Regional measurements of PCDD/PCDF concentrations in Korean atmosphere and comparison with gas–particle partitioning models. *Chemosphere* **2002**, *49*, 755–764. [CrossRef]
35. Koukoulakis, K.; Chrysohou, E.; Kanellopoulos, P.; Karavoltsos, S.; Katsouras, G.; Dassenakis, M.; Nikolelis, D.; Bakeas, V.B. Trace elements bound to airborne PM10 in a heavily industrialized site nearby Athens: Seasonal patterns, emission sources, health implications. *Atmos. Pollut. Res.* **2019**, *10*, 1347–1356. [CrossRef]
36. Kanellopoulos, P.G.; Verouti, E.; Chrysochou, E.; Koukoulakis, K.; Bakeas, V.B. Primary and secondary organic aerosol in an urban/industrial site: Sources, health implications and the role of plastic enriched waste burning. *J. Environ. Sci.* **2021**, *99*, 222–238. [CrossRef]
37. Kanellopoulos, P.G.; Chrysochou, E.; Koukoulakis, K.; Vasileiadou, E.; Kizas, C.; Savvides, C.; Bakeas, V.B. Polar organic compounds in PM10 and PM2.5 atmospheric aerosols from a background Eastern Mediterranean site during the winter period: Secondary formation, distribution and source apportionment. *Atmos. Environ.* **2020**, *237*, 117622. [CrossRef]
38. Li, X.; Li, Y.; Zhang, Q.; Wang, P.; Yang, H.; Jiang, G.; Wei, F. Evaluation of atmospheric sources of PCDD/Fs, PCBs and PBDEs around a steel industrial complex in northeast China using passive air samplers. *Chemosphere* **2011**, *84*, 957–963. [CrossRef] [PubMed]
39. Ercan, Ö.; Dincer, F. Atmospheric concentrations of PCDD/Fs, PAHs, and metals in the vicinity of a cement plant in Istanbul. *Air Qual. Atmos. Health* **2015**, *9*, 159–172. [CrossRef]
40. Jung, K.H.; Yan, B.; Chillrud, S.N.; Perera, F.P.; Whyatt, R.M.; Camann, D.; Kinney, P.L.; Miller, R.K. Assessment of Benzo(a)pyrene-equivalent Carcinogenicity and Mutagenicity of Residential Indoor versus Outdoor Polycyclic Aromatic Hydrocarbons Exposing Young Children in New York City. *Int. J. Environ. Res. Public Health* **2010**, *7*, 1889–1900. [CrossRef] [PubMed]
41. Pereira, G.M.; Teinilä, K.; Custódio, D.; Santos, A.G.; Xian, H.; Hillamo, R.; Alves, C.; De Andrade, J.B.; Da Rocha, G.O.; Kumar, P.; et al. Particulate pollutants in the Brazilian city of São Paulo: 1-year investigation for the chemical composition and source apportionment. *Atmos. Chem. Phys. Discuss.* **2017**, *17*, 11943–11969. [CrossRef]
42. Manoli, E.; Kouras, A.; Karagkiozidou, O.; Argyropoulos, G.; Voutsa, D.; Samara, C. Polycyclic aromatic hydrocarbons (PAHs) at traffic and urban background sites of northern Greece: Source apportionment of ambient PAH levels and PAH-induced lung cancer risk. *Environ. Sci. Pollut. Res.* **2016**, *23*, 3556–3568. [CrossRef] [PubMed]
43. US EPA. United States Environmental Protection Agency (USEPA) Risk-Based Concentration Table 2012. Available online: http://www.epa.gov/reg3hwmd/risk/human/index.htm (accessed on 28 October 2020).
44. US EPA. *Human Health Risk Assessment Protocol for Hazardous Waste Combustion Facilities*; U.S. Environmental Protection Agency: Washington, DC, USA, 1998.
45. Čupr, P.; Flegrová, Z.; Franců, J.; Landlová, L.; Klánová, J. Mineralogical, chemical and toxicological characterization of urban air particles. *Environ. Int.* **2013**, *54*, 26–34. [CrossRef]
46. OEHHA. *Air Toxics Hot Spots Program Risk Assessment Guidelines. Part II: Technical Support Document for Describing Available Cancer Potency Factors*; Office of Environmental Health Hazard Assessment: Sacramento, CA, USA, 2002.
47. Mantis, J.; Chaloulakou, A.; Samara, C. PM10-bound polycyclic aromatic hydrocarbons (PAHs) in the Greater Area of Athens, Greece. *Chemosphere* **2005**, *59*, 593–604. [CrossRef]
48. Manoli, E.; Chelioti-Chatzidimitriou, A.; Karageorgou, K.; Kouras, A.; Voutsa, D.; Samara, C.; Kampanos, I. Polycyclic aromatic hydrocarbons and trace elements bounded to airborne PM10 in the harbor of Volos, Greece: Implications for the impact of harbor activities. *Atmos. Environ.* **2017**, *167*, 61–72. [CrossRef]
49. Vasilakos, C.; Veros, D.; Michopoulos, J.; Maggos, T.; O'Connor, C. Estimation of selected heavy metals and arsenic in PM10 aerosols in the ambient air of the Greater Athens Area, Greece. *J. Hazard. Mater.* **2007**, *140*, 389–398. [CrossRef]
50. Directive, C. 107/EC of the European Parliament and of the Council of 15 December 2004: Relating to arsenic, cadmium, mercury, nickel and polycyclic aromatic hydrocarbons in ambient air (Fourth Daughter Directive). *Off. J. Eur. Communities* **2004**, *26*, 3–6.
51. Directive, E. 50/EC of the European Parliament and of the Council of 21 May 2008 on ambient air quality and cleaner air for Europe. *Off. J. Eur. Union* **2008**, *11*, 1–44.

52. Cheng, J.; Yuan, T.; Wu, Q.; Zhao, W.; Xie, H.; Ma, Y.; Ma, J.; Wang, W. PM10-bound Polycyclic Aromatic Hydrocarbons (PAHs) and Cancer Risk Estimation in the Atmosphere Surrounding an Industrial Area of Shanghai, China. *Water Air Soil Pollut.* **2007**, *183*, 437–446. [CrossRef]
53. Kouimtzis, T.; Samara, C.; Voutsa, D.; Balafoutis, C.; Müller, L. PCDD/Fs and PCBs in airborne particulate matter of the greater Thessaloniki area, N. Greece. *Chemosphere* **2002**, *47*, 193–205. [CrossRef]
54. Menichini, E.; Iacovella, N.; Monfredini, F.; Turrio-Baldassarri, L. Atmospheric pollution by PAHs, PCDD/Fs and PCBs simultaneously collected at a regional background site in central Italy and at an urban site in Rome. *Chemosphere* **2007**, *69*, 422–434. [CrossRef] [PubMed]
55. Die, Q.; Nie, Z.; Liu, F.; Tian, Y.; Fang, Y.; Gao, H.; Tian, S.; He, J.; Huang, Q. Seasonal variations in atmospheric concentrations and gas-particle partitioning of PCDD/Fs and dioxin-like PCBs around industrial sites in Shanghai, China. *Atmos. Environ.* **2015**, *119*, 220–227. [CrossRef]
56. Degrendele, C.; Okonski, K.; Melymuk, L.; Landlová, L.; Kukučka, P.; Čupr, P.; Klánová, J. Size specific distribution of the atmospheric particulate PCDD/Fs, dl-PCBs and PAHs on a seasonal scale: Implications for cancer risks from inhalation. *Atmos. Environ.* **2014**, *98*, 410–416. [CrossRef]
57. Muñoz-Arnanz, J.; Roscales, J.L.; Vicente, A.; Ros, M.; Barrios, L.; Morales, L.; Abad, E.; Jiménez, B. Assessment of POPs in air from Spain using passive sampling from 2008 to 2015. Part II: Spatial and temporal observations of PCDD/Fs and dl-PCBs. *Sci. Total Environ.* **2018**, *634*, 1669–1679. [CrossRef] [PubMed]
58. Lohmann, R.; Jones, K.C. Dioxins and furans in air and deposition: A review of levels, behaviour and processes. *Sci. Total Environ.* **1998**, *219*, 53–81. [CrossRef]
59. Government of Germany Report of the Federal Immission Control Committee (in German). 2004. Available online: https://www.lai-immissionsschutz.de/servlet/is/20170/LAI_Schutz_Publikum_Schalleinwirkungen_Anhaenge.pdf?command=downloadContent&filename=LAI_Schutz_Publikum_Schalleinwirkungen_Anhaenge.pdf (accessed on 28 October 2020).
60. Government of Japan. Information Brochure Dioxins. 2012. Available online: https://www.env.go.jp/en/chemi/dioxins/brochure2012.pdf (accessed on 28 October 2020).
61. Choi, S.D.; Baek, S.Y.; Chang, Y.S. Atmospheric levels and distribution of dioxin-like polychlorinated biphenyls (PCBs) and polybrominated diphenyl ethers (PBDEs) in the vicinity of an iron and steel making plant. *Atmos. Environ.* **2008**, *42*, 2479–2488. [CrossRef]
62. Trinh, M.M.; Tsai, C.L.; Hien, T.T.; Thuan, N.T.; Chi, K.H.; Lien, C.G.; Chang, M.B. Atmospheric concentrations and gas-particle partitioning of PCDD/Fs and dioxin-like PCBs around Hochiminh city. *Chemosphere* **2018**, *202*, 246–254. [CrossRef] [PubMed]
63. Coutinho, M.; Pereira, M.; Borrego, C. Monitoring of ambient air PCDD/F levels in Portugal. *Chemosphere* **2007**, *67*, 1715–1721. [CrossRef] [PubMed]
64. Min, Y.; Lee, M.; Kim, D.; Heo, J. Annual and seasonal variations in atmospheric PCDDs/PCDFs and dioxin-like PCBs levels in satellite cities of Seoul, Korea during 2003–2009. *Atmos. Environ.* **2013**, *77*, 222–230. [CrossRef]
65. Finardi, S.; Radice, P.; Cecinato, A.; Gariazzo, C.; Gherardi, M.; Romagnoli, P. Seasonal variation of PAHs concentration and source attribution through diagnostic ratios analysis. *Urban Clim.* **2017**, *22*, 19–34. [CrossRef]
66. Paterakis, S.; Fameli, K.M.; Assimakopoulos, V.; Bougiatioti, A.; Maggos, T.; Mihalopoulos, N. Levels, sources and health risk of PM2.5 and PM1-bound PAHs across the Greater Athens Area: The role of the type of environment and the meteorology. *Atmosphere* **2019**, *10*, 622. [CrossRef]
67. Katsoyiannis, A.; Sweetman, A.J.; Jones, K.C. PAH molecular diagnostic ratios applied to atmospheric sources: A critical evaluation using two decades of source inventory and air concentration data from the UK. *Environ. Sci. Technol.* **2011**, *45*, 8897–8906. [CrossRef]
68. Tobiszewski, M.; Namieśnik, J. PAH diagnostic ratios for the identification of pollution emission sources. *Environ. Pollut.* **2012**, *162*, 110–119. [CrossRef]
69. Duarte-Davidson, R.; Stewart, A.; Alcock, R.E.; Cousins, I.T.; Jones, K.C. Exploring the balance between sources, deposition, and the environmental burden of PCDD/Fs in the U.K. terrestrial environment: An aid to identifying uncertainties and research needs. *Environ. Sci. Technol.* **1997**, *31*, 1–11. [CrossRef]

70. Motelay-Massei, A.; Harner, T.; Shoeib, M.; Diamond, M.; Stern, G.; Rosenberg, B. Using passive air samplers to assess urban–rural trends for persistent organic pollutants and polycyclic aromatic hydrocarbons. 2. Seasonal trends for PAHs, PCBs, and organochlorine pesticides. *Environ. Sci. Technol.* **2005**, *39*, 5763–5773. [CrossRef]
71. Baek, S.Y.; Choi, S.D.; Park, H.; Kang, J.H.; Chang, Y.S. Spatial and seasonal distribution of polychlorinated biphenyls (PCBs) in the vicinity of an iron and steel making plant. *Environ. Sci. Technol.* **2010**, *44*, 3035–3040. [CrossRef]
72. Cleverly, D.; Ferrario, J.; Byrne, C.; Riggs, K.; Joseph, D.; Hartford, P. A general indication of the contemporary background levels of PCDDs, PCDFs, and coplanar PCBs in the ambient air over rural and remote areas of the United States. *Environ. Sci. Technol.* **2007**, *41*, 1537–1544. [CrossRef] [PubMed]
73. Guo, H.; Lee, S.C.; Ho, K.F.; Wang, X.M.; Zou, S.C. Particle-associated polycyclic aromatic hydrocarbons in urban air of Hong Kong. *Atmos. Environ.* **2003**, *37*, 5307–5317. [CrossRef]
74. Sienra, M.D.R.; Rosazza, N.G.; Préndez, M. Polycyclic aromatic hydrocarbons and their molecular diagnostic ratios in urban atmospheric respirable particulate matter. *Atmos. Res.* **2005**, *75*, 267–281. [CrossRef]
75. Xu, M.X.; Yan, J.H.; Lu, S.Y.; Li, X.D.; Chen, T.; Ni, M.J.; Dai, H.F.; Wang, F.; Cen, K.F. Concentrations, profiles, and sources of atmospheric PCDD/Fs near a municipal solid waste incinerator in Eastern China. *Environ. Sci. Technol.* **2009**, *43*, 1023–1029. [CrossRef]
76. Yang, H.H.; Chen, C.M. Emission inventory and sources of polycyclic aromatic hydrocarbons in the atmosphere at a suburban area in Taiwan. *Chemosphere* **2004**, *56*, 879–887. [CrossRef] [PubMed]
77. Masiol, M.; Hofer, A.; Squizzato, S.; Piazza, R.; Rampazzo, G.; Pavoni, B. Carcinogenic and mutagenic risk associated to airborne particle-phase polycyclic aromatic hydrocarbons: A source apportionment. *Atmos. Environ.* **2012**, *60*, 375–382. [CrossRef]
78. Sarigiannis, D.A.; Karakitsios, S.P.; Zikopoulos, D.; Nikolaki, S.; Kermenidou, M. Lung cancer risk from PAHs emitted from biomass combustion. *Environ. Res.* **2015**, *137*, 147–156. [CrossRef]

Publisher's Note: MDPI stays neutral with regard to jurisdictional claims in published maps and institutional affiliations.

© 2020 by the authors. Licensee MDPI, Basel, Switzerland. This article is an open access article distributed under the terms and conditions of the Creative Commons Attribution (CC BY) license (http://creativecommons.org/licenses/by/4.0/).

Communication

Towards Improving Transparency of Count Data Regression Models for Health Impacts of Air Pollution

John F. Joseph [1],*, Chad Furl [1], Hatim O. Sharif [1], Thankam Sunil [2] and Charles G. Macias [3]

[1] Department of Civil and Environmental Engineering, University of Texas at San Antonio, One UTSA Circle, San Antonio, TX 78249, USA; chad.furl@utsa.edu (C.F.); Hatim.sharif@utsa.edu (H.O.S.)
[2] Department of Public Health, University of Tennessee, Knoxville, 1914 Andy Holt Ave., Knoxville, TN 37996, USA; tsunil@utk.edu
[3] Center for Clinical Effectiveness and Evidence-Based Outcome Center, Baylor College of Medicine/Texas Children's Hospital, 6621 Fannin St., Houston, TX 77030, USA; Charles.Macias@UHhospitals.org
* Correspondence: john.joseph@utsa.edu

Citation: Joseph, J.F.; Furl, C.; Sharif, H.O.; Sunil, T.; Macias, C.G. Towards Improving Transparency of Count Data Regression Models for Health Impacts of Air Pollution. *Appl. Sci.* **2021**, *11*, 3375. https://doi.org/10.3390/app11083375

Academic Editor: Thomas Maggos

Received: 20 February 2021
Accepted: 7 April 2021
Published: 9 April 2021

Publisher's Note: MDPI stays neutral with regard to jurisdictional claims in published maps and institutional affiliations.

Copyright: © 2021 by the authors. Licensee MDPI, Basel, Switzerland. This article is an open access article distributed under the terms and conditions of the Creative Commons Attribution (CC BY) license (https:// creativecommons.org/licenses/by/ 4.0/).

Abstract: In studies on the health impacts of air pollution, regression analysis continues to advance far beyond classical linear regression, which many scientists may have become familiar with in an introductory statistics course. With each new level of complexity, regression analysis may become less transparent, even to the analyst working with the data. This may be especially true in count data regression models, where the response variable (typically given the symbol y) is count data (i.e., takes on values of 0, 1, 2, ...). In such models, the normal distribution (the familiar bell-shaped curve) for the residuals (i.e., the differences between the observed values and the values predicted by the regression model) no longer applies. Unless care is taken to correctly specify just how those residuals are distributed, the tendency to accept untrue hypotheses may be greatly increased. The aim of this paper is to present a simple histogram of predicted and observed count values (POCH), which, while rarely found in the environmental literature but presented in authoritative statistical texts, can dramatically reduce the risk of accepting untrue hypotheses. POCH can also increase the transparency of count data regression models to analysts themselves and to the scientific community in general.

Keywords: count data; correlation; regression models

1. Introduction

In count data regression analysis, the response variable takes on count values (i.e., 0, 1, 2, ...). The consequences of this property of the response variable can be understood by comparison with classical linear regression analysis.

In classical linear regression analysis, for a set of n datapoints, the predicted value of the response variable $\hat{y}_i$ may be given by

$$\hat{y}_i = \hat{\beta}_0 + \hat{\beta}_1 x_{1i} + \hat{\beta}_2 x_{2i} + \ldots + \hat{\beta}_m x_{mi} \quad for\ i = 1,\ 2,\ \ldots,\ n \tag{1}$$

where $x_1, \ldots, x_m$ are the covariates, $\hat{\beta}_0, \ldots, \hat{\beta}_m$ are the parameters, and $\hat{y}_i$ is the predicted value of the response variable. $\hat{y}_i$ is also the estimate of the expected value of the response variable given the covariate values. Hence, (1) is referred to as the conditional mean model (CMM).

The CMM residuals, i.e., the differences between $\hat{y}_i$ and observed values y_i, are distributed about the conditional mean according to the normal probability density function (pdf):

$$f(res_i) = \frac{1}{\sigma\sqrt{2\pi}} \exp\left(-\frac{1}{2}\left(\frac{res_i}{\sigma}\right)^2\right) \tag{2}$$

where $res_i = y_i - \hat{y}_i$, the residual for the i-th observed value, and σ is the standard deviation of the residuals. The closer res_i is to 0, the higher the value of $f(res_i)$. If the residuals are also identically distributed (i.e., come from the same, vast, imaginary pool of residuals) and independently distributed (i.e., one residual is not useful in predicting the value of another), then the pdfs may be multiplied together to form the normal-based likelihood function:

$$\begin{aligned}\mathcal{L}_{normal} &= \prod_{i=1}^{n} \frac{1}{\sigma\sqrt{2\pi}} \exp\left(-\frac{1}{2}\left(\frac{res_i}{\sigma}\right)^2\right) \\ &= \prod_{i=1}^{n} \frac{1}{\sigma\sqrt{2\pi}} \exp\left(-\frac{1}{2}\left(\frac{y_i - (\hat{\beta}_0 + \hat{\beta}_1 x_{1i} + \hat{\beta}_2 x_{2i} + \ldots + \hat{\beta}_m x_{mi})}{\sigma}\right)^2\right)\end{aligned} \quad (3)$$

The best estimates of CMM parameters may be found by adjusting them until they maximize this normal-based likelihood function.

Several properties of this likelihood function allow classical regression analysis to be transparent, both to the analyst working with the data and to the general audience reviewing the published results. Maximizing the likelihood function corresponds to minimizing the sum of the squares of the residuals, and thus a plot of the resulting conditional mean shows it passing more-or-less through the middle of the scattering of observed values. One senses that shifting or rotating that best-fit line would not improve the fit. Also, the relatively simple R^2, which varies from 0 to 1, and is a measure of the portion of the variation in the response variable accounted for by the conditional mean model, is visually represented in the plot.

In classical linear regression, the standard deviation appearing in the likelihood function can be estimated directly from the residuals to give a fairly reasonable representation of the spread of the data, even if the residuals are not exactly normally distributed. This, in turn, allows for p-values that tend to be relatively trustworthy. There is still a risk that a covariate that is not truly associated with the response variable will have a low p-value due to mere chance. This risk increases as the number of covariates under consideration for inclusion in the CMM increases. Overfitting of the CMM (i.e., the inclusion of covariates or other complexities that represent merely random effects rather than actual associations) may then occur. However, the dataset can be divided into two subsets—training data (to build the CMM) and testing data. The training data can be further subdivided for k-fold cross-validation, with reductions in R^2 or other simple measures to help detect the presence of inappropriate covariates. Finally, because such covariates may elude even k-fold cross-validation, the final CMM is applied to the testing data, and, again, reductions in R^2 or other simple measures will further aid in detecting false inference and overfitting.

Unfortunately, many of the above desirable features are not available in count data regression analysis. To begin with, the CMM immediately becomes more complex with the right side typically being exponentiated:

$$\hat{y}_i = e^{\hat{\beta}_0 + \hat{\beta}_1 x_{1i} + \hat{\beta}_2 x_{2i} + \ldots + \hat{\beta}_m x_{mi}} \quad (4)$$

If one plots the conditional mean through the scattering of observed values, the correct placement of the line may now seem counter-intuitive because non-linearity in the CMM, along with other factors, mean that the distribution of the observed values will not likely be symmetric about the best fit line. Casual assessment of the goodness-of-fit by eye is difficult, as can be seen, for example, in Figure 1, which shows candidates of best-fit lines for childhood asthma data in Houston, Texas (Figure 1 will be discussed in more detail in the next section). Furthermore, the normal pdf will now need to be replaced by any one of dozens of probability mass functions (pmfs) to build the likelihood function. Incorrect pmf selection can lead to underestimation of the spread of the data, resulting in falsely low p-values [1], false inference, and overfitting. Worse still, there is no longer a simple, universally recognized R^2 or other intuitively appealing measures of goodness-of-fit that can be conveniently used in k-fold cross-validation or in application to test data to help warn against overfitting. There are only various forms of the more difficult to interpret pseudo-R^2, and other measures, depending on the representation of the residuals [2]. This may explain why authoritative "how-to" guides on data analysis in R may demonstrate k-fold cross-validation for various model types but not for count data regression [3,4]. In our literature review of the impact of air quality on respiratory health, we found k-fold cross-validation and application of testing data was used [5], but never for a count data response variable in a CMM.

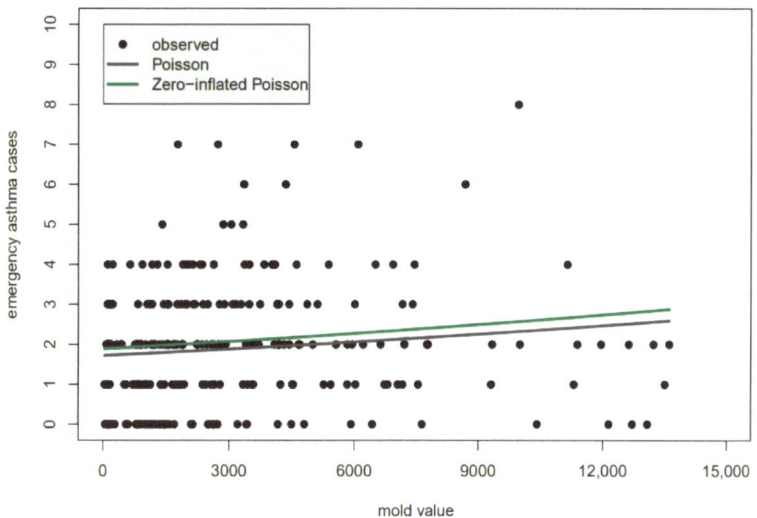

Figure 1. Emergency department childhood asthma arrivals in response to mold during summers of 2003–2011 in Houston, Texas.

Addressing all the ramifications of misspecification of the pmf in count data regression analysis is beyond the scope of this brief commentary. The impact of misspecification on p-values for covariate parameter estimates, and a simple strategy to reduce the tendency for the underestimation to occur, are illustrated in the following sections.

2. Illustration of False Inference and Overfitting Due to pmf Misspecification

The consequences of misspecifying the pmf in count data regression analysis can be seen in our own analysis of the relationship between air quality and childhood asthma in Houston, Texas, during the summers of 2003–2011. Concentrations of aeroallergens (mold and pollen) and anthropogenic contaminants (butane, nitrous oxide, ozone, sulfur dioxide, and particulates) were initially included in the model as covariates. The number of children arriving per day at particular hospital emergency departments for asthma was the response variable. We initially assumed the Poisson distribution for the pmf. With this pmf, a strong association between the response variable and the mold concentration was found, with p-value $< 10^{-15}$.

However, the most appropriate CMM and pmf among those being considered may be identified as that which yields the lowest Akaike information criteria (AIC) value [6]

$$AIC = -2 \cdot ln(\mathcal{L}) + 2 \cdot k \qquad (5)$$

where $\mathcal{L}$ is the likelihood function value for the selected pmf, and k is the number of parameters that may be adjusted to increase $\mathcal{L}$. The second term is thus a way of penalizing the inclusion of parameters, as including an additional adjustable parameter will always increase $\mathcal{L}$, even if the parameter is not truly representative of actual statistical relationships. Variations of the AIC may also be used. We use the original AIC here because it is commonly available in software packages. The chooseDist() function of the R gamlss package [7] runs through dozens of pmfs for building likelihood functions, adjusts parameters to maximize each, and then identifies the one with the lowest AIC. By using this process, dozens of pmfs were found, which yielded a lower AIC than did the Poisson distribution. An alternative pmf, the zero-inflated Poisson, which allows for a higher number of zeros than would be expected for the Poisson and thus, in turn, has a substantially broader spread than the Poisson would show for our dataset, was found to yield the lowest AIC among the dozens of available pmfs. The resulting p-value for the mold covariate was now 0.051, a p-value increase of many orders of magnitude compared to that provided by the Poisson pmf, leading the mold covariate to be accepted as statistically significant only under far less strict criteria.

Figure 1 shows a plot of the best-fit line through the data based on the Poisson pmf (gray) and the zero-inflated Poisson pmf (green). Due to the non-linearity of the CMM and other factors, one would be hard-pressed to say whether either of the lines fits the data well, let alone which fits the data better to justify the use of one CMM or pmf over the other. Indeed, as we will see in the following discussion of the generation and analysis of synthetic data, radically different pmfs may yield essentially identical CMMs, completely eliminating the usefulness of plots, such as in Figure 1, in determining which pmf is superior.

To show that the impact of pmf misspecification on p-values is not unique to peculiarities of the somewhat small air quality and childhood asthma dataset we ourselves are working with, we developed a synthetic dataset that readers are free to view, re-generate with parameters of their choice, and re-test through the link provided in the data availability statement below. Figure 2 shows how we generated the synthetic dataset and how the reader could use the code to generate their own. The three blocks forming the left column of the schematic are all the reader would need to select to build the synthetic dataset.

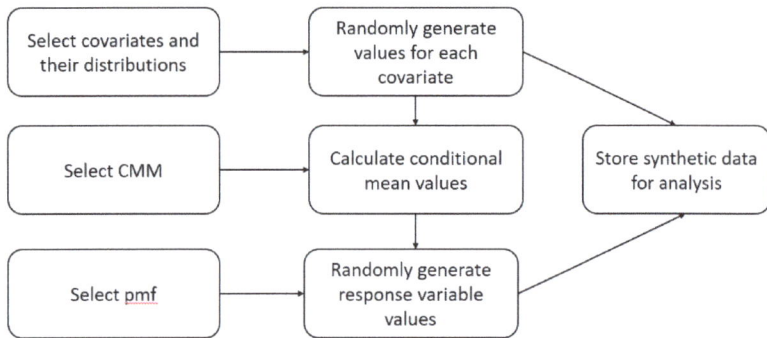

Figure 2. Schematic for the generation of synthetic data with count data as the response variable. pmf: probability mass functions; CMM: conditional mean model.

For our synthetic dataset, which we analyzed in Table 1 below, the code provided through the link was applied in R version 4.0.0 [8] to generate 1000 values for each of three covariates, x_1, x_2, and x_3, from the normal distribution with the mean $\mu = 10$ and standard deviation $\sigma = 1$. Parameter values were then assigned to create 1000 conditional mean values as follows, with x_3 excluded:

$$\hat{y}_i = e^{1+0.1x_{1i}+0.1x_{2i}} \text{ for } i = 1, 2, \ldots, 1000 \quad (6)$$

Observed y_i values were distributed about these $\hat{y}_i$ values according to the negative binomial pmf, which has a standard deviation of $\sigma_i = \sqrt{\hat{y}_i + \alpha \hat{y}_i^2}$. (This is in contrast with the Poisson distribution, which is less spread out, with $\sigma_i = \sqrt{\hat{y}_i}$.) A value of 0.5 was chosen for α, the dispersion parameter.

Table 1. Results for regression analysis using negative binomial (Neg. Bin.) pmf ($\mathcal{L}_{optimal}$ columns) and Poisson pmf ($\mathcal{L}_{Poisson}$ columns).

	$\hat{y}=e^{\hat{\beta}_0+\hat{\beta}_1 x_1}$ as a Conditional Mean Model		$\hat{y}=e^{\hat{\beta}_0+\hat{\beta}_1 x_1+\hat{\beta}_2 x}$ as a Conditional Mean Model		$\hat{y}=e^{\hat{\beta}_0+\hat{\beta}_1 x_1+\hat{\beta}_2 x+\hat{\beta}_3 x_3}$ as a Conditional Mean Model	
	$\mathcal{L}_{Poisson}$	$\mathcal{L}_{optimal}$	$\mathcal{L}_{Poisson}$	$\mathcal{L}_{optimal}$	$\mathcal{L}_{Poisson}$	$\mathcal{L}_{optimal}$
pmf	Poisson	Neg. bin.	Poisson	Neg. bin.	Poisson	Neg. bin.
σ_i	$\sqrt{\hat{y}_i}$	$\sqrt{\hat{y}_i + \alpha \hat{y}_i^2}$	$\sqrt{\hat{y}_i}$	$\sqrt{\hat{y}_i + \alpha \hat{y}_i^2}$	$\sqrt{\hat{y}_i}$	$\sqrt{\hat{y}_i + \alpha \hat{y}_i^2}$
α	NA	0.515	NA	0.512	NA	0.511
AIC	15,054.0	7905.6	14,973.3	7900.9	14,960.1	7901.3
$\hat{\beta}_0$ (p-value)	1.52 ($< 2 \times 10^{-16}$)	1.47 (7.6×10^{-9})	0.85 (7.2×10^{-16})	0.84 (0.017)	1.15 ($< 2 \times 10^{-16}$)	1.17 (0.0072)
$\hat{\beta}_1$ (p-value)	0.15 ($< 2 \times 10^{-16}$)	0.15 (1.2×10^{-9})	0.15 ($< 2 \times 10^{-16}$)	0.15 (1.1×10^{-9})	0.15 ($< 2 \times 10^{-16}$)	0.15 (9.3×10^{-10})
$\hat{\beta}_2$ (p-value)	NA	NA	0.065 ($< 2 \times 10^{-16}$)	0.063 (0.0099)	0.064 ($< 2 \times 10^{-16}$)	0.063 (0.011)
$\hat{\beta}_3$ (p-value)	NA	NA	NA	NA	−0.029 (0.00010)	−0.033 (0.20)

The results for each of the three CMMs are shown in Table 1 $\mathcal{L}_{optimal}$ columns. In each case, the optimal pmf is, not surprisingly, the same one used to generate the data. In some cases, adding covariates may cause a switch to a pmf with a less spread structure [1]. As expected, x_3, which was not used to generate the response variable, has a coefficient with a p-value well above 0.05, and slightly increases the AIC. It is to be excluded from the CMM.

For comparison, results for the Poisson pmf, often used in the literature, appear in the $\mathcal{L}_{Poisson}$ columns. The p-values are now falsely low, sometimes by several orders of magnitude. The false inference would now lead to including x_3. The lowering of the AIC value by including x_3 shows that the AIC is inadequate for preventing CMM overfitting.

Hilbe, an author of more than 10 books on statistical modeling, has cautioned that "Many analysts have been deceived into thinking that they have developed a well-fitted model" because the spread of the residuals was greater than represented in their count data regression model [1]. In our own dataset of childhood asthma and air quality in Houston, the distribution of the daily arrivals to the emergency department appears to be zero-inflated, i.e., there is an inexplicably high number of days with zero arrivals if the observed values are assumed to be Poisson distributed about the conditional mean. The zero-inflated Poisson pmf accounted for what is in effect an increase in the spread of the residuals, thereby giving a more realistic p-value (0.051), which is many orders of magnitude higher than that suggested by the Poisson pmf ($< 10^{-15}$).

Utilizing the most appropriate pmf can dramatically reduce the risk of false inference and overfitting. However, it must be noted that the AIC and related criteria do not establish appropriateness in any absolute sense but only identify the best choice among a set of choices. It could be that none of the choices is ultimately appropriate. In recognition of this limitation of such criteria, and in recognition of limitations among various software packages to select the most appropriate pmf, and to provide an intuitively appealing visual check on the selected pmf and CMM, a predicted-and-observed count histogram (POCH) is discussed in the following section.

3. The Predicted-And-Observed Count Histogram

Figures 3 and 4 are predicted-and-observed count histograms (POCH), similar to what is presented but not formally named in authoritative count data regression analysis texts [1,2]. Black dots and other markers are where the tops of the more traditional vertical histogram bars would be to represent the number of times that the response variable takes on the count value. The black dots represent the number of occurrences of the observed response variable values, while the green and gray markers represent the number of occurrences predicted by the models. In Figure 3, for example, the black dot at the count of 0 indicates that there were 0 childhood asthma emergency department arrivals on 57 of the summer days, while the gray square indicates that the Poisson pmf anticipates 0 arrivals to occur on only 40 of the summer days. Figure 3 shows that while the Poisson and zero-inflated Poisson had similar performance in predicting the number of days for which three or more arrivals occur, the zero-inflated Poisson pmf was a substantial improvement overall for the lower arrival numbers. In Figure 4, the model having two covariates and using the negative binomial pmf was clearly a better fit than was the three-covariate model with the Poisson pmf. Though not shown in either POCH, the analyst may generate predicted values from the pmf that best fits the observed histogram directly, i.e., without a CMM, note the resulting AIC value, and thus have a baseline AIC value from which to develop the CMM.

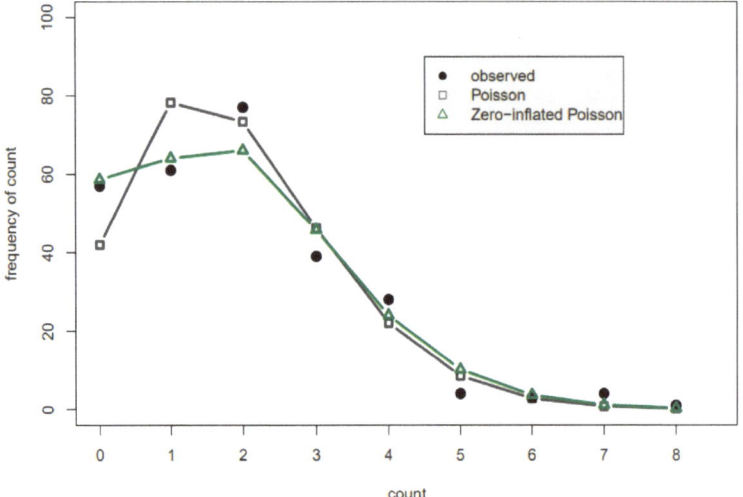

Figure 3. Predicted-and-Observed Count Histogram for modeling of emergency department arrivals with mold as the covariate, for summers from 2003–2011 in Houston, Texas.

In both Figures 3 and 4, the POCH shows Poisson pmfs (as opposed to the zero-inflated and negative binomial pmfs) have a more narrow distribution than does the actual data. It thus clearly warns that *p*-values with the Poisson pmf for these particular datasets will be falsely low. Such charts immediately provide transparency of the complicated count data regression analysis to the analyst working with the data and to the broader audience.

The POCH is easily generated for even the most complex count data regression analysis models, including ones that incorporate smoothing splines, autoregressive parameters, etc., as in generalized additive models and models in which the count data is binary, such as in case-crossover studies. The POCH merely requires a predicted response variable and a representation of the distribution of residuals, and so can be developed even for a quasi-likelihood method [9].

A POCH helps assess the correctness of the pmf not only in regards to spread, but also in regard to skewness, zero-inflation, hurdles, and other potentially important features. A POCH will not entirely address every violation of statistical assumptions. For example, one still needs to check for autocorrelation among residuals. However, where the POCH does not directly address them, it may provide a solid starting point. For example, testing for autocorrelation in count models requires standardizing the residuals before plotting the autocorrelation function [2,10]. The POCH can help identify the correct pmf for the standardization. Once the final model is selected, perhaps including autoregressive parameters, the POCH should be re-generated to re-confirm the appropriateness of the model.

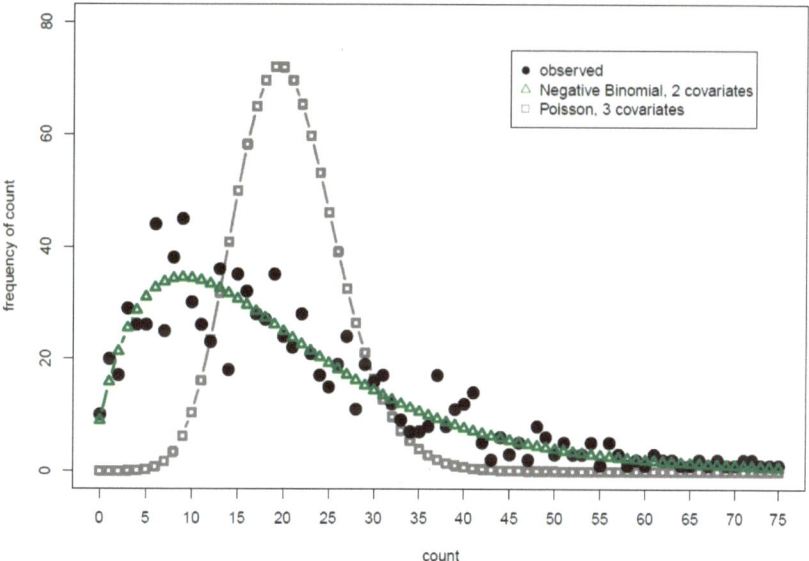

Figure 4. Predicted-and-Observed Count Histogram for the synthetic dataset.

In our literature review of the impact of air quality on health, we found no histogram such as a POCH, or any explicit evidence that the most appropriate pmf was used. This absence even among excellent articles [11–18] suggests a systemic issue extending beyond individual authors. We recommend that publishers require a POCH for articles involving count data regression models.

4. Conclusions

The complexity of count data regression models can lead to false inference and overfitting. A remedy is a predicted-and-observed count histogram POCH, which makes the analysis more transparent to analysts themselves and to the scientific community in general.

Author Contributions: J.F.J. and H.O.S. guided this research and contributed significantly to preparing the manuscript for publication. H.O.S., C.G.M. and T.S. participated in development the research methodology. J.F.J. and C.F. developed the scripts used in the analysis. J.F.J. and C.F. performed the data analysis with contribution from H.O.S. C.G.M. compiled that data. J.F.J. prepared the first draft. J.F.J., H.O.S., C.G.M. and T.S. performed the final overall proof reading of the manuscript. All authors have read and agreed to the published version of the manuscript.

Funding: This research received no external funding.

Informed Consent Statement: Not applicable.

Data Availability Statement: The synthetic data and R code used to generate and analyze it is available at the Open Science Framework website at https://osf.io/rjtkz (accessed on 1 February 2021). Access may require going to https://osf.io first and then searching for public profile rjtkz.

Acknowledgments: This work has been supported in part through the Robert Wood Johnson Demonstration Project (grant #043506) for Texas Emergency Department Asthma Surveillance.

Conflicts of Interest: The authors declare no conflict of interest.

References

1. Hilbe, J.M. *Modeling Count Data*; Cambridge University Press: New York, NY, USA, 2014.
2. Cameron, A.C.; Trivedi, P.K. *Regression Analysis of Count Data*, 2nd ed.; Cambridge University Press: New York, NY, USA, 2013.
3. Kabacoff, R.I. *R in Action: Data Analysis and Graphics with R*; Manning Publications Co.: Shelter Island, NY, USA, 2015.
4. Rigby, R.A.; Stasinopoulos, D.M.; Heller, G.Z.; De Bastiani, F. *Distributions for Modeling Location, Scale, and Shape: Using Gamlss in R*; CRC Press: Boca Raton, FL, USA; Taylor & Francis Group: Boca Raton, FL, USA, 2020.
5. Vitolo, C.; Scutari, M.; Ghalaieny, M.; Tucker, A.; Russell, A. Modeling air pollution, climate, and health data using Bayesian networks: A case study of the English regions. *Earth Space Sci.* **2018**, *5*, 76–88. [CrossRef]
6. Akaike, H. Information Theory and an Extension of the Maximum Likelihood Principle. In *Proceedings of the Second International Symposium on Information Theory, Tsahkadsor, Armenia, 2–8 September 1971*; Petrov, B.N., Caski, F., Eds.; Akademiai Kiado: Budapest, Hungary, 1973; pp. 267–281.
7. Rigby, R.A.; Stasinopoulos, D.M. Generalized additive models for location, scale and shape (with discussion). *Appl. Stat.* **2005**, *54*, 507–554. [CrossRef]
8. R Core Team. *R: A Language and Environment for Statistical Computing*; R Foundation for Statistical Computing: Vienna, Austria, 2020. Available online: https://www.R-project.org/ (accessed on 1 February 2021).
9. Wedderburn, R.W.M. Quasi-likelihood functions, generalized linear models, and the Gauss-Newton method. *Biometrika* **1974**, *61*, 439–447.
10. Li, W.K. Testing model adequacy for some Markov regression models for time series. *Biometrika* **1991**, *78*, 83–89. [CrossRef]
11. Choi, M.; Curriero, F.C.; Johantgen, M.; Mills, M.E.C.; Sattler, B.; Lipscomb, J. Association between ozone and emergency department visits: An ecological study. *Int. J. Environ. Health Res.* **2011**, *21*, 201–221. [CrossRef] [PubMed]
12. Hyrkas-Palmu, H.; Ikäheimo, T.M.; Laatikainen, T.; Jousilahti, P.; Jaakkola, M.S.; Jaakkola, J.J.K. Cold weather increases respiratory symptoms and functional disability especially among patients with asthma and allergic rhinitis. *Sci. Rep.* **2018**, *8*, 10131. [CrossRef] [PubMed]
13. Lam, H.C.; Li, A.M.; Chan, E.Y.; Goggins, W.B., III. The short-term association between asthma hospitalisations, ambient temperature, other meteorological factors and air pollutants in Hong Kong: A time-series study. *Thorax* **2016**, *71*, 1097–1109. [CrossRef] [PubMed]
14. Lin, Y.; Chang, S.; Lin, C.; Chen, Y.; Wang, Y. Comparing ozone metrics on associations with outpatient visits for respiratory diseases in Taipei Metropolitan area. *Environ. Pollut.* **2013**, *177*, 177–184. [CrossRef] [PubMed]
15. O'Lenick, C.R.; Winquist, A.; Chang, H.H.; Kramer, M.R.; Mulholland, J.A.; Grundstein, A.; Sarnat, S.E. Evaluation of individual and area-level factors as modifiers of the association between warm-season temperature and pediatric asthma morbidity in Atlanta, GA. *Environ. Res.* **2017**, *156*, 132–144. [CrossRef] [PubMed]
16. Rublee, C.S.; Sorensen, C.J.; Lemery, J.; Wade, T.J.; Sams, E.A.; Hilborn, E.D.; Crooks, J.L. Associations between dust storms and intensive care unit admissions in the United States, 2000–2015. *GeoHealth* **2020**, *3*, e2020GH000260. [CrossRef] [PubMed]
17. Xu, Z.; Huang, C.; Su, H.; Turner, L.R.; Qiao, Z.; Tong, S. Diurnal temperature range and childhood asthma: A time-series study. *Environ. Health* **2013**, *12*, 12. Available online: http://www.ehjournal.net/content/12/1/12 (accessed on 1 February 2021).
18. Zhang, H.; Liu, S.; Chen, Z.; Zu, B.; Zhao, Y. Effects of variations in meteorological factors on daily hospital visits for asthma: A time-series study. *Environ. Res.* **2020**, *182*, 109115. [CrossRef] [PubMed]

MDPI
St. Alban-Anlage 66
4052 Basel
Switzerland
Tel. +41 61 683 77 34
Fax +41 61 302 89 18
www.mdpi.com

Applied Sciences Editorial Office
E-mail: applsci@mdpi.com
www.mdpi.com/journal/applsci

www.ingramcontent.com/pod-product-compliance
Lightning Source LLC
LaVergne TN
LVHW072331090526
838202LV00019B/2401